Mathematical Cognition and Learning

Acquisition of Complex Arithmetic Skills and Higher-Order Mathematics Concepts

Mathematical Cognition and Learning

Series Editors

Daniel B. Berch
David C. Geary
Kathleen Mann Koepke

VOLUME 3

Acquisition of Complex Arithmetic Skills and Higher-Order Mathematics Concepts

Volume Editors

David C. Geary
Daniel B. Berch
Robert J. Ochsendorf
Kathleen Mann Koepke

Mathematical Cognition and Learning

Acquisition of Complex Arithmetic Skills and Higher-Order Mathematics Concepts

Edited by

David C. Geary
Psychological Sciences
University of Missouri
Columbia, MO, United States

Daniel B. Berch
Curry School of Education
University of Virginia
Charlottesville, VA, United States

Robert J. Ochsendorf
Directorate for Education and Human Resources
National Science Foundation
Arlington, VA, United States

Kathleen Mann Koepke
Eunice Kennedy Shriver National Institute of Child Health and Human Development (NICHD)
National Institutes of Health (NIH)
Bethesda, MD, United States

ELSEVIER

ACADEMIC PRESS
An imprint of Elsevier
elsevier.com

Academic Press is an imprint of Elsevier
125 London Wall, London EC2Y 5AS, United Kingdom
525 B Street, Suite 1800, San Diego, CA 92101-4495, United States
50 Hampshire Street, 5th Floor, Cambridge, MA 02139, United States
The Boulevard, Langford Lane, Kidlington, Oxford OX5 1GB, United Kingdom

Library of Congress Cataloging-in-Publication Data
A catalog record for this book is available from the Library of Congress

British Library Cataloguing-in-Publication Data
A catalogue record for this book is available from the British Library

ISBN: 978-0-12-805086-6

For information on all Academic Press publications
visit our website at https://www.elsevier.com/

Publisher: Nikki Levy
Acquisition Editor: Emily Ekle
Editorial Project Manager: Barbara Makinster
Production Project Manager: Caroline Johnson
Designer: Greg Harris

Typeset by Thomson Digital

Contents

Part II
Rational Number Processing

6. Fraction Development in Children: Importance of Building Numerical Magnitude Understanding

Nancy C. Jordan, Jessica Rodrigues, Nicole Hansen, Ilyse Resnick

7. Numbers as Mathematical Models: Modeling Relations and Magnitudes with Fractions and Decimals

Melissa DeWolf, Miriam Bassok, Keith J. Holyoak

Part III
Algebraic, Geometric, and Trigonometric Concepts

Contributors

Christina Barbieri, University of Delaware, Newark, DE, United States

Marcia A. Barnes, Department of Special Education & Meadows Center for Preventing Educational Risk, University of Texas, Austin, TX, United States

Miriam Bassok, Department of Psychology, University of Washington, Seattle, WA, United States

Kreshnik N. Begolli, Temple University, Philadelphia, PA, United States

Daniel B. Berch, Curry School of Education, University of Virginia, Charlottesville, VA, United States

Julie L. Booth, Temple University, Philadelphia, PA, United States

Rebecca Bull, National Institute of Education, Singapore, Singapore

Sara Caviola, Department of Psychology, University of Cambridge, Cambridge, United Kingdom

Briana Chang, Temple University, Philadelphia, PA, United States

Jodi L. Davenport, WestEd, San Francisco, CA, United States

Melissa DeWolf, Department of Psychology, University of California, Los Angeles, CA, United States

Kelley Durkin, Peabody Research Institute, Vanderbilt University, Nashville, TN, United States

Mary Wagner Fuhs, University of Dayton, Dayton, OH, United States

David C. Geary, Psychological Sciences, University of Missouri, Columbia, MO, United States

David Giofrè, Liverpool John Moores University, Natural Sciences and Psychology, Liverpool, United Kingdom

Nicole Hansen, Fairleigh Dickinson University, Teaneck, NJ, United States

Keith J. Holyoak, Department of Psychology, University of California, Los Angeles, CA, United States

Caroline Byrd Hornburg, University of Notre Dame, Notre Dame, IN, United States

Nancy C. Jordan, University of Delaware, Newark, DE, United States

Kathleen Mann Koepke, *Eunice Kennedy Shriver*, National Institute of Child Health and Human Development (NICHD), National Institutes of Health (NIH), Bethesda, MD, United States

Kerry Lee, National Institute of Education, Singapore, Singapore

Irene C. Mammarella, Department of Developmental and Social Psychology, University of Padova, Padova, Italy

James L. McClelland, Department of Psychology, Stanford University, Stanford, CA, United States

Kelly M. McGinn, Temple University, Philadelphia, PA, United States

Nicole M. McNeil, University of Notre Dame, Notre Dame, IN, United States

Kevin W. Mickey, Department of Psychology, Stanford University, Stanford, CA, United States

Dana Miller-Cotto, Temple University, Philadelphia, PA, United States

Swee Fong Ng, National Institute of Education, Singapore, Singapore

Connor D. O'Rear, University of Notre Dame, Notre Dame, IN, United States

Robert J. Ochsendorf, Directorate for Education and Human Resources, National Science Foundation, Arlington, VA, United States

Kimberly P. Raghubar, Department of Pediatrics, Baylor College of Medicine and Psychology Service, Texas Children's Hospital & Department of Psychology, University of Houston, Houston, TX, United States

Ilyse Resnick, University of Delaware, Newark, DE, United States

Bethany Rittle-Johnson, Department of Psychology and Human Development, Vanderbilt University, Nashville, TN, United States

Katherine M. Robinson, Department of Psychology, Campion College, University of Regina, Regina, SK, Canada

Jessica Rodrigues, University of Delaware, Newark, DE, United States

Robert S. Siegler, Carnegie-Mellon University, Pittsburgh, PA, United States

Jon R. Star, Graduate School of Education, Harvard University, Cambridge, MA, United States

Catherine Thevenot, University of Lausanne, Institute of Psychology, Lausanne, Switzerland

Xenia Vamvakoussi, University of Ioannina, Ioannina, Epirus, Greece

Wim Van Dooren, KU Leuven, Leuven, Flanders, Belgium

Jo Van Hoof, KU Leuven, Leuven, Flanders, Belgium

Lieven Verschaffel, KU Leuven, Leuven, Flanders, Belgium

Laura K. Young, Temple University, Philadelphia, PA, United States

Foreword: Build It and They Will Come

Robert S. Siegler
Carnegie-Mellon University, Pittsburgh, PA, United States

In 1978, I was a member of the review panel for the first math learning grants competition of the National Institute of Education (NIE). Panel members were told that there was enough money to fund 10 grants; the only problem was that there were only 2 proposals that clearly merited funding and 2 others where opinion was mixed. Once the reviews were presented and discussed, the conversation turned to whether the panel had to recommend 10 grants for funding, because that would mean funding at least 6 proposals for which there was no enthusiasm. Most panel members, me included, argued against funding them. However, Susan Chipman, who was Assistant Director of NIE and in charge of the review competition, stated that giving out all of the money was essential. Although acknowledging that the applications were far from stellar, she argued that once the word got out that there was substantial funding for math learning research, better proposals would soon follow. The panel reluctantly agreed and voted to fund the 10 least dirty shirts in the dirty laundry basket.

How things have changed! A similar competition today would elicit at least 20 worthy proposals. The challenge would be to distinguish the really excellent ones from the merely good or very good ones. There would be passionate debates, just as before, but now they would stem from panel members arguing that their favorite proposals just had to be funded no matter what.

A web search of the program from the most recent Society for Research in Child Development meeting provides quantitative evidence of the prominence that research in mathematics development has attained. "Math" was a keyword or appeared in the abstract of 216 presentations at the 2015 SRCD Conference. This number of mentions exceeded that for other popular areas of developmental psychology such as "perception" (117), "attention" (166), "memory" (141), "reasoning" (67), "space" or "spatial" (87), "moral" (110), "reading" (91), and "executive function" (180). Susan Chipman was right, and her vision deserves recognition: NIE built it, and they did come. The excellent chapters in the present volume attest to the theoretical and practical importance of the area that Susan and the NIE helped create; the chapters also say a lot about where the area is today and where it seems to be going.

Probably the most striking feature of this volume is the range of topic areas and age groups that are addressed. Before roughly 2010, the overwhelming focus of mathematical development research was on whole numbers and on children from birth to early elementary school. There were many studies of the early development of nonsymbolic numbers, counting, whole number arithmetic, number conservation, and symbolic magnitude representations, but not many on more advanced mathematical topics or with older children and adolescents. Starting around 2010, the focus of math development research widened to include rational numbers: fractions, decimals, percentages, and negatives. The latest and most exciting developments in the study of rational numbers are well represented in this volume, in particular in the chapters of Jordan on improving fraction instruction; of Rittle-Johnson, Star, and Durkin on use of comparison to improve learning of rational numbers; of van Hoof, Van Dooren, Vamvakoussi, and Verschaffel on the developmental transition from natural to rational numbers; and of DeWolf, Bassok, and Holyoak on the importance of relational reasoning with rational numbers for algebra performance. Most of the participants in these studies ranged from late elementary school through the end of middle school, though adults' knowledge of rational numbers also received a fair amount of attention.

Rational numbers have received a great deal of research attention since 2010, so the focus on them in this volume was not too surprising. More surprising was the emphasis on topics beyond rational numbers: on learning of algebra in the chapters of Booth et al.; of Lee, Ng, and Bull; of Rittle-Johnson et al.; and of DeWolf et al.; on geometry in the chapter of Mammarella, Giofre, and Caviola; and on trigonometry in the chapter by Mickey and McClelland. Of course, all of these topics have received some research attention for many years, but the emphasis on them in this volume is striking. I believe that the focus on rational numbers, algebra, geometry, and trigonometry represents an important part of where research on mathematical development is going. Together with prior research on whole numbers, this research will build the database necessary to construct more encompassing theories of mathematical development than have heretofore been possible.

The integrated theory of numerical development, which I have formulated in recent years (Siegler & Braithwaite, in press; Siegler, 2016; Siegler, Thompson, & Schneider, 2011), provides a means for bringing together acquisition of whole and rational number knowledge within a single framework. Within that framework, numerical development presents two central challenges. One challenge is to understand that all real numbers share the property of representing magnitudes that can be located and ordered on number lines. The other challenge is to understand that many other properties of natural numbers do not consistently characterize rational numbers. Properties that apply to natural but not rational numbers include being represented by a single, unique symbol; having a unique predecessor and a unique successor; never decreasing with multiplication or increasing with division; and so on. It was gratifying for me personally that this theory was used in productive ways by many of the authors in the present volume.

A challenge now is to integrate the new work that is emerging on algebra, geometry, and trigonometry with current understanding of whole and rational numbers. We already know that just as precision of whole numbers has proved predictive of precision of representations of rational numbers (e.g., Bailey, Siegler, & Geary, 2014; Mou et al., 2016), so has precision of representations of rational numbers proved predictive of algebraic proficiency (e.g., Booth, Newton, & Twiss-Garrity, 2014; Mou et al., 2016). These relations are present even after many relevant variables are statistically controlled: IQ, working memory, executive functioning, parental income and education, etc.

Largely unknown, however, are the causal pathways that lead to these predictive relations. Several chapters in this volume, in particular those of Booth et al., Rittle-Johnson et al., and DeWolf et al., provide promising leads to how the integrative process operates, but a lot of work remains to be done before an integrative theory of *mathematical development*, as opposed to numerical development, will be possible. Formulating such an integrative theory of mathematical development clearly deserves high priority.

Another striking feature of the present volume is the amount of research that focuses on conceptual understanding of mathematical procedures. Although considerable progress has been made in understanding conceptual bases of very early numerical procedures, for example the work of Gelman and Gallistel (1978) on counting, much less research has been devoted to conceptual understanding of whole number arithmetic procedures, much less rational number ones. To be clear, there has been considerable documentation of *misconceptions* in both whole number arithmetic (e.g., Brown & VanLehn, 1980 and rational number arithmetic (e.g., Resnick & Omanson, 1987), but these studies have left unclear the *conceptions* on which the flawed procedures are based. To illustrate, consider the well-known long subtraction error of inverting the top and bottom digits in a column when the top digit is smaller (e.g., treating 145-108 as if it were 148-105). More than 35 years after the Brown and VanLehn studies that convincingly documented this error pattern, we still do not know whether this and other "misconceptions" stem from (1) belief that the erroneous procedure is correct, (2) belief that the erroneous procedure is one of a few possibilities that might be correct, or (3) belief that the erroneous procedure is erroneous, but use of it anyway due to not knowing a better alternative. Simply put, we still do not understand the conception on which the "misconception" is based.

Thus, it was refreshing to see a great deal of attention in this volume to the conceptual underpinnings of relatively advanced mathematical procedures. Chapters that particularly emphasized these conceptual underpinnings included Robinson's work on understanding of multiplication and division,

McNeil et al.'s chapter on understanding of mathematical equivalence, DeWolf et al.'s and Booth et al.'s chapters on algebra, Rittle-Johnson et al.'s chapter on the role of comparison processes, Thevenot's chapter on arithmetic word problems, and Mammarella et al.'s chapter on geometry.

This focus on conceptual understanding of mathematical procedures is welcome for the same reason as the focus on algebra, geometry, and trigonometry. The chapters and related work provide crucial empirical data for identifying the conceptual understanding that influences, and is influenced by, knowledge of mathematical procedures. At the same time, these expansions pose the same type of challenge for formulating an encompassing theory of mathematical development as the expansions of research into more advanced mathematical topics. One sign of the challenge of expanding our theories to include conceptual as well as procedural knowledge is that our descriptions of the development of mathematical procedures are almost invariably far more concrete and specific than our descriptions regarding conceptual understanding of the procedures.

Recently, Hugo Lortie-Forgues and I also became intrigued by the conceptual underpinnings of mathematical procedures, in particular, rational number arithmetic procedures. In a pair of studies, one on understanding of fraction arithmetic procedures (Siegler & Lortie-Forgues, 2015) and one on understanding of decimal arithmetic procedures (Lortie-Forgues & Siegler, 2015), we found that large majorities of both middle school students and preservice teachers believed that multiplication of pairs of numbers between 0 and 1 invariably led to answers greater than either multiplicand and that division of pairs of numbers between 0 and 1 invariably led to answers smaller than the number being divided. When asked why they believed that, participants most often answered that multiplication always makes numbers bigger and that division makes them smaller. Such explanations characterized most participants whose multiplication and division of rational numbers was flawless, as well as ones who did not know the procedures as well. This is a striking example of failing to learn from experience, given the hundreds if not thousands of fraction and decimal arithmetic problems that college students especially would have solved during their lives. The findings suggest that studying conceptual understanding of more advanced mathematical procedures is also likely to reveal more misconceptions than correct conceptions, many bits and pieces of knowledge that are only loosely connected to the classes of problems for which they are appropriate, and too often no understanding at all of why mathematical procedures are justified.

A third major theme of this volume is the intensifying effort to use findings from cognitive science research to improve math learning. This is most evident in areas where a substantial empirical base is available to guide the instructional efforts, such as mathematical equivalence, arithmetic word problems, and rational numbers, as indicated in the chapters by McNeil et al., Thevenot, Barnes and Raghubar, and Jordan et al. However, it also is evident in areas with a smaller empirical base, in particular in the chapters on algebra by Booth et al. and by Rittle-Johnson et al. These instructional efforts promise to yield practical benefits for educating students; they also promise to yield theoretical benefits for understanding mathematical development.

I'll close with a few comments about the editors of this volume, Dave Geary, Dan Berch, Rob Ochsendorf, and Kathy Mann Koepke. They were the ones who selected the authors, and they probably anticipated the topics that the authors would emphasize. The extensive coverage in this volume of older children's mathematical development, of more advanced areas of mathematics, and of conceptual understanding of procedures does not just reflect trends in the field. It also reflects the wisdom of the editors in choosing these authors to contribute chapters to this volume. This wisdom is also evident in their insightful summary chapter. I encourage readers to think hard about the findings reported in this volume, to consider their implications for theory and practice, and to build on them to move toward a fully integrative theory of mathematical development.

REFERENCES

Bailey, D. H., Siegler, R. S., & Geary, D. C. (2014). First grade predictors of middle school fraction knowledge. *Developmental Science*, *17*, 775–785.

Booth, J. L., Newton, K. J., & Twiss-Garrity, L. (2014). The impact of fraction magnitude knowledge on algebra performance and learning. *Journal of Experimental Child Psychology*, *118*, 110–118.

Brown, J. S., & VanLehn, K. (1980). Repair theory: A generative theory of bugs in procedural skills. *Cognitive Science*, *4*, 379–426.

Gelman, R., & Gallistel, C. R. (1978). *The child's understanding of number*. Cambridge, MA: Harvard University Press.

Lortie-Forgues, H., & Siegler, R. S. (2015). Conceptual knowledge of decimal arithmetic. *Journal of Educational Psychology*. Published online ahead of print, doi: http://dx.doi.org/10.1037/edu0000148

Mou, Y., Li, Y., Hoard, M. K., Nugent, L., Chu, F., Rouder, J., & Geary, D. C. (2016). Developmental foundations of children's fraction magnitude knowledge. *Cognitive Development*, *39*, 141–153.

Resnick, L. B., & Omanson, S. F. (1987). Learning to understand arithmetic. In R. Glaser (Ed.), *Advances in instructional psychology* (pp. 41–95). (Vol. 3). Hillsdale, NJ: Erlbaum.

Siegler, R. S. (2016). Continuity and change in the field of cognitive development and in the perspectives of one cognitive developmentalist. *Child Development Perspectives*, *10*, 128–133.

Siegler, R. S., & Braithwaite, D. W. (in press). Numerical development. *Annual Review of Psychology*.

Siegler, R. S., & Lortie-Forgues, H. (2015). Conceptual knowledge of fraction arithmetic. *Journal of Educational Psychology*, *107*, 909–918.

Siegler, R. S., Thompson, C. A., & Schneider, M. (2011). An integrated theory of whole number and fractions development. *Cognitive Psychology*, *62*, 273–296.

Preface

With this edition, we have now launched three of the five volumes of the Mathematical Cognition and Learning series. The first two broadly covered the biological foundations and correlates of mathematical cognition and learning, focusing first on evolutionary continuities in the system for representing relative quantity and carrying out operations on these representations (e.g., addition) and how this intuitive number sense contributes to young children's emerging understanding of aspects of formal and symbolic mathematics. The second volume covered the rapidly growing neuroscience approaches to mathematical cognition, focusing largely on the brain systems underlying our inherent number sense and children's early symbolic quantitative competencies. This second volume also provided an overview of genetic and environmental influences on individual differences in mathematics achievement and specific aspects of mathematical competence, as well as discussion of how certain types of genetic disorders can compromise the development of mathematical competencies.

With this third volume, we move to cognitive science approaches to mathematical cognition and learning and to the more complex domains of formal mathematics that are taught in elementary grades through high school. The volume centers on the interface between cognitive processing (e.g., prior knowledge, working memory), and mathematical learning highlighting recent findings concerning this interface and moves into new territory (e.g., trigonometry). To put this volume in context, in the 1970s cognitive psychologists' early groundbreaking studies of formal mathematics was primarily targeted on the processes and representations underlying adults' competence with simple arithmetic, with a few exceptions. Over the next two decades, these studies expanded to include children's developing arithmetic competencies, but the focus remained fairly narrow. As the reader will see, this is no longer the case. Cognitive scientists are now better engaged with mathematics educators and study the full range of school-taught mathematics, from simple arithmetic to trigonometry.

As with the previous volumes, we anticipate this volume will be of interest to researchers, graduate students, and undergraduates specializing in cognitive development, cognitive neuroscience, educational psychology, special education, and many other disciplines. More so than the first two volumes, the current volume should be especially interesting for researchers and practitioners in mathematics education. The chapters herein provide cutting-edge reviews of what cognitive scientists have discovered about students' mathematical understanding and learning, with important implications for future research and for practice. This volume could be used as a textbook for several kinds of courses taught in psychology (e.g., educational psychology) and education (e.g., mathematics education, instruction, and learning).

We have organized the volume into four sections. In the first are chapters that focus on aspects of arithmetic that were not as well studied in the first wave of cognitive science research. Across these chapters, the reader is provided with comprehensive and insightful reviews of students' understanding of core arithmetic concepts, such as associativity, and the influence of prior mathematical knowledge and everyday experiences on skill at solving arithmetical word problems. The section closes with a review of cognitive studies of the arithmetical learning of children with neurological disorders. In the second section, we move to the critical topic of rational numbers (i.e., fractions and decimals). The reader will be provided with incisive reviews of how and why students' knowledge of whole numbers can sometimes facilitate and sometimes interfere with their understanding of rational numbers, and why people better understand fractions in relational contexts (e.g., three out of the four children were girls) and decimals in contexts that involve continuous quantities (e.g., proportion of two liquids in a mixture). The third section moves to the critical but understudied topics of the factors that influence students' understanding of mathematical equivalence and algebra, geometry, and trigonometry. The volume concludes with a duo of chapters on recent intervention and instructional studies that are focused on improving students' mathematics learning and especially their learning of algebra, a critical gateway topic in high school mathematics.

We thank the Child Development and Behavior Branch of the *Eunice Kennedy Shriver* National Institute of Child Health and Human Development, NIH for the primary funding of the conference on which this volume is based and Dr. Joan McLaughlin, Commissioner of the National Center for Special Education Research (NCSER), U.S. Department of Education for the Department's supplementary funding of the conference. We also thank Dr. Ann Steffen from the Department of Psychology at the University of Missouri at St. Louis for her support and assistance in organizing the annual conference on which this volume is based.

David C. Geary
Daniel B. Berch
Robert J. Ochsendorf
Kathleen Mann Koepke

Chapter 1

Insights from Cognitive Science on Mathematical Learning*

David C. Geary*, Daniel B. Berch, Robert J. Ochsendorf†, Kathleen Mann Koepke‡**

**Psychological Sciences, University of Missouri, Columbia, MO, United States; **Curry School of Education, University of Virginia, Charlottesville, VA, United States; †Directorate for Education and Human Resources, National Science Foundation, Arlington, VA, United States;* ‡Eunice Kennedy Shriver, *National Institute of Child Health and Human Development (NICHD), National Institutes of Health (NIH), Bethesda, MD, United States*

For the most part, Volumes 1 and 2 of this series focused on basic nonsymbolic quantitative processing and symbolic numerical processing, including their evolutionary origins, early development, neural substrates, and genetic influences. As such, generally the most complex quantitative processing treated in the series thus far has been concerned with single-digit arithmetic. In contrast, the present volume covers higher-level mathematical processing and operations, including the development of multidigit arithmetic skills, learning to understand fractions and other rational number concepts and procedures, engaging in algebraic reasoning, acquiring geometric principles and competencies, and visualizing trigonometric expressions. As a consequence, the theoretical underpinnings of this work are more complex, multilayered, and wide-ranging in nature, drawing not only on some of the more rudimentary foundations of natural number processing, but also other domain-specific and domain-general skills needed for processing higher-order quantitative relations among multidigit Arabic numerals, rational numbers, number words, spatial relations, and abstract mathematical symbols.

In this first section, we provide the reader with the foundational components of cognitive theories that undergird research in these areas, and also preview

*This chapter is based, in part, upon work supported by (while Robert Ochsendorf was serving at) the National Science Foundation. Any opinion, findings, and conclusions or recommendations expressed in this material are those of the authors and do not necessarily reflect the views of the National Science Foundation.

Acquisition of Complex Arithmetic Skills and Higher-Order Mathematics Concepts
http://dx.doi.org/10.1016/B978-0-12-805086-6.00001-1

the kinds of theoretical accounts our authors advance when interpreting the evidence they review. In the second and third respective sections, we overview how basic research in cognitive science informs studies of mathematical learning and how a few core principles and general empirical findings help to tie together the basic questions being addressed with the research described across the chapters of these volumes, and how the answers to these questions set the stage for future basic and applied (e.g., design of interventions) research.

ON THE NATURE OF THEORIES AND MODELS IN COGNITIVE PSYCHOLOGY

Theories in the cognitive and behavioral sciences are composed of any more or less formalized sets of propositions (Marx, 1963). *Theoretical propositions* are statements—in the form of sentences or equations—of relations between *constructs*. Constructs, in turn, comprise symbolic representations of relations among *variables*, which consist of classes of objects, events, or their properties. For example, the construct of "memory" is frequently defined as the encoding (mental representation), storage, and retrieval of information supplied by the environment (e.g., phone numbers, names, facial expressions, or multiplication facts), and as theoretical constructs are by definition abstract, they are sometimes instantiated in the form of concrete models, such as familiar physical entities, as a way of characterizing mental processes in a more immediately recognizable manner. Roediger (1979) compiled a lengthy list of these kinds of models for illustrating that the construct of memory has been considered analogous to receptacles or spaces that contain objects, such as a house (James, 1890), stores (Atkinson & Shiffrin, 1968), and a library (Broadbent, 1971). Through the use of other metaphors that likewise possess spatial features, memory has also been likened to a wax tablet (Plato, Aristotle), a tape recorder (Posner & Warren, 1972), and a conveyor belt (Murdock, 1974).

As "memory" is considered a kind of overarching construct in cognitive psychology, it actually comprises several subconstructs, such as short-term memory, working memory, and long-term memory. *Working memory* refers to the capacity to temporarily store information in the service of performing more complex cognitive tasks (Baddeley, 1986). For the purposes of the present volume, one might generate a theoretical proposition hypothesizing that the constructs "working memory capacity" and "mathematics learning" are related in the following manner: *The greater one's working memory capacity, the better his or her mathematics learning*. Indeed, in Chapter 4, Barnes and Raghubar discuss this very proposition and the kinds of evidence supporting it.

The Role of Theories in Cognitive Psychology

What functions do theories actually serve in cognitive psychological research? There are primarily two: (1) a *tool*, in that they can possess heuristic value (i.e., furthering research); and (2) a *goal*, that is, to explain the mechanisms that give rise to various phenomena. The chapters in this volume make ample use of theories for both of these purposes. For example, with respect to theories serving as both a tool and a goal, in Chapter 5, van Hoof et al. specifically acknowledge how they make use of a combination of two complementary theories of the origin and development of the natural number bias (Ni & Zhou, 2005)—*the framework approach to conceptual change theory* and *the dual process theory of reasoning*—to "increase our understanding of learners' transition from natural to rational number understanding, and the role of the natural number bias in this transition." Similarly, in Chapter 6, Jordan et al. inform the reader that one of the main aims of their paper is to situate their own research, along with studies conducted by others, within the *integrated theory of number development* (Siegler, Thompson, & Schneider, 2011). That is, they describe studies designed to test predictions growing out of this theory, thus using it as a tool for designing studies to generate testable hypotheses, and also as a goal for explaining how the building of magnitude understanding can improve the development of fraction learning in children.

As another example, in Chapter 8, McNeil et al. describe how the *change-resistance theory* (McNeil & Alibali, 2005) can account for the difficulties children experience with acquiring an understanding of mathematical equivalence—affording the goal of explaining this phenomenon. Furthermore, these authors also discuss how they have "used the change-resistance account as a framework for designing a more comprehensive intervention" (i.e., as a tool). Finally, although Booth et al. begin their chapter by informing the reader that they will review the evidence for eight cognitive *principles* that are considered to be particularly promising for improving mathematics instruction, it soon becomes clear that the ideas from which these principles were originally derived were *theories*, including Vygotsky's (1978) *sociocultural theory of cognitive development* (principle of scaffolding), Sweller's (1988) *Cognitive Load Theory* (worked examples principle), and Festinger's (1957) *cognitive dissonance theory* (error reflection principle), among others.

Furthermore, a number of other theories are treated in these chapters that cover a rather broad range of topics, some of which were originally developed to explain cognitive development, learning, and related phenomena in general, including—*Piaget's theory of cognitive development* (Piaget, 1952, 1980; Piaget & Inhelder, 1967; Piaget, Inhelder, & Szeminska, 1960; Piaget & Szeminska, 1995; see Booth et al., Mammarella, Giofrè, and Caviola, McNeil et al., and Robinson chapters), Geary's (2008) *evolutionary educational psychology theory* (see Van Hoof et al. chapter), *Vosniadou's conceptual change theory* (Christou & Vosniadou, 2005, 2012; Stafylidou & Vosniadou, 2004; Vamvakoussi & Vosniadou, 2010, 2012; Vamvakoussi, Vosniadou, & Van Dooren, 2013; Vosniadou, 1994, 2003; Vosniadou & Vamvakoussi, 2006; Vosniadou, Vamvakoussi, & Skopeliti, 2008; Vosniadou & Verschaffel, 2004; see Booth et al., DeWolf et al., Jordan et al., Thevenot, and Van Hoof et al. chapters), *schema theory* (Kintsch & Greeno, 1985: see DeWolf et al. and Thevenot chapters), and *mental models* (Johnson-Laird, 1983; see Thevenot chapter). Also tending toward the more general form of theories of cognitive processing are: *working memory* theories (see Barnes et al., Booth et al., Jordan et al., Lee et al., Mammarella et al., McNeil et al., Rittle-Johnson et al., Robinson, Thevenot, and Van Hoof et al. chapters), *core conceptual structure* (Case, 1996; see Mickey & McClelland chapter), and *analogical reasoning theory* (Gentner, 1983; Gick & Holyoak, 1983; see Booth et al., DeWolf, Bassok, and Holyoak, and Rittle-Johnson, Star, and Durkin chapters). In contrast, other authors describe research that makes use of or tests more circumscribed theories, including van Hiele's (1986) *theory of geometric development*, and *core implicit knowledge of geometry* (Dehaene, Izard, Pica, & Spelke, 2006; see Mammarella et al. chapter).

Theory Testing and Validation

Readers of this volume will encounter several chapters that discuss alternative theories of a given phenomenon. Although space does not permit a detailed treatment of all the factors that are considered in the process of evaluating the validity of theories, it is nonetheless important to at least briefly describe some of the issues associated with how theories can be corroborated, if not proven. Namely, it is important to recognize that theories cannot be "logically verified," because false as well as true theoretical premises or assumptions can give rise to true deductions. Relatedly, unlike logical arguments, one cannot directly "prove" a theory to be true, and even though empirical confirmation of a theoretical proposition certainly provides "support" for a theory, it does not by itself directly increase the truth value of the theory (due to the logical fallacy known as *asserting the consequent*), because one cannot eliminate all possible alternative interpretations of the findings. Finally, in order to be of any scientific value, theories must be falsifiable (Popper, 1959), that is, the theoretical propositions must be made explicit enough that empirical tests could in principle disprove them. That said, theories in cognitive psychology are rarely disproved. Rather, those that are increasingly unsubstantiated by the data are gradually replaced by theories that not only do a better job of yielding predictions supported by evidence, but that are also able to account for a comparatively greater range of empirical phenomena while possessing fewer theoretical primitives or assumptions. In the next section, we illustrate how a few core principles and general empirical findings help to tie together the basic questions being addressed with the research described across the chapters of these volumes (including the first two volumes of the series), and how the answers to these questions set the stage for future basic and applied (e.g., design of interventions) research.

METHODOLOGICAL CONSIDERATIONS

At some level, all of the chapters in this volume and in the cognition literature more generally grapple with the issues of what to measure, when to measure it, how to measure it, who to measure, and of course why measure it at all. These issues are further complicated for developmental and educational research because the answer to many of these questions can change over time, either as a result of maturation or grade-related differences in mathematical (or whatever the topic of interest) content, and, in fact, it is even more complicated than this, as the rate of developmental or education-related change (e.g., due to curriculum differences across countries) can vary from one seemingly related topic to the next (e.g., Fennell et al., 2008; Siegler, 1983), adding the further issue of how often the domain of interest needs to be measured (Siegler & Crowley, 1992).

Why?

At one level, the question of why measure students' understanding of arithmetic, rational numbers, algebra, and so forth is straightforward: These are core topics in the standard mathematics curriculum throughout the developed and developing world (Fennell et al., 2008). The broader question concerns whether competence in these areas will influence the long-term wellbeing of individual students and ultimately the communities within which they reside (National Mathematics Advisory Panel, 2008). In multiple domains of life, these competencies matter greatly. For instance, adults who have difficulty solving word problems that require whole number arithmetic, fractions, simple algebra, and measurement have compromised employment prospects (Bynner, 1997; Parsons & Bynner, 1997; Rivera-Batiz, 1992), and they have difficulties in other areas of life that require quantitative knowledge (e.g., Reyna, Nelson, Han, & Dieckmann, 2009), independent of intelligence, reading ability, and demographic factors.

Whereas there is consensus regarding the importance of these numeracy-related competencies, there is debate regarding the importance of much of high school mathematics (Hacker, 2016). Hacker argued that standard high school algebra courses (and beyond, except for statistics) are an unnecessary burden for most students, even some students in traditional STEM fields (science, technology, engineering, and mathematics), such as computer science. For the sake of argument, let's assume that this position is correct; that most people only need numeracy-related competencies and some understanding of statistics. Beyond academic and career success, managing everyday life, including monetary transactions, finances, and health affairs requires basic fraction, algebra and probability knowledge (Reyna et al., 2009; Rivera-Batiz, 1992). The questions then become how do we instill these competencies in students and how do we ensure these competencies are maintained throughout adulthood?

The latter question has not been well addressed in the literature, but the study of Bahrick and Hall (1991) for the long-term retention of algebraic concepts and procedures is suggestive; "even in the absence of further rehearsal activities, individuals who take college-level mathematics courses at or above the level of calculus have minimal losses of high school algebra for half a century" (Bahrick & Hall, 1991, p. 20). In other words, lifelong maintenance of these competencies was found for people who practiced algebra in the context of more complex mathematics that involves considerable use of algebra. The implication is that the long-term retention of numeracy-related competencies may be facilitated if these competencies are practiced in the context of high school mathematics. Of course, all but a rudimentary understanding of statistics requires some basic algebra and geometry skills, and calculus for a deeper understanding of statistics.

What, When, How, and Who?

Fig. 1.1 organizes our discussion of the remaining issues, with a focus on themes that are common across the chapters of this volume. The "what" is represented by the mathematical content of the respective chapters, and is important for the reasons noted previously. These reasons however, do not capture the core contributions of a cognitive science analysis of mathematical cognition and learning. These contributions are, in part, represented by the prior knowledge and cognitive biases and abilities components of Fig. 1.1, and get at the "how" experiences influence mathematical cognition and learning. We'll call these the conceptual how to contrast them with the methodological how (below).

The importance of prior knowledge is well illustrated by McNeil et al.'s (this volume) demonstration that limited experiences in how arithmetic equations are solved (e.g., 3 + 7 = ?, and not 10 = 7 + ?) leads to a misunderstanding of mathematical equivalence, and by Van Hoof et al.'s (this volume) demonstration that children's understanding of natural numbers (i.e., "1, 2, 3, 4 …") leads to systematic conceptual misunderstandings and procedural errors when learning about rational numbers (Ni & Zhou, 2005). Thevenot (this volume) shows us that it's not simply prior mathematical knowledge that influences subsequent mathematical learning, but also experiences outside of the classroom; this is true for other areas of academic learning as well (Clement, 1982; Shtulman, 2006). Of course, prior knowledge, mathematical, or otherwise, often facilitates subsequent learning, as illustrated by Thevenot and other chapters in the volume (e.g., Jordan et al., this volume).

All of these effects, for better or worse, reflect a general cognitive bias; that is, if a conceptual or procedural solution to the issue at hand is not forthcoming, fall back on related existing knowledge (and hope for the best) (Siegler, 1983). One critical feature of this existing knowledge is that it can be used analogically to help understand new concepts or reason through novel problems (Holyoak, 2012; for discussion of other biases see Booth et al., this volume). One component of this process is to retrieve the most appropriate analogy from long-term memory and in this circumstance prior knowledge can lead one astray. As Thevenot (this volume) and others have described, the semantic relations between real-world objects influence, again for better or worse, how students set up the mathematical equations used to solve word problems (Bassok, Chase, & Martin, 1998; Martin & Bassok, 2005). Addition problems

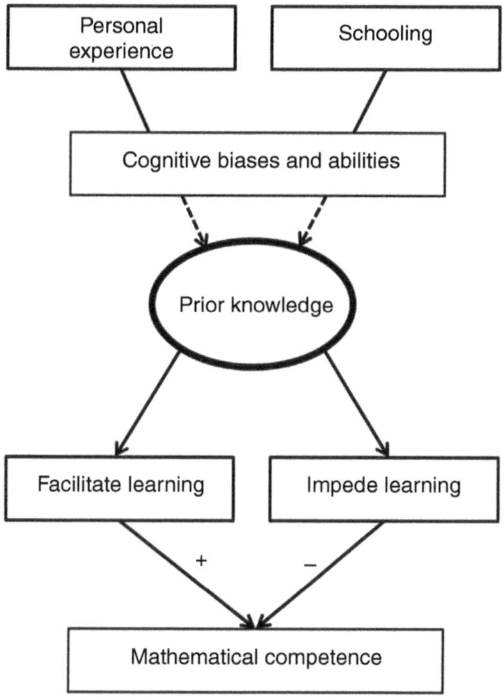

FIGURE 1.1 A common theme across the chapters of this volume and in cognitive science research more generally is that priori knowledge can facilitate or impede learning. Students' priori knowledge comes from personal out-of-school experiences and school-related experiences (including the design of textbooks) that foster domain specific expertise. These knowledge bases are filtered by cognitive biases (e.g., in how analogies are drawn between related concepts) and their use in learning and problem solving contexts is dependent on several specific cognitive abilities (e.g., working memory).

are typically composed of categorically similar items, such that translating statements containing sparrows and blue jays into an addition problem (to find the total number of birds) is easier than statements containing dissimilar objects, such as sparrows and trees. Similarly the translation of division problems is easier for functionally related objects, such as the number of eggs distributed across nests. Bassok et al. (1998) found that these types of problems are very common in mathematics textbooks and influenced students' construction and translation of word problems. The wording of problems can trigger real-world functional relations that sometimes facilitate and sometimes detract from mathematical problem solving (Thevenot, this volume).

There are also deep analogies within mathematics that are critical to students' comprehension of the field, but these connections are not always made, as illustrated by Robinson's (this volume) discussion of the lag between students' understanding of the inverse relation between addition and subtraction and the analogous relation between multiplication and division. Making the connection between these types of analogous relations does not always occur and in fact often does not occur without explicitly directing students' attention to the relation. It may seem odd that students readily draw (often times implicitly) on analogies based on everyday experiences during the solving of some types of mathematics problems, but often fail to do so for clear analogous relations within mathematics itself. At one level, the differences are likely related to level of mathematical expertise (Holyoak, 2012), but at a deeper level may reflect the difficulty of learning abstract, evolutionarily novel information (Geary & Berch, 2016). Learning in these domains is difficult because it is highly dependent on working memory and explicit problem solving (Geary, 2005; Sweller, 2015). This argument is in keeping with the conclusions of Lee, Ng, and Bull (this volume) and with considerable prior research on the relation between working memory, problem solving abilities, and mathematical achievement (Deary, Strand, Smith, & Fernandes, 2007; Friso-van den Bos, van der Ven, Kroesbergen, & van Luit, 2013; Geary, 2011), that is, the cognitive abilities component in Fig. 1.1, and with research on reasoning and problem solving in novel domains more generally (Stanovich & West, 2000).

There are several potentially important implications for the study of mathematical cognition and learning. The first addresses, in part, the "when" different types of prior knowledge are important; specifically, the level of mathematical expertise will influence when every day experiences and mathematics-related experiences will influence performance. During the early phases of learning, we would assume that the influence of prior knowledge largely comes from everyday experiences and that previously learned mathematical knowledge will become increasingly important over time. Our assumption here is in keeping with findings of two recent longitudinal studies documenting across-grade increases in the relative importance of domain-specific mathematical knowledge (vs. domain-general abilities) on subsequent mathematics achievement (Geary, Nicholas, Li, & Sun, in press; Lee & Bull, 2016). In this view, the search for potential detrimental effects of prior knowledge on mathematical learning and problem solving would focus on potential analogies in everyday experiences of naïve learners (Thevenot, this volume) and, for the more knowledgeable learner, inappropriate analogies drawn from well-learned prior mathematical knowledge (Van Hoof et al., this volume) or incorrect inferences based on prior inadequate mathematics instruction (McNeil et al., this volume). Our overall point is that cognitive science studies of mathematical learning can be improved with a more consistent consideration of students' prior knowledge, including the eventual development of measures of prior knowledge that can be used by teachers in classroom settings.

The second implication concerns the cognitive abilities that facilitate the development of mathematical expertise. As noted, learning in evolutionarily novel domains, including much of academic mathematics is predicted to be dependent on working memory capacity and competence at explicit problem solving. One component of the former is the ability to inhibit prior knowledge from influencing abstract problem solving (Stanovich & West, 2000). Although the inhibition component of working memory is not as consistently related to mathematics achievement as the updating component (see Bull & Lee, 2014), it has not been thoroughly examined with respect to the potential disrupting effects of everyday experiences and prior mathematical knowledge on error patterns during mathematical problem solving. Van Hoof et al.'s (this volume) dual process model is based on the same argument, although not from an evolutionary perspective, and their description of incongruency effects—where knowledge of whole number properties leads to incorrect inferences about rational number properties—is consistent with a role for inhibition as a moderator of the influence of prior knowledge on aspects of mathematical learning and problem solving. In other words, we should consider that different components of working memory may influence different aspect of mathematical cognition and learning (not just focus on performance on mathematics achievement tests) (Miyake et al., 2000).

The importance of working memory and explicit problem solving point to these as critical individual differences variables—part of the "who" (see Barnes, this volume; Mammarella et al., this volume)—and perhaps more critically as factors to be considered in the design of instructional materials and approaches (e.g., pacing of instruction), as noted by Rittle-Johnson et al. (this volume) and Booth et al. (this volume). The importance of these competencies is in keeping with Sweller's cognitive load theory (Sweller, 1988; Sweller & Cooper, 1985), his recent integration of this theory into an evolutionary perspective (Sweller, 2015), and with respect to children's education, the focus on the development of domain-specific expertise (Tricot & Sweller, 2013), mathematics in this case. Our point is that studies of students' mathematical cogition and learning would be more complete if they included measures of working memory and competence at explicit problem solving (e.g., fluid intelligence) either as covariates or potential moderators.

Psychologists and other cognitive scientists have at their disposal a rich array of techniques—the methodological "how"—with which to study mathematical cognition and learning. As shown in Fig. 1.2, the influence of prior knowledge on learning can be implicit (occurring without students' awareness of the influence) or explicit (making a direct link in working memory between prior knowledge and current learning or problem solving). If prior knowledge implicitly facilitates current learning, the solving of problems on newly learned materials will be faster (reaction times) and more accurate (error patterns). If prior knowledge implicitly impedes current learning, problem solving will be slower and more prone to errors. Any such effects will be exaggerated by priming prior knowledge; for instance, by asking students to solve problems based on this prior knowledge before assessing their new knowledge (e.g., McNeil et al, this volume; Thevenot, this volume).

As Van Hoof et al. (this volume) note, these common measures can also be used in very subtle ways. The natural number bias might not result in more

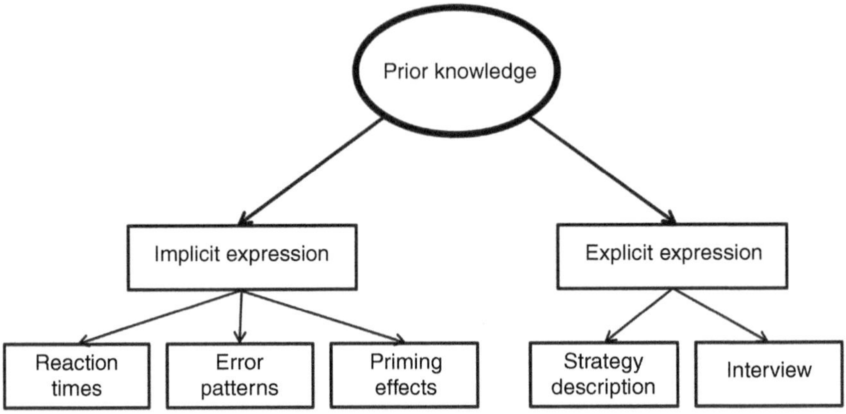

FIGURE 1.2 Students' prior knowledge can have implicit (i.e., outside of their awareness) and explicit (e.g., their beliefs about the topic) influences on learning and problem solving. These in turn are reflected in a variety of methods for studying cognition.

errors when solving superficially similar rational number problems, but reactions times are often longer. For instance, knowing that 4 > 3 leads to a common assumption that 1/4 > 1/3 in novices but not students who understand rational numbers. The combination of knowledge of natural numbers and knowledge of rational numbers leads to two potential answers, that 1/4 > 1/3 and 1/4 < 1/3. The conflict can trigger either engagement of the explicit problem solving system that adjudicates between these alternatives or inhibition of the answer based on natural number knowledge. The result is the correct response but with longer problem solving times. Tellingly, the natural number bias sometimes disappears for expert mathematicians, depending on how they solve the problems (Obersteiner, Hoof, Verschaffel, & Dooren, 2016; Obersteiner, Van Dooren, Van Hoof, & Verschaffel, 2013).

The overall pattern suggests that inappropriate reliance on prior knowledge can result in fast but incorrect problem solving during the initial phases of learning new material, at least for problems in which prior knowledge is inconsistent with current learning goals; the errors themselves provide clues as to what prior knowledge is being used as a fallback. Paradoxically, one early indication that students are learning the new material is they become slower, but they are also becoming more accurate at problem solving. Slower problem solving results from the competition between the new knowledge and inconsistent prior knowledge. As students gain expertise, their speed of problem solving increases, because they largely inhibit interference from prior knowledge, having integrated the new knowledge into a conceptual framework that is consistent with prior knowledge (e.g., representing fractions as magnitudes rather than a composite of two whole numbers) (Siegler et al., 2011), or they solve the problems in ways that do not elicit prior knowledge. Whatever the specifics, the combination of reaction times, along with error rates, and error patterns—potentially facilitated by use of primary experiments (e.g., DeWolf et al., this volume; McNeil et al., this volume)—provides considerable information on which prior knowledge and the extent to which this knowledge is influencing current mathematical learning and problem solving.

For some types of problem solving, students' explicit descriptions of how they solved the problems or their rationale for their answer can also provide insights into the influence of prior knowledge (Siegler & Shrager, 1984; Obersteiner et al., 2016; Van Hoof et al., this volume). This type of information is probably less sensitive to subtle influences of prior knowledge than reaction time and error pattern analyses but can be useful nevertheless, as long as there is convergent information (e.g., reaction times) consistent with the descriptions (Siegler, 1987). However, it is likely that these methods are useful for assessing only some types of student knowledge; specifically, knowledge that can be represented in working memory.

All of these methods tap the immediate or short-term influence of prior knowledge on mathematical learning and problem solving. These are often sufficient for the study of reasonably well-educated college students, but must be incorporated into broader designs to capture maturational or grade-related changes in these influences. These broader designs include cross-sectional, longitudinal, and cross-sequential (mix of cross-sectional and longitudinal designs) methods (Lee & Bull, 2016). Even within the context of these broader designs, the issue of the rate with which the knowledge of interest is changing needs to be considered. For conceptual or procedural knowledge that changes slowly, once a year assessments might be sufficient but if change is occurring more rapidly, more frequent measures will be needed. As Siegler and Crowley (1991) noted, even frequent "snapshots" might not be sufficient to capture the moment of conceptual insight or procedural discovery; this requires the embedding of microgenetic studies into the broader designs. The goal is to narrow the time-frame of change—but beware that this can vary considerably across contexts (e.g., nationally) for academic learning (e.g., Geary, Bow-Thomas, Liu, & Siegler, 1996)—such that the proximate factors, including prior knowledge, that promote or interfere with this change can be better identified and ultimately change with instructional interventions.

CHALLENGES FOR INSTRUCTION

As described in the Booth et al. chapter (this volume), several key principles that have emerged from cognitive science research over the last few decades could have important implications for the design and delivery of mathematics instruction. Booth and her colleagues review evidence from laboratory-based studies in addition to evidence from classroom-based studies for a number of powerful learning principles including scaffolding, worked examples, interleaved problem types, and analogical comparison. While the classroom-based body of evidence for these principles has been growing in recent years, there remains little evidence that such principles are being systematically and widely incorporated into mathematics instruction in the classroom via curriculum materials or textbooks. This is somewhat surprising given that (1) the empirical evidence in support is often fairly strong and convincing and (2) many of the principles would be quite simple to incorporate into instruction.

As an example, a recent national research and development effort was funded by the Institute of Education Sciences (IES) to revise and then evaluate in a large randomized trial the efficacy of *Connected Mathematics Project* (*CMP*) after incorporating principles of cognitive science. *CMP* was revised to explicitly include (1) graphics with verbal descriptions to promote the integration of concepts, (2) interleaved worked examples with new problem types, (3) spaced learning of critical content over time, and (4) focused feedback on quizzes and homework to promote learning. The revised *CMP* materials are being experimentally contrasted against the original *CMP* materials. The findings from this center are still being disseminated but preliminary results seem promising with the revised materials demonstrating improved learning in three of the four *CMP* units (Davenport, Kao, & Schneider, 2013).

There are recent examples where principles of cognitive science have been explicitly incorporated into mathematics interventions for students with learning disabilities. Fuchs, Malone, Schumacher, Namkung, and Wang (in press) developed one such intervention for a fourth grade fractions curriculum. The program builds on Siegler et al.'s (2011) *integrated theory of number development* to explicitly emphasize the magnitude representation of fractions and regularly provides students the opportunity to experience and compare rational numbers situated on a number line, just as whole numbers are commonly represented. This approach to fractions instruction has proven to be quite effective, with students consistently outperforming students in business-as-usual conditions. The authors conclude that when intervention is delivered in small groups and relies on systematic, explicit instructional principles and state-of-the-art thinking about the content area (i.e., fractions magnitude understanding), struggling students' performance improves on complex curricular content beyond what is achieved for matched control group students who receive school-based classroom and more traditional interventions.

This research by Fuchs and colleagues highlights the possibility for more explicit cognitive strategy use for students who struggle and may have, for example, poor language comprehension and/or limited working memory. For many of these students, typical classroom instruction is not sufficiently meeting their needs particularly in mathematics where the content sequentially builds and becomes increasingly layered (Fuchs et al., 2015). If students with limited working memory capacity are struggling to master arithmetic operations, such as single-digit addition and subtraction, when fractions or more complicated concepts are introduced, then mastery of such new content becomes that much more difficult. Thus, as mentioned earlier, a better understanding of students' cognitive profiles (e.g., working memory or attention) could help provide clues for more or less effective intervention approaches (see also Sweller, 1988). For example, Fuchs et al. (2014) demonstrated that students with low working memory capacity performed better on fraction number lines in a conceptual practice condition (e.g., self-explanation similar to "worked-examples") as compared to a fluency condition (e.g., speeded practice). Conversely, the fluency condition seemed to benefit students with higher working memory capacity. Focusing such research and development efforts on interventions for students with learning disabilities who often exhibit memory or executive function deficits may prove particularly fruitful given that many of the approaches described in this volume help to minimize the working memory or cognitive load demands of problem solving. These cognitive principles, perhaps when coupled with more explicit forms of instruction, could be quite powerful for improving the education outcomes of struggling learners. More research is needed with these populations to better understand when and to what extent students may benefit from more explicit use of these potentially powerful cognitive science principles.

It is worth considering the role that such cognitive research could play in influencing teacher preparation and practice. For example, a principle, such as interleaved problem types demonstrates fairly robust evidence in laboratory and classroom based studies showing that mathematics learning improves as students develop stronger associations between a given problem type and its corresponding strategy (Rohrer, Dedrick, & Stershic, 2015; Rohrer, Dedrick, & Burgess, 2014). Interleaved practice is a relatively simple strategy that could be quite easily implemented with little training or support in nearly every mathematics classroom in the country. However, it is not clear if and how such a principle is being included as part of preservice teacher preparation or in-service teacher professional development. As principles of cognitive science increasingly inform education research, more research is necessary to better understand how such new knowledge is being utilized both by teacher preparation programs and by school districts providing learning opportunities for in-service teachers.

CONCLUSIONS AND FUTURE DIRECTIONS

The United States' lag behind other industrial countries in mathematical achievement portends significant socioeconomic (National Mathematics Advisory Panel, 2008) and quality of life (Department for Education and Skills, 2003) challenges for current and rising generations. Identifying impediments to learning as well as facilitators of successful acquisition, retention, and transfer of mathematical competencies are critical goals. Better understanding is needed in areas such as instructional sequencing and timing so as to minimize prior learning interference, as well as means to enable conceptual shifts (e.g., from natural to rational numbers), and effective or preferred symbolic representations to facilitate initial learning (e.g., 0.25 versus ¼ versus "one quarter"), to name just a few. Cognitive psychology, its theories, methods, and findings, as articulated in the following chapters, has provided vital keys to effective learning and intervention for those who struggle to learn quantitative skills.

Though cognitive psychology and cognitive science more generally has perhaps loosened the Gordian knot, it is not yet fully unraveled. Many questions remain, and the contributions of other sciences to our understanding of how we learn mathematics and develop quantitative reasoning must be incorporated into innovative new cognitive science models and new tools developed to enable these approaches (see chapters in Berch, Geary, & Mann Koepke, 2016). Ultimately, this basic science and the theories that derive from it must successfully make their way into learning environments and effectively enable learners at all stages of arithmetic and complex mathematics development. Working in partnership, U.S. federal funding agencies, such as the *Eunice Kennedy Shriver* National Institute of Child Health and Human Development (NICHD), the U.S. Department of Education's Institute of Education Sciences (IES), and the National Science Foundation (NSF) remain committed to advancing this research in order to promote and improve mathematics learning and its concomitant prosperity and wellness across our entire population.

REFERENCES

Atkinson, R. C., & Shiffrin, R. M. (1968). Human memory: A proposed system and its control processes. In K. W. Spence, & J. T. Spence (Eds.), *The psychology of learning and motivation* (pp. 89–195). (Vol. 2). New York: Academic Press.

Baddeley, A. D. (1986). *Working Memory*. Oxford: Clarendon Press.

Bahrick, H. P., & Hall, L. K. (1991). Lifetime maintenance of high school mathematics content. *Journal of Experimental Psychology: General, 120*, 20–33.

Bassok, M., Chase, V. M., & Martin, S. A. (1998). Adding apples and oranges: alignment of semantic and formal knowledge. *Cognitive Psychology, 35*, 99–134.

Berch, D. B., Geary, D. C., & Mann Koepke, K. (Eds.). (2016). *Development of mathematical cognition: Neural substrates and genetic influences*. San Diego, CA: Elsevier/Academic Press.

Broadbent, D. E. (1971). *Decision and stress*. New York: Academic Press.

Bull, R., & Lee, K. (2014). Executive functioning and mathematics achievement. *Child Development Perspectives, 8*, 36–41.

Bynner, J. (1997). Basic skills in adolescents' occupational preparation. *Career Development Quarterly, 45*, 305–321.

Case, R. (1996). Introduction: reconceptualizing the nature of children's conceptual structures and their development in middle childhood. *Monographs of the Society for Research in Child Development, 61*(1–2), 1–26.

Clement, J. (1982). Students' preconceptions in introductory mechanics. *American Journal of Physics, 50*(66–71).

Christou, K. P., & Vosniadou, S. (2005). How students interpret literal symbols in algebra: A conceptual change approach. In B. G. Bara, L. Barsalou, & M. Bucciarelli (Eds.), *Proceedings of the XXVII Annual Conference of the Cognitive Science Society* (pp. 453–458). Stresa, Italy: Erlbaum.

Christou, K. P., & Vosniadou, S. (2012). What kinds of numbers do students assign to literal symbols? Aspects of the transition from arithmetic to algebra. *Mathematical Thinking and Learning, 14*, 1–27.

Davenport, J.L., Kao, Y.S., & Schneider, S.A. (2013). *Integrating cognitive principles to redesign a middle school math curriculum*. Available from: http://iesmathcenter.org/research/2013_Davenport.pdf

Deary, I. J., Strand, S., Smith, P., & Fernandes, C. (2007). Intelligence and educational achievement. *Intelligence*, *35*, 13–21.

Dehaene, S., Izard, V., Pica, P., & Spelke, E. (2006). Core knowledge of geometry in an Amazonian indigene group. *Science*, *311*, 381–384.

Department for Education and Skills (DfES). (2003). *Skills for life the national strategy for improving adult literacy and numeracy skills: focus on delivery to 2007*. London: DfES.

Fennell, F. S., Faulkner, L. R., Ma, L., Schmid, W., Stotsky, S., Wu, H. H., & Flawn, T. (2008). Report of the task group on conceptual knowledge and skills. In: *National Mathematics Advisory Panel, Reports of the task groups and subcommittees* (pp. 3-i–3-65). Washington, DC: United States Department of Education.

Festinger, L. (1957). *A theory of cognitive dissonance*. Stanford, CA: Stanford University Press.

Friso-van den Bos, I., van der Ven, S. H., Kroesbergen, E. H., & van Luit, J. E. (2013). Working memory and mathematics in primary school children: A meta-analysis. *Educational Research Review*, *10*, 29–44.

Fuchs, L.S., Malone, A.S., Schumacher, R.F., Namkung, J., & Wang, A. (in press). Fraction intervention for students with mathematics difficulties: lessons learned from five randomized control trials. *Journal of Learning Disabilities*.

Fuchs, L. S., Fuchs, D., Compton, D. L., Wehby, J., Schumacher, R. F., Gersten, R., & Jordan, N. (2015). Inclusion versus specialized intervention for very-low-performing students: What does access mean in an era of academic challenge? *Exceptional Children*, *81*, 134–157.

Fuchs, L. S., Schumacher, R. F., Sterba, S. K., Long, J., Namkung, J., Malone, A., Hamlett, C. L., Jordan, N. C., Gersten, R., Siegler, R. S., & Changas, P. (2014). Does working memory moderate the effects of fraction intervention? An aptitude-treatment interaction. *Journal of Educational Psychology*, *106*, 499–514.

Geary, D. C. (2005). *The origin of mind: Evolution of brain, cognition, and general intelligence*. Washington, DC: American Psychological Association.

Geary, D. (2008). An evolutionarily informed education science. *Educational Psychologist*, *43*, 179–195.

Geary, D. C. (2011). Cognitive predictors of individual differences in achievement growth in mathematics: a five year longitudinal study. *Developmental Psychology*, *47*, 1539–1552.

Geary, D. C., & Berch, D. B. (2016). Evolution and children's cognitive and academic development. In D. C. Geary, & D. B. Berch (Eds.), *Evolutionary perspectives on child development and education* (pp. 217–249). New York: Springer.

Geary, D. C., Bow-Thomas, C. C., Liu, F., & Siegler, R. S. (1996). Development of arithmetical competencies in Chinese and American children: Influence of age, language, and schooling. *Child Development*, *67*, 2022–2044.

Geary, D.C., Nicholas, A., Li, Y., & Sun, J. (in press). Developmental change in the influence of domain-general abilities and domain-specific knowledge on mathematics achievement: An eight-year longitudinal study. *Journal of Educational Psychology*.

Gentner, D. (1983). Structure-mapping: A theoretical framework for analogy. *Cognitive Science*, *7*, 155–170.

Gick, M. L., & Holyoak, K. J. (1983). Schema induction and analogical transfer. *Cognitive Psychology*, *15*, 1–38.

Hacker, A. (2016). *The math myth: And other STEM delusions*. New York: New Press.

Holyoak, K. J. (2012). Analogy and relational reasoning. In K. J. Holyoak, & R. G. Morrison (Eds.), *The Oxford handbook of thinking and reasoning* (pp. 234–259). New York: Oxford University Press.

James, W. (1890). *Principles of psychology*. New York: Holt.

Johnson-Laird, P. N. (1983). *Mental models: Towards a cognitive science of language, inference, and consciousness*. Cambridge, MA: Harvard University Press.

Kintsch, W., & Greeno, J. G. (1985). Understanding and solving word arithmetic problems. *Psychological Review*, *92*, 109–129.

Lee, K., & Bull, R. (2016). Developmental changes in working memory, updating, and math achievement. *Journal of Educational Psychology*, *108*, 869–882.

Martin, S. A., & Bassok, M. (2005). Effects of semantic cues on mathematical modeling: Evidence from word-problem solving and equation construction tasks. *Memory & Cognition*, *33*, 471–478.

Marx, M. H. (1963). The general nature of theory construction. In M. H. Marx (Ed.), *Theories in contemporary psychology* (pp. 4–46). New York: Macmillan.

McNeil, N. M., & Alibali, M. W. (2005). Why won't you change your mind? Knowledge of operational patterns hinders learning and performance on equations. *Child Development*, *76*, 883–899.

Miyake, A., Friedman, N. P., Emerson, M. J., Witzki, A. H., Howerter, A., & Wager, T. D. (2000). The unity and diversity of executive functions and their contributions to complex "frontal lobe" tasks: A latent variable analysis. *Cognitive Psychology*, *41*, 49–100.

Murdock, B. B. (1974). *Human memory: Theory and data*. Hillsdale, NJ: Erlbaum, 1974.

National Mathematics Advisory Panel (2008). *Foundations for success: Final report of the national mathematics advisory panel*. Washington, DC: United States Department of Education.

Ni, Y., & Zhou, Y. D. (2005). Teaching and learning fraction and rational numbers: The origins and implications of whole number bias. *Educational Psychologist*, *40*, 27–52.

Obersteiner, A., Van Dooren, W., Van Hoof, J., & Verschaffel, L. (2013). The natural number bias and magnitude representation in fraction comparison by expert mathematicians. *Learning and Instruction*, *28*, 64–72.

Obersteiner, A., Hoof, J. V., Verschaffel, L., & Dooren, W. V. (2016). Who can escape the natural number bias in rational number tasks? A study involving students and experts. *British Journal of Psychology*.

Parsons, S., & Bynner, J. (1997). Numeracy and employment. *Education and Training*, *39*, 43–51.

Piaget, J. (1952). *The child's conception of number*. London, UK: Routledge & Kegan Paul.

Piaget, J. (1980). *Experiments in contradictions*. Chicago: University of Chicago Press.

Piaget, J., & Inhelder, B. (1967). *The child's conception of space*. New York, NY: W.W. Norton.

Piaget, J., Inhelder, B., & Szeminska, A. (1960). *The child's conception of geometry*. London, UK: Routledge.

Piaget, J., & Szeminska, A. (1995). The child's conception of number (Gattegno, C. & Hadgson, F. M. Trans.). In H. E. Gruber & J. J. Voneche (Eds.), *The essential Piaget* (pp. 298–341). Northvale, NJ: Jason Aronson. (Original work published 1941).

Popper, K. (1959). *The logic of scientific discovery*. New York, NY: Basic Books.

Posner, M. I., & Warren, R. E. (1972). Traces, concepts, and conscious constructions. In A. W. Melton, & E. Martin (Eds.), *Coding processes in human memory*. Washington, DC: Winston.

Reyna, V. F., Nelson, W. L., Han, P. K., & Dieckmann, N. F. (2009). How numeracy influences risk comprehension and medical decision making. *Psychological Bulletin*, *135*, 943–973.

Rivera-Batiz, F. (1992). Quantitative literacy and the likelihood of employment among young adults in the United States. *Journal of Human Resources*, *27*, 313–328.

Roediger, H. L. (1979). Implicit and explicit memory models. *Bulletin of the Psychonomic Society*, *13*, 339–342.

Rohrer, D., Dedrick, R. F., & Burgess, K. (2014). The benefit of interleaved mathematics practice is not limited to superficially similar kinds of problems. *Psychonomic Bulletin and Review, 21*, 1323–1330.

Rohrer, D., Dedrick, R. F., & Stershic, S. (2015). Interleaved practice improves mathematics learning. *Journal of Educational Psychology, 107*, 900–908.

Shtulman, A. (2006). Qualitative differences between naïve and scientific theories of evolution. *Cognitive Psychology, 52*, 170–194.

Siegler, R. S. (1983). Five generalizations about cognitive development. *American Psychologist, 38*, 263.

Siegler, R. S. (1987). The perils of averaging data over strategies: an example from children's addition. *Journal of Experimental Psychology: General, 116*, 250–264.

Siegler, R. S., & Crowley, K. (1991). The microgenetic method: A direct means for studying cognitive development. *American Psychologist, 46*, 606–620.

Siegler, R. S., & Shrager, J. (1984). Strategy choice in addition and subtraction: how do children know what to do? In C. Sophian (Ed.), *Origins of cognitive skills* (pp. 229–293). Hillsdale, NJ: Erlbaum.

Siegler, R. S., Thompson, C., & Schneider, M. (2011). An integrated theory of whole number and fractions development. *Cognitive Psychology, 62*, 273–296.

Stanovich, K. E., & West, R. F. (2000). Individual differences in reasoning: Implications for the rationality debate? *Behavioral and Brain Sciences, 23*, 645–726.

Stafylidou, S., & Vosniadou, S. (2004). The development of student's understanding of the numerical value of fractions. *Learning and Instruction, 14*, 508–518.

Sweller, J. (1988). Cognitive load during problem solving: Effects on learning. *Cognitive Science, 12*, 257–285.

Sweller, J. (2015). In academe, what is learned, and how is it learned? *Current Directions in Psychological Science, 24*, 190–194.

Sweller, J., & Cooper, G. A. (1985). The use of worked examples as a substitute for problem solving in learning algebra. *Cognition and Instruction, 2*, 59–89.

Tricot, A., & Sweller, J. (2013). Domain-specific knowledge and why teaching generic skills does not work. *Educational Psychology Review, 26*, 265–283.

Vamvakoussi, X., & Vosniadou, S. (2010). Understanding the structure of the set of rational numbers: A conceptual change approach. *Learning and Instruction, 14*, 453–467.

Vamvakoussi, X., & Vosniadou, S. (2012). Bridging the gap between the dense and the discrete: The number line and the "rubber line" bridging analogy. *Mathematical Thinking and Learning, 14*, 265–284.

Vamvakoussi, X., Vosniadou, S., & Van Dooren, W. (2013). The framework theory approach applied to mathematics learning. In S. Vosniadou (Ed.), *International handbook of research on conceptual change* (2nd ed., pp. 305–321). Routledge: New York, NY.

Van Hiele, P. M. (1986). *Structure and insight: A theory of mathematics education*. Orlando, FL: Academic Press.

Vosniadou, S. (1994). Capturing and modelling the process of conceptual change. *Learning and Instruction, 4*, 45–69.

Vosniadou, S. (2003). Exploring the relationships between conceptual change and intentional learning. In G. M. Sinatra, & P. R. Pintrich (Eds.), *Intentional conceptual change* (pp. 377–406). London: Lawrence Erlbaum Associates.

Vosniadou, S., & Vamvakoussi, X. (2006). Examining mathematics learning from a conceptual change point of view: Implications for the design of learning environments. In L. Verschaffel, F. Dochy, M. Boekaerts, & S. Vosniadou (Eds.), *Instructional psychology: past, present and future trends. Sixteen essays in honour of Erik De Corte* (pp. 55–70). Oxford, UK: Elsevier Science Ltd.

Vosniadou, S., Vamvakoussi, X., & Skopeliti, I. (2008). The framework theory approach to conceptual change. In S. Vosniadou (Ed.), *International handbook of research on conceptual change* (pp. 3–34). Mahwah, NJ: Erlbaum.

Vosniadou, S., & Verschaffel, L. (2004). Extending the conceptual change approach to mathematics learning and teaching. *Learning and Instruction*, *14*, 445–451.

Vygotsky, L. S. (1978). *Mind in society: The development of higher psychological processes.* Cambridge, MA: Harvard University Press.

Part I

Complex Arithmetic Processing

Chapter 2

The Understanding of Additive and Multiplicative Arithmetic Concepts

Katherine M. Robinson
Department of Psychology, Campion College, University of Regina, Regina, SK, Canada

INTRODUCTION

In a report requested by the U.S. National Research Council on the state of research and knowledge on children's mathematics learning, Kilpatrick, Swafford, and Findell (2001) concluded that: "The more mathematical concepts they understand, the more sensible mathematics becomes" (p. 131). According to Kilpatrick et al., conceptual understanding helps organize mathematical information into a coherent whole, helps relate new knowledge to old, enhances retention of mathematical knowledge and information, and therefore provides students with the ability to connect concepts and procedures. Conceptual knowledge can also facilitate learning because it enables children to identify similarities across problem solving situations even when the situations appear to be, at least on the surface, unrelated. Children with good conceptual knowledge have deeply organized their knowledge into a strong network of facts and principles (Kilpatrick et al., 2001).

One of the first studies in psychology to examine children's understanding of an arithmetic concept was reported by Starkey and Gelman (1982). Starkey and Gelman based their study on Piaget's (1952) tenet that children can only truly understand addition and subtraction if they understand that the two operations are inversely related to one another. Piaget was building on his idea of how children acquire conservation—that a property or object remains fundamentally unchanged despite some irrelevant transformation. In the classic conservation of liquid task, Piaget showed children two short and wide glasses both filled with the same amount of liquid and then poured the contents of one of them into a tall narrow glass. Piaget noted that a child who understood conservation could explain why the amount of liquid in both final containers—one short and

Acquisition of Complex Arithmetic Skills and Higher-Order Mathematics Concepts
http://dx.doi.org/10.1016/B978-0-12-805086-6.00002-3

wide and the other tall and narrow—was the same, even though the glasses containing the same amounts of liquid were themselves different, by using the inversion explanation. If the liquid in the new tall and thin container was poured back into its original short and wide container then this would be the reverse or inverse of the original action and the amount of liquid would be demonstrably the same. Piaget applied this notion of conservation or quantitative identity and how an amount stayed unchanged despite an irrelevant physical transformation to several different types of tasks, such as conservation of mass and conservation of length. He also extended it into his examination of children's concepts about number. His work about how children understand mathematics is considered integral for educators and researchers concerned with how children's performance in mathematics is hindered when an understanding of mathematics is not an integral part of their learning (Fuson, 2009).

Using Piaget's ideas, Starkey and Gelman (1982) tested whether children understood the concept of inversion as it pertained to addition and subtraction by asking 3-, 4-, and 5-year-olds to solve problems, such as 1 + 2 – 2 using concrete objects. If children understood that addition and subtraction are inversely related then they would not need to perform any calculations to arrive at the correct answer of 1 as the same number was both added and subtracted. Starkey and Gelman found children's accuracy on these problems was moderate to high across development and hypothesized that solving the problems accurately could mean that a child understood the concept of inversion as it applies to the operations of addition and subtraction. Although not without some methodological issues, this study has served as an important cornerstone for subsequent research on children's understanding of arithmetic concepts.

WHAT IS CONCEPTUAL KNOWLEDGE OF ARITHMETIC?

Over the years, there have been a number of definitions proposed for conceptual knowledge of arithmetic and therein lies one of the difficulties of conducting research in this area (Crooks & Alibali, 2014). A second difficulty is finding measures of knowledge—a point that will be discussed later. With respect to the first point, the issue is that if there is no consensus about the definition of conceptual knowledge, then how can studies of conceptual knowledge from different labs using different measures or tasks be compared?

A number of definitions have been proposed over the last 30 years or so. Hiebert and Lefevre (1986) defined conceptual knowledge as an interconnected web of knowledge that includes how information within a domain, such as mathematics is related. Kilpatrick et al. (2001) proposed that conceptual knowledge is the understanding of mathematical concepts, operations, and relations. Similarly, Baroody and Ginsburg (1986) considered conceptual knowledge to be implicit or explicit understanding of concepts or principles. This definition has been used by a large number of researchers across a wide variety of mathematical concepts (e.g., Rittle-Johnson & Alibali, 1999). Bisanz and LeFevre

(1990) further expanded on Baroody and Ginsburg's definition to propose that conceptual knowledge of arithmetic can be thought of as "the principles that reflect the underlying structure of mathematics and that can be inferred from the selective use of effective procedures under conditions where those principles apply" (p. 216).

Although still a matter of debate regarding which definition(s) of conceptual knowledge should be used in guiding research (Crooks & Alibali, 2014), a matter that will probably not be resolved in the near future given the complexities and variety of definitions and measures of conceptual knowledge. The research discussed in this chapter is based on Bisanz and LeFevre's (1990) definition of conceptual knowledge, and includes two of the three most common definitions identified by Crooks and Alibali (2014).

THE IMPORTANCE OF CONCEPTUAL KNOWLEDGE

Despite the lack of consensus on what is meant by conceptual knowledge, there does appear to be strong agreement on why parents, teachers, and researchers should care about children's understanding of arithmetic concepts. The National Mathematics Advisory Panel (2008) proposed that "conceptual understanding is critical for children's ability to identify and correct errors, for appropriately transferring algorithms to solve novel problems, and for understanding novel problems in general (pp.4–14)." Conceptual knowledge may also be an important basis for how children discover new problem solving procedures (Baroody, 2003; Rittle-Johnson, Schneider, & Star, 2015). Knowledge of concepts can provide shortcuts during problem solving procedures (Jordan, Hanich, & Uberti, 2003) and therefore result in faster problem solving and more accurate answers (Lai, Baroody, & Johnson, 2008).

Even though there is an almost universal acknowledgement about how important conceptual knowledge of mathematics is—not only for current but future mathematics learning—research on this knowledge has lagged far behind research on the other two types of knowledge (procedural and factual). Bisanz (1999) noted that the role of conceptual understanding of how children's knowledge and skills in mathematics develop was underresearched. Interestingly, although the development of children's conceptual knowledge is attracting more attention from researchers (Crooks & Alibali, 2014), much remains to be learned about this development (Gilmore & Papadatou-Pastou, 2009).

A BRIEF HISTORY OF RESEARCH ON CONCEPTUAL KNOWLEDGE

Research on conceptual knowledge has been hindered not only by definitional ambiguity, but also by how to best measure this knowledge (Prather & Alibali, 2009). Most researchers have concluded conceptual knowledge is best measured indirectly (Bisanz & LeFevre, 1990; Canobi, 2009; Carpenter, 1986;

Dixon & Boncoddo, 2009). The most commonly used method is to present children with arithmetic problems that are most easily and quickly solved if children have knowledge of the underlying concepts, principles, or relations. For example, if children understand that addition and subtraction are inversely related operations, even when presented with a problem, such as 354297 + 8638298 − 8638298, they should be able to quickly and accurately solve the problem by stating the first number. This approach is typically called the inversion shortcut (Bisanz & LeFevre, 1990). If, however, they do not understand that addition and subtraction are inversely related and they do not have a calculator or even pencil and paper handy, this problem will take most children a long time to solve and they are likely to make a calculation error.

Simply asking children what they know, for example, about the relation between addition and subtraction often does not yield information relating to children's knowledge of the inverse relation between addition and subtraction. Instead, researchers have used novel tasks, such as the inversion problem (Starkey & Gelman, 1982). When students encounter new or novel problems, they can make use of their conceptual knowledge to generate novel problem solving procedures (Schneider, Rittle-Johnson, & Star, 2011). The generation of these procedures suggests that the students at least implicitly understand the associated concepts. If children spontaneously report using conceptual knowledge to solve problems then they are demonstrating explicit knowledge which is typically considered a more stringent measure of their conceptual understanding (Canobi, 2009) as compared to, when they are taught conceptually-based problem solving procedures (Carpenter, 1986).

So, when measuring conceptual knowledge through the use of problem solving procedures—a notion integral in Bisanz and LeFevre's (1990) definition of conceptual knowledge, the use of novel problems is important. Novel problems mean that children must spontaneously generate a new problem solving procedure or transfer a known procedure from a conceptually similar but superficially different problem. In this way, there is no possibility that children are using the rote application of a previously learned procedure. Application of such a rotely learned procedure would mean that children are not required to understand the concepts or principles being assessed in order to solve the problem successfully. Moreover, the necessity of understanding a specific concept or principle in order to generate a problem solving procedure yields important information about the strategies that children can spontaneously use (Dixon & Moore, 1996) so it helps researchers and educators to not only know more about children's conceptual knowledge of arithmetic and, consequently also provides insights into children's procedural knowledge.

It so happens that, for most children, the inversion problems originally formulated by Starkey and Gelman (1982) are novel to them. In school, particularly in the early years, they spend most of their time dealing only with two-term problems (e.g., 3 × 4) or three-term problems involving the same operations (e.g., 2 + 5 + 13). Three-term (or more) problems only become more common in the late middle school years when children start learning algebra but in those problems, children are more focused on solving for the unknown. That said, researchers have proposed that understanding of the inverse relation between not only addition and subtraction but also between multiplication and division is integral for learning algebra (Nunes et al., 2008). In this way, inversion problems (e.g., 2 + 23 – 23) are ideal for assessing children's understanding of the inverse relation between addition and subtraction and multiplication and division (Gilmore & Papadatou-Pastou, 2009). Bisanz and LeFevre (1990) played a key role in taking Starkey and Gelman's inversion problems and developing a design to effectively assess conceptual knowledge.

Bisanz and LeFevre (1990) not only assessed children's accuracy on inversion problems (e.g., 6 + 3 – 3 = __), they also asked children to provide a verbal report of their problem solving procedure. In this way, Bisanz and LeFevre could differentiate between children who were calculating the answer by adding the first two numbers (e.g., 6 + 3) and then subtracting the third number (e.g., 9 – 3) from those children who were applying their understanding of the inversion concept. That is, the children reported that they did not add or subtract because adding and subtracting 3 left the first number, 6, unaffected—they applied the conceptually-based inversion shortcut. Additionally, Bisanz and LeFevre included a second type of problem, which they called the *standard problem*, as a further check for whether children applied their conceptual knowledge of the inverse relation between addition and subtraction. While inversion problems are of the form $a + b - b$, standard problems are of the form $a + b - c$ (e.g., 6 + 5 – 2). The use of standard problems cleverly served two purposes. First, on standard problems, unlike inversion problems, calculation is required. Therefore, slower problem solving and more errors should occur and children should verbally report using a calculation strategy on standard problems than when using the inversion shortcut on inversion problems.

Second, a problem size effect can also be assessed with standard problems. Ashcraft (1982) first investigated the problem size effect on two-term problems and found that larger numbers in a problem were associated with more errors and slower problem solving. A problem, such as 2 + 4 would tend to be solved accurately, and relatively quickly, even by young children, compared to a problem, such as 265 + 382 or even 8 + 9. By manipulating the size of the numbers included in both the inversion and standard problems, Bisanz and LeFevre posited that the problem size effect would be found for standard problems that required calculations but would not be found for inversion problems if children were solving these based on their conceptual knowledge. This is the pattern they found for 6-, 9-, and 11-year-olds as well as adults. On inversion problems, children's accuracy did not vary with the size of the numbers in the problem, whereas on standard problems, as the size of the numbers in the problem increased, so did the errors. With their innovative inclusion of the standard problem and assessment of the problem size effect, Bisanz and LeFevre laid out the methodological groundwork for all subsequent work on the inversion concept.

Overall, the bulk of research on conceptual knowledge has tended to focus on a fairly restricted set of concepts. As Crooks and Alibali (2014) note, equivalence and inversion are two of the most thoroughly researched arithmetic concepts. Equivalence is the focus of the chapter by McNeil and coworkers in Chapter 8 so it will not be considered any further here. Other concepts, such as commutativity (e.g., Baroody, Ginsburg, & Waxman, 1983; Canobi, 2005; Cowan & Renton, 1996), that is if $a + b = c$ then $b + c = a$, have been investigated, but as they have not received as much research attention it is more difficult to draw strong conclusions from them compared to the concepts of inversion and equivalence (Crooks & Alibali, 2014; Gilmore & Papadatou-Pastou, 2009; Prather & Alibali, 2009). Also, concepts, such as commutativity are usually explicitly taught to children (Canobi & Bethune, 2008) so, unlike novel problems, such as inversion and equivalence problems, it is not clear whether children are applying their conceptual knowledge when solving these problems or applying a procedure that they were taught and the conceptual basis of which they may not understand (Cowan, 2003).

THE IMPORTANCE OF MULTIPLICATIVE CONCEPTS AND THE STATE OF CURRENT RESEARCH

What is notable about most of the research on arithmetic concepts is that almost all researchers have focused exclusively on additive concepts, that is, concepts involving the operations of addition and/or subtraction. Research on children's understanding of multiplicative concepts, that is, concepts involving the operations of multiplication and/or division, is relatively sparse, despite its importance for more complex and advanced mathematical skills (Cowan & Renton, 1996; NMAP, 2008; Nunes et al., 2009). This is potentially problematic for a number of reasons. First, multiplication and division are typically considered to be more difficult operations for children to understand and master, particularly division (Dixon, Deets, & Bangert, 2001). As a result, concepts about the relationship between multiplication and division may also be more difficult than concepts about the relationship between addition and subtraction (Fuson, 1988). This leads to the second reason; specifically, just because a child understands a concept on an additive problem (e.g., $2 + 3 + 8 = 2 + 11$), it does not mean that they will understand the same concept on a multiplicative problem (e.g., $2 \times 3 \times 8 = 2 \times 24$) (Nunes, Bryant, Hallett, Bell, & Evans, 2009). Third, there is evidence that children may develop an implicit understanding of the inverse relation between addition and subtraction (Klein & Bisanz, 2000), but coming to understand the inverse relation between multiplication and division is likely to require formal instruction.

ADDITIVE VERSUS MULTIPLICATIVE CONCEPTS

The two arithmetic concepts of inversion and associativity have been investigated using both additive and multiplicative problems. Both concepts involve understanding the relations among arithmetic operations—an understanding that is critical to the development of children's mathematical skills (Bisanz, Watchorn, Piatt, & Sherman, 2009; National Council of Teachers of Mathematics, 2000; National Council of Teachers in Mathematics, 2014). As described earlier, the use of these conceptually-based problem solving procedures or shortcuts has long been used as a measure of conceptual knowledge by various researchers (e.g., Canobi & Bethune, 2008; Gilmore & Bryant, 2006; Lai et al., 2008; Rasmussen, Ho, & Bisanz, 2003; Robinson & Dubé, 2012; Siegler & Stern, 1998; Vilette, 2002). Conceptually-based shortcut use can be determined through accuracy measures, solution times, and/or verbal reports of problem solving procedures, although using at least two of the three measures is recommended in order to obtain corroborating evidence. For example, Robinson and Dubé (2009a) asked 2nd–4th graders to report their problem solving procedures on inversion and associativity problems and also collected accuracy and solution time data. Students who reported using conceptually-based shortcuts made fewer errors and had shorter solution times than students who reported using a left-to-right problem solving procedure indicating that verbal reports were veridical.

THE INVERSION CONCEPT

The inversion concept has been studied extensively from the preschool years to adulthood using addition and subtraction inversion problems of the form $a + b - b$ (Bisanz & LeFevre, 1990; Robinson & Ninowski, 2003; Sherman & Bisanz, 2007). Multiplication and division inversion problems of the form $d \times e \div e$ (e.g., $4 \times 7 \div 7$) have also been investigated with a more restricted developmental range of late middle childhood to adulthood (Dubé & Robinson, 2010a; Robinson, Ninowski, & Gray, 2006). Younger children are generally unsuccessful on these types of problems, which is not surprising given that, for most children, multiplication is first formally introduced in the 3rd grade followed by division in the 4th grade. Grade 4 and 5 children's knowledge of division facts is much weaker than it is for multiplication (Robinson et al., 2006; Siegler, 1988). It is only in the 5th grade that most children begin to rely on their knowledge of multiplication facts to solve associated division problems (Robinson et al., 2006).

THE ASSOCIATIVITY CONCEPT

The associativity concept refers to the idea that numbers can be, for example, added in any order and the result will be the same (e.g., $3 + 4 + 5$ can be solved correctly by adding 3 and 4 and then adding 5 or by adding 4 and 5 together and then adding 3). Most of the research on associativity has focused on one operation at a time and, typically, the operation of addition or subtraction (e.g., Shumway, 1974), but associativity can also be applied to the standard problems devised by Bisanz and LeFevre (1990). These standard problems can also be used to assess conceptual knowledge so are often called associativity problems

(Klein & Bisanz, 2000). Associativity problems are of the form $a + b - c$ and $d \times e \div f$ (e.g., $4 + 9 - 3$ and $4 \times 9 \div 3$) and can often be more quickly and accurately solved if the second and third numbers are dealt with first (Klein & Bisanz, 2000; Robinson & Ninowski, 2003) (i.e., $9 - 3 = 6 + 4$ and $9 \div 3 = 3 \times 4$). The advantage of using the associativity shortcut becomes obvious on problems, such as $453 + 987 - 972$ or $23 \times 39 \div 13$. If these problems have to be solved using mental calculation, it is much faster and would result in fewer errors if 972 was subtracted from 987 before the addition of 453 and if 39 was divided by 13 before multiplying by 23.

There are two problem characteristics that can discourage the application of the associativity concept via the associativity shortcut. First, unlike inversion problems there is no clear visual pattern (i.e., $b - b$ or $e \div e$) that occurs in these associativity problems. Thus, applying the associativity concept requires more reflection on the part of the child in order to notice that there is an easier way to solve the problem. Second, some associativity problems are more conducive to the use of the associativity shortcut than others. On a problem, such as $6 + 23 - 21$, subtracting the third number from the second results in a positive number whereas, on a problem, such as $7 + 25 - 29$, subtracting the third number from the second results in a negative number. Negative numbers are more challenging for children to understand and use during problem solving (Young & Booth, 2015) and therefore may prevent children from applying their conceptual knowledge of associativity even if they do notice the associativity shortcut can be used. This issue is further compounded on multiplicative problems. Using the associativity shortcut to solve $6 \times 18 \div 2$ is much more straightforward then using the shortcut to solve $4 \times 9 \div 27$; this is because the latter would involve a fraction and children are often not comfortable dealing with even simple fractions (see also Van Hoof et al., Chapter 5; as well as Hallett, Nunes, Bryant, & Thorpe, 2012; Siegler & Lortie-Forgues, 2015). Finally, using the associativity shortcut requires calculation, just as the left-to-right problem solving procedure does. If problem solvers are seeking to use the most efficient problem solving procedures (Siegler & Araya, 2005) then some students may not realize or deem the associativity shortcut to be sufficiently more efficacious than the left-to-right approach.

ARE ADDITIVE AND MULTIPLICATIVE CONCEPTS THE SAME?

Robinson and LeFevre (2012) explored the issue of whether children understand additive and multiplicative concepts, specifically the concepts of inversion and associativity, in the same way. If this is the case, then children who understand the additive version of the concept should also understand the multiplicative version. Nunes et al. (2008) pointed out that, for at least the concept of inversion, understanding the relation between addition and subtraction should be a precursor for understanding the relation between multiplication and division and this point could be extrapolated to the concept of associativity as well. Enough studies have been conducted that either directly compare additive and multiplicative versions of inversion and associativity or have used similar research designs to address these issues.

Inversion

Formal schooling does not seem to be required for children to understand the inversion concept on addition and subtraction inversion problems (Klein & Bisanz, 2000; Rasmussen et al., 2003). However, in most of these studies on early understanding, concrete objects (e.g., XX + XXXX – XXXX) rather than symbolic versions of inversion problems (e.g., 2 + 4 – 4) have been used. When symbolic problems are used, however, the inversion concept is applied on only about a third of problems in 2nd grade (Robinson & Dubé, 2009a, 2013), increasing to almost two-thirds in 8th grade (Robinson et al., 2006), and then on almost all problems for undergraduate university students (Robinson & Ninowski, 2003).

The inversion shortcut is used less frequently on multiplication and division problems. Overall, 6th graders use the inversion shortcut on less than a quarter of the multiplication and division problems (Robinson & Dubé, 2009b; Robinson & Ninowski, 2003), increasing to about half the problems in 9th grade (Dubé, 2014), but over two thirds of problems in undergraduates (Dubé, 2014; Robinson & Ninowski, 2003). Across a series of studies there is a clear pattern that emerges: both children and adults are less likely to use inversion shortcuts on multiplication and division inversion problems.

Two studies provide further insights into the difference in conceptual knowledge of inversion between additive and multiplicative versions of the concept. Siegler and Stern (1998) conducted a microgenetic study in which 2nd graders who did not spontaneously use the inversion shortcut on addition and subtraction problems were presented with a set of inversion problems once a week for several weeks. By the end of the study, every student had discovered and applied the inversion shortcut during problem solving. Robinson and Dubé (2009c) used the same design with 6th graders who did not spontaneously use the inversion shortcut on multiplication and division problems. By the end of their study, over a third of participants had still not discovered and applied the inversion shortcut. It appears that it is easier for children to induce the inverse relation between addition and subtraction than that between multiplication and division.

Only one study with children has directly compared performance on addition and subtraction inversion problems with performance on multiplication and division inversion problems. Robinson et al. (2006) had 6th and 8th graders solve both types of problems but half of the children solved the addition and subtraction problems first and the other half the multiplication and division problems first. Not surprisingly, inversion shortcut use was higher on the addition and subtraction problems. The key findings related to Robinson et al.'s hypothesis that because knowledge of the additive inversion concept develops even before formal schooling, solving the addition and subtraction problems first might increase inversion shortcut use on the multiplication and division problems. However, contrary to this hypothesis, no transfer effects were found. Robinson and Ninowski (2003) found a similar lack of transfer with undergraduates. The students who solved the addition and subtraction inversion problems first were no more likely to subsequently use the inversion shortcut on the multiplication and division inversion problems than those who had solved the multiplication and division problems first. This is consistent with the studies described earlier and provides support for the implication that there is a disconnect between the additive and multiplicative versions of the inversion concept. The results of these studies as well as the two microgenetic studies provide further evidence that children and adults have a relatively weaker understanding of the inversion concept when it is being assessed with the operations of multiplication and division compared to the operations of addition and subtraction.

Associativity

Klein and Bisanz (2000), in a study of 4-year-olds' solving of concrete problems, were the first to notice that standard problems of the form $a + b - c$ can be solved using conceptual knowledge. They found that although almost a third of 4-year-olds used the associativity shortcut at least once, overall use was only 5%. Other research has shown that second graders' use of the associativity shortcut ranges from 10% to just under 15% of the problems, from 10% to 30% in 3rd grade, from under 10% to just over 25% in 4th grade, and to 25% by 5th grade (Robinson & Dubé, 2009a, 2012, 2013), reflecting a tendency for the number of students who use the shortcut to increase across grade rather than more students using it occasionally as they get older. One study has examined associativity shortcut use on addition and subtraction problems (in this study called the "right-to-left" strategy) in older students with about 10% use in 6th grade and less than 25% in 8th grade (Robinson et al., 2006). The difference between this study and studies with higher rates of shortcut use may be due to the inclusion in this study of standard problems that were not conducive to associativity shortcut use. For some of the problems in Robinson et al. (2006), subtracting the third from the second number resulted in a negative number; so this study may underestimate children's understanding of the associativity concept on addition and subtraction problems.

Overall, however, associativity shortcut use seems to develop very slowly across childhood. Even undergraduates only use the shortcut to solve about 60% of the problems on which it could be used (Robinson & Ninowski, 2003), but once again this study included problems that were not as conducive to use of the associativity shortcut. To examine whether associativity shortcut use is impacted by problem format, LeFevre and Robinson (2015, unpublished manuscript) separated the standard problems into problems that were "conducive" to associativity shortcut use (e.g., 6 + 28 – 22) and problems that were "nonconducive" (e.g., 9 + 25 – 29). They found almost 75% associativity shortcut use on conducive problems and just over 65% use on nonconducive problems, suggesting that problem format does impact application of conceptual knowledge.

On multiplication and division problems, associativity shortcut use drops even further. Robinson and Ninowski (2003) found that no 6th graders reported using the shortcut, even on problems that were more conducive to it, and 8th graders used the shortcut on less than 5% of problems. More recently, Dubé (2014) used only conducive problems and found that associativity shortcut use ranged from just under 15% in 7th grade, to over 25% in 9th grade, and to almost 40% in 11th grade. When presented with only conducive problems, undergraduate students' use of the associativity shortcut ranged from 45% to 55% (Dubé, 2014; Dubé & Robinson, 2010a). Robinson and Ninowski (2003) used a mix of conducive and nonconducive problems and found that combined associativity shortcut use was 30%. However, LeFevre and Robinson (2015, unpublished manuscript) found that associativity shortcut use was 30% on nonconducive problems but rose to more than 75% on conducive problems that may be explained by the study's inclusion of a large proportion of university students with very strong mathematical skills.

To sum up, the use of the associativity shortcut starts low and remains low across childhood and even among university students it is not the predominant strategy, particularly when the provisional result is a negative number or a fraction. Moreover, as with the inversion shortcut, associativity shortcut use is less frequent on multiplication and division problems compared to addition and subtraction problems.

Inversion Versus Associativity

A strong pattern emerges from the studies reported previously: students use the inversion shortcut more frequently than the associativity shortcut. This suggests that conceptual knowledge of inversion is stronger than that of associativity; but it is also possible that students are simply less likely to apply their knowledge of associativity than their knowledge of inversion through the use of shortcuts.

The first piece of evidence supporting the idea that students do have conceptual knowledge of associativity but are not always willing to apply it comes from the work of LeFevre and Robinson (2015, unpublished manuscript). As previously mentioned, they found that, on associativity problems that yielded a negative number or a fraction when dealing with the second and third numbers, associativity shortcut use dropped significantly on both additive and multiplicative problems. The second piece of evidence comes from the format of the problems themselves. On inversion problems there is always an obvious visual pattern; every inversion problem contains either $b - b$ or $e \div e$ so even a quick glance at the problem will cue the participant to use the inversion shortcut (Robinson & LeFevre, 2012). On associativity problems there is no consistent visual pattern across problems.

The third piece of evidence comes from the performance advantage of the inversion shortcut compared to the associativity shortcut. For the inversion shortcut, no calculations are required therefore very few errors occur on these problems (if they do it is usually because the participant states the *b* term by mistake) and problem solving solution times are very quick. In contrast, the associativity shortcut still requires calculation and therefore more errors and longer solution times will occur. For example, in a study of addition and subtraction concepts with 2nd–4th graders, when the inversion shortcut was used on inversion problems overall accuracy (98.7% vs. 55.4%) and solution times (6044 vs. 13,355 ms) were better than when a left-to-right calculation strategy was used (Robinson & Dubé, 2009a). In contrast, when the associativity shortcut was used on the associativity problems, overall accuracy (68.3% vs. 48.5% and solution times (13,194 vs. 17.656 ms) were still better for the shortcut as compared to when a left-to-right calculation strategy was used, but the gains were smaller than for inversion problems. Robinson, Dubé, and Beatch (2015) found a similar pattern for multiplication and division problems with 6th–8th graders. When the inversion shortcut was used accuracy (90.2% vs. 44.7%) and solution times (1818 vs. 2848 ms) were better than when a left-to-right strategy was used. When the associativity shortcut was used, once again the shortcut was associated with better performance (51.4% vs. 36.4% for accuracy and 2469 vs. 2993 ms for solution times) compared to a left-to-right strategy but not by as great a margin. In particular, note that the solution time advantage of the associativity shortcut compared to using a left-to-right strategy is even smaller on multiplication and division inversion problems than on addition and subtraction problems (0.5 vs. 4.5 s, respectively) which may provide another reason why associativity shortcut use is particularly infrequent on multiplication and division problems—the performance advantage may not be a strong enough incentive to abandon well-learned and practiced left-to-right approaches to problem solving (Lemaire & Lecacheur, 2011; McNeil, Rittle-Johnson, Hattikudur, & Peterson, 2010).

There is also evidence that students' conceptual knowledge of associativity is actually weaker than that of inversion. In most studies of the inversion and associativity concept, researchers have relied on the use of shortcuts to infer conceptual knowledge. However, other measures can be used (Bisanz et al., 2009; Prather & Alibali, 2009). One approach is to provide participants with a demonstration of the shortcuts and ask them to evaluate them. If children understand the concept behind the shortcuts they should provide positive evaluations of the shortcut and explanations for why the shortcuts are a good way to solve the inversion or associativity problems (Bisanz & LeFevre, 1992; Bisanz et al., 2009). This task is most commonly referred to as the "evaluation of procedures task" and has been used to investigate a range of mathematical concepts and tap into conceptual knowledge that children may not have been able to explicitly use or verbalize during problem solving (Crooks & Alibali, 2014).

In studies using both the problem solving task and the evaluation of procedures task, an interesting pattern of results has emerged and suggests that children have better conceptual knowledge of the concepts of inversion and associativity than they demonstrate during problem solving. Robinson and Dubé (2009a) found that almost all 2nd–4th graders (97%) evaluated the inversion shortcut as being a good problem solving procedure for solving additive problems, and most students (78%) thought the inversion shortcut was better than a left-to-right problem solving procedure. Both of these were much higher than the 46% use of the inversion shortcut during the problem solving task. During this task, associativity shortcut use was 22% and yet endorsement of the associativity shortcut during the evaluation of procedures task was at 89% and preference for the shortcut was only marginally higher (59%) compared to preference for a left-to-right problem solving procedure. Robinson and Dubé (2012) replicated these results and also extended them to slighter older children (3rd–4th graders) (Robinson & Dubé, 2013).

In a study of 6th–8th graders, Robinson and Dubé (2009b) used the evaluation of procedures task in conjunction with the usual problem solving task on multiplication and division inversion problems. In the problem solving task, use of the inversion shortcut was low (16%) and yet 83% of the students approved of the inversion shortcut while only 51% preferred the shortcut compared to a left-to-right strategy. Robinson et al. (2016) investigated the evaluation of both the inversion and associativity concepts by students in the same age range but focused only on the preference measure on the evaluation of procedures task as a more stringent assessment of conceptual knowledge. Inversion shortcut use was higher (29%) on the problem solving task but again the evaluation of procedures task suggested that children have a better conceptual understanding of the inverse relation between multiplication and division than they can express during actual problem solving; preference for the shortcut was 75%. Associativity shortcut use (16%) was lower than inversion shortcut use in the problem solving task and preference for the associativity shortcut in the evaluation of procedures task was only 52%.

Overall, the pattern is consistent regardless of which task is used to assess conceptual knowledge: children appear to have a better understanding of the inversion concept than the associativity concept. As a final note on this point, in all of the years that these studies have been conducted (not all of which are discussed in this chapter) and amongst the hundreds of participants involved, only one participant to date has used the associativity shortcut and not used the inversion shortcut during the problem solving task. Every other associativity shortcut user, regardless of additive or multiplicative versions, has always used the inversion shortcut.

INDIVIDUAL DIFFERENCES AND FACTORS IN THE USE OF CONCEPTUALLY-BASED SHORTCUTS

To date, this discussion of arithmetic concepts has focused on the difference between additive and multiplicative concepts, between the concepts of inversion and associativity, and also outlined the pattern of developmental change in these concepts but no mention has been made of individual differences and the factors that relate to conceptual knowledge, two areas of investigation which have recently been gaining interest (Gilmore & Papadatou-Pastou, 2009). Individual differences are considered first, followed by the examination of the factors relating to the use of conceptually-based shortcuts during problem solving.

INDIVIDUAL DIFFERENCES

Up until this point in the chapter, my focus has centered primarily on the use of the inversion shortcut during problem solving as a good measure of conceptual knowledge (Bisanz & LeFevre, 1990; Crooks & Alibali, 2014; Prather & Alibali, 2009). However, a more lenient measure can also be used. Bisanz and LeFevre (1990) noticed a third approach to solving inversion problems, which they termed the "negation strategy." With this strategy, students use a mix of a left-to-right calculation procedure and the inversion shortcut. So, on a problem, such as $896 \times 5193 \div 5193$, a participant might calculate the first part of the problem by multiplying 896 and 5193 together and then realize that dividing by the third number, 5193, negates the number they just multiplied leaving them the number they started with, 896. These participants have at least some understanding of the inversion concept because they understand the inverse or reverse relationship between multiplying and dividing (or adding and subtracting) the second and third numbers. In Siegler and Stern's (1998) microgenetic study of the development of the inversion shortcut on addition and subtraction problems, they found that many students moved from using a left-to-right problem solving procedure, to using the negation strategy, to using the inversion shortcut. On multiplication and division inversion problems, the same pattern was found for some individuals but other individuals switched directly from using a left-to-right procedure to the inversion shortcut (Robinson & Dubé, 2009c).

Gilmore and Bryant (2006) were the first to use cluster analyses as a means to investigate individual differences in children's understanding of arithmetic concepts. Cluster analysis is a statistical technique that sorts individuals into subgroups based on their performance on a range of tasks or measures. Robinson and Dubé (2009a) used cluster analyses to investigate individual differences in children's understanding of both inversion and associativity on additive problems. In this instance, they only examined the use of the inversion and associativity shortcuts as measures of conceptual knowledge. They found three distinct clusters. The Dual Concept cluster, the smallest cluster included only 18% of the students who used both the inversion and associativity shortcuts frequently (87 and 80%, respectively). The Inversion Concept cluster included 40% of the students who used only the inversion shortcut frequently (69 and 16%). The No Concept cluster included 42% of the students who rarely used either the inversion or the associativity shortcuts (4 and 1%). Interestingly, grade did not predict cluster membership such that, for example, the 2nd graders were just as likely as the 3rd and 4th graders to be in the Dual Concept cluster.

In a follow up study, Robinson and Dubé (2013) investigated individual differences in more detail with 3rd, 4th, and 5th graders by adding negation strategy use to the two shortcuts in their cluster analysis and a 4th cluster was identified: the Negation cluster. Once again, students in the Dual Concept cluster (16% of the participants) had frequent use of the inversion and associativity shortcuts and fairly infrequent use of the negation strategy (84, 74, and 4%, respectively). Students in the Inversion Concept cluster (19% of the participants) only used the inversion shortcut frequently and used the associativity shortcut and the negation strategy infrequently (70, 4, and 14%). Students in the Negation cluster (28% of the participants) rarely used the inversion or associativity shortcuts but frequently used the negation strategy (17, 6, and 56%). Finally, students in the No Concept cluster showed little evidence of conceptual understanding (3, 2, and 11%). Once again, grade did not predict cluster membership suggesting that the development of both the inversion and associativity concepts may not be as influenced by years of education as analysis considering only frequency of shortcut use might suggest.

For multiplication and division problems, a more detailed investigation of individual differences has been conducted. Robinson and Dubé (2015) drew on the data from several studies to conduct a cluster analysis based on the data from 540 participants in Grades 6, 7, and 8. This time, five clusters were identified. Students in the No Concept cluster (45% of participants) used the inversion and associativity shortcuts as well as the negation strategy infrequently (4, 2, and 7%, respectively). Students in the Negation Concept cluster (27% of participants) only used the negation strategy frequently (6, 1, and 59% for inversion, associativity, and negation, respectively), whereas, students in the Inversion Concept cluster (14% of participants) only used the inversion shortcut frequently (78, 3, and 11%). In this metaanalysis, the Dual Concept cluster identified in previous studies (both with additive and multiplicative problems) further subdivided into two further subgroups: the Low Dual Concept and the High Dual Concept clusters. These are both small clusters (4 and 10% for the Low and High Dual Concept clusters, respectively) which is presumably why they had not been identified as separate subgroups in previous studies. Students in the Low Dual Concept cluster used both shortcuts but only moderately so (34, 49, and 8%) whereas, students in the High Dual Concept clusters used both shortcuts frequently (89, 87, and 2%). Interestingly, the metaanalysis revealed that grade predicted cluster membership with 8th graders more likely to be in one of the Dual Concept cluster and for 7th graders more likely to be in the Negation cluster. This latter finding fits with the findings of Robinson and Dubé (2009b) who found that although overall inversion shortcut use and use of the negation strategy did not differ across grade in the problem solving task, 7th graders had the lowest preference for the shortcut compared to a left-to-right solution procedure (39% vs. 56% and 59% for 6th and 8th graders, respectively).

The original three clusters have also been identified in studies of adults (Robinson & LeFevre, 2012). That is, the No Concept cluster (which is identified as the Weak Concept cluster because overall use of both shortcuts tends to be higher in all adults compared to almost all children), the Inversion Concept cluster, and the Dual Concept cluster were all found. That these same three clusters have been found with younger children on additive problems, with older children on multiplicative problems, and with adults on both additive and multiplicative problems, suggests that these clusters represent patterns of individual differences that may be continuous across development.

FACTORS RELATING TO CONCEPTUALLY-BASED SHORTCUT USE

Computational Skills and Age

A number of researchers have been interested in the factors that predict to the use of conceptually-based shortcuts. Most of the research has been with addition and subtraction inversion problems. Sherman and Bisanz (2007) looked at whether counting skills and knowledge of number names was related to performance on concrete inversion problems but only found weak correlations with arithmetic skills. Gilmore and Papadatou-Pastou (2009) conducted a metaanalysis of addition and subtraction inversion studies that included assessments of symbolic arithmetic skills and identified three clusters of individuals confirming previous research with the same clusters (Gilmore & Bryant, 2006). One cluster included children with both a strong understanding of the inversion concept and good calculation skills. A second cluster included children with a poor understanding of inversion and weak calculation skills. Finally, a third cluster included children with a good understanding of inversion but weak calculation skills. Interestingly, as has been found in other analysis of individual differences on the additive concepts, no grade differences were found in cluster membership and Gilmore and Papadatou-Pastou (2009) therefore concluded that although arithmetic skills generally increase across age or grade, the understanding of inversion does not. Watchorn et al. (2014) found a similar pattern of results with the same three clusters as well as a fourth cluster of participants with good arithmetic skills but low use of the inversion shortcut.

Only one study using a slightly different approach has examined whether computational skills are related to inversion shortcut use on multiplication and division problems and no research at all has been conducted for use of the associativity shortcut. Robinson and Dubé (2009b) identified three clusters of students comprised of 6th–8th graders. Students in the Inversion cluster used the inversion shortcut frequently (64%) and had the highest computational skills. Students in the Negation cluster used the negation strategy frequently (58%) and the inversion shortcut infrequently (5%) and their computational skills did not differ from students in the latter Computation cluster. These students rarely used the inversion shortcut (1%) or the negation strategy (3%) and instead relied on a left-to-right computational approach to solve the inversion problems.

Overall, it appears that computation skills may be related—but not for all individuals—to the understanding of the inverse relation between pairs of operations. This suggests that more than computational skills or age/grade account for the large individual differences in the understanding of arithmetic concepts.

Working Memory

Domain-general cognitive abilities have also been explored as potential contributors to individual differences in conceptual knowledge. Rasmussen et al. (2003) investigated the role of working memory in their study of the inversion concept on addition and subtraction problems with preschool and 1st graders. Corsi span, a measure of visual-spatial working memory, correlated with preschoolers' but not 1st graders' use of the inversion shortcut. Digit span, a measure of the phonological loop component of working memory, did not correlate with inversion shortcut use either in preschool or 1st grade students. Dubé and Robinson (2010b) investigated working memory and inversion shortcut use on multiplication and division problems with 6th and 8th grade students. In both grades, two measures of working memory, including the backward digit span, correlated with inversion shortcut use. Based on their findings they proposed that working memory might play a role in the use of the inversion shortcut because it helps direct attention towards the right side of an inversion problem (i.e., to the $e \div e$ part of the problem). Watchorn et al. (2014) in their study on addition and subtraction inversion problems found that children with greater attention skills were more likely to use the inversion shortcut, particularly if they also had strong computation skills.

Inhibition and Attention

Dubé and Robinson (2010b) also hypothesized that the inhibition component of working memory may be particularly important for use of the inversion shortcut. Siegler and Araya (2005) proposed that to apply the inversion shortcut (at least on addition and subtraction problems), children must not only pay attention to the right side of the problem but also inhibit their well-established tendency to solve problems from left-to-right, and instead direct attention to the right side of the problem. Robinson and Dubé (2013) investigated the role of inhibition in children's use of the inversion and associativity shortcuts on addition and subtraction problems. In this study, as previously reported, four clusters of individuals were found: the No Concept, Negation, Inversion Concept, and Dual Concept clusters. Children in the two clusters with the highest use of conceptually-based shortcuts, the Inversion and Dual Concept clusters, also scored highest on the Stop-Signal task, a standard measure of inhibitory abilities. This finding suggests that these participants were able to inhibit their tendency to rotely solve problems from left-to-right and thereby process all of the presented numbers before executing the shortcut strategy.

No research to date has been conducted on the role of working memory or attention on the use of shortcuts when solving multiplication and division problems. However, a recent study by Dubé (2014) sheds some light on Siegler and Araya's (2005) proposal that applying the inversion shortcut requires the interruption of the tendency to solve problems from left-to-right. Using multiplication and division inversion and associativity problems, Dubé (2014) had 7th, 9th, and 11th grade students, as well as undergraduate students solve problems using a task that checked whether they had calculated the product of the first two numbers. Calculation of the first two numbers on an inversion problem would mean that they did not inhibit or interrupt the left-to-right procedure to use the inversion shortcut which requires focusing on the right side of the problem. The same holds for the associativity shortcut which also requires focusing on the right side of the problem by dividing the second number by the third and then multiplying the result by the first number. Dubé (2014) found compelling evidence that inversion and associativity shortcut users had indeed inhibited their tendency to use left-to-right problem solving procedures in order to implement their use of the shortcuts. These findings suggest that inhibition is likely to also be involved in the use of conceptually-based shortcuts on multiplication and division problems and for associativity problems as well as inversion problems, at least on problems typically used in these studies (cf., $43 - 43 + 78$ or $64 \div 32 \times 21$). In sum, working memory, attention, and inhibition all need to be considered as important cognitive factors in the development of arithmetic concepts as well as in theories of mathematical development (Cragg & Gilmore, 2014).

Attitudes

Although less research has been conducted in this area, there is evidence that children's attitudes can influence their use of conceptually-based shortcuts. In these instances, children may have the necessary conceptual knowledge but may refuse to use it during problem solving. This is potentially problematic given that researchers have long relied on the use of conceptually-based shortcuts as one of the best ways, if not the best way, to measure conceptual knowledge (Bisanz et al., 2009; Crooks & Alibali, 2014; Gilmore & Papadatou-Pastou, 2009; Prather & Alibali, 2009). In early studies on 3rd–8th graders and using the evaluation of procedures task (Robinson & Dubé, 2009a, 2009b), it became apparent that some children had strong attitudes as to whether the inversion and/or associativity shortcuts were "good" or appropriate problem solving procedures. Some students, when told about how a fictional child had used the inversion or associativity shortcut to solve a problem, reacted with strong approval and when asked for their reasoning often expressed admiration for the cleverness of the fictional child, noted how efficient or smart the shortcut was, and some even wished that they had thought of the shortcut themselves when they had solved the inversion and associativity problems earlier in the session.

In contrast, other students reacted with strong disapproval. Some students stated that when solving problems, no steps should be omitted or done out of order therefore the shortcuts were not appropriate, and some stated that when solving math problems calculations always had to be performed. Other students questioned the abilities of the fictional child's teacher and others accused the fictional child of cheating. These positive and negative responses were found with children in all grades, for both inversion and associativity, and on both additive and multiplicative problems.

Kilpatrick et al. (2001) and others (e.g., Ellis, 1997; Martin, Anderson, Bobis, Vellar, & Way, 2012) have recognized how attitudes toward mathematics can impact success in mathematics. Researchers have found that noncognitive factors, such as mathematics anxiety negatively affect mathematics performance (e.g., Ashcraft & Rudig, 2012; Ramirez, Chang, Maloney, Levine, & Beilock, 2016) but little research has specifically examined attitudes and how they relate to conceptual knowledge of arithmetic. Two recent studies have helped fill this gap. Robinson and Dubé (2013) had 2nd–4th grade students solve two sets of addition and subtraction inversion and associativity problems. Between the two sets of problems, students did the evaluation of procedures task in which the inversion and associativity shortcuts were demonstrated along with a left-to-right procedure and students were asked whether they preferred the shortcut or the left-to-right procedure. The students then solved the second set of problems. For both inversion and associativity, overall shortcut use was higher for students in the second set. In particular, associativity shortcut use increased significantly more for the participants who had preferred the shortcut. This suggests two possibilities: that the evaluation of procedures task promoted understanding of the associativity concept or that students felt that, because of the demonstration of the shortcut, it was appropriate for them to now start using the associativity shortcut. However, as there was no control group it is also possible that more exposure to the problems alone promoted shortcut discovery though this would not be consistent Robinson and Dubé's (2009c) microgenetic study.

In a more recent study, Robinson et al. (2016) investigated how 6th–8th grade students' attitudes about the inversion and associativity shortcuts influenced their use of the shortcuts during multiplication and division problem solving. Half of the students solved two sets of problems consecutively while the other half completed the evaluation of procedures task between the two sets of problems. On inversion problems, inversion shortcut use increased across problem set for both groups but the increase was greater for the group that completed the evaluation of procedures task between the problem sets. Conversely, on associativity problems, associativity shortcut use only increased across problem set for the group that completed the evaluation of procedures task between the problems sets. Preference for a shortcut versus a left-to-right procedure also influenced their use. For this, only data from the group that completed the evaluation of procedures task between the two problem sets was analyzed. For both inversion and associativity problems, students with a preference for the inversion shortcut used it more frequently than other students. Therefore, there is evidence that the evaluation of the procedures task itself lead to an increase in conceptual knowledge or at least an increased willingness to apply conceptual knowledge, but there is also evidence that children's attitudes towards the demonstrated shortcuts in the evaluation of procedures task influenced their subsequent shortcut use. The impact of attitudes on conceptually-based shortcut use is concerning but is consistent with recent findings that mathematics anxiety prevents some children from using more advanced problem solving strategies (Ramirez et al., 2016). Overall, the current research on attitudes adds further complexity to the variables involved in the development of conceptual knowledge as assessed through the use of conceptually-based shortcuts.

Educational Experiences

One final factor needs to be briefly discussed. Contradictory evidence exists on the role of age or grade on the use of conceptually-based shortcuts as some studies have found that use of the inversion and/or associativity increases across grade while others have not (e.g., Gilmore & Papadatou-Pastou, 2009; Robinson & Dubé, 2015). By adulthood, inversion and associativity shortcut use are at their highest (e.g., Dubé, 2014), suggesting that conceptual knowledge has developed since childhood and adolescence, but within tighter time spans (e.g., from 2nd to 4th grade), often there is little conceptual change (e.g., the finding that grade often does not predict cluster membership). Therefore, instruction is probably important in the development of conceptual knowledge but for some students, this does not appear to be sufficient for them to understand the concepts of inversion and associativity. Other educational factors, however, may impact mathematical performance.

Harold Stevenson, in his examination of cross-cultural differences in academic achievement consistently found that American children underperformed in mathematics in comparison to East Asian children. Stevenson et al. (1990) attributed Americans' underperformance to children's lower motivation toward mathematics, parents' lower expectations for their children in mathematics, teachers' lower levels of interest in teaching mathematics, and a less rigorous mathematical curriculum. Laski and Yu (2014) concluded that one of the reasons that Chinese outperform Chinese-American Kindergarten children on arithmetic tasks was due to Chinese teachers spending more class time discussing and explaining numerical concepts. Ma (1999) has proposed that one of the most marked differences between Chinese and American teachers relates to their conceptual understanding of elementary mathematics. A number of researchers have examined cross-cultural differences in conceptual knowledge (Ho & Fuson, 1998; Miller, Major, Shu, & Zhang, 2000; Torbeyns, Schneider, Xin, & Siegler, 2015) but only one has focused on the concepts of inversion and associativity. Robinson and Beatch (2015) had Canadian-educated adults and adults who had received all of their elementary and high school education in Asia (mostly in China) solve both additive and multiplicative problems to assess their understanding of the inversion and associativity concepts. Asian participants used more shortcuts overall and were more likely to be placed into the equivalent of the Dual Concept cluster discussed earlier. Findings from this study support the notion that cross-cultural differences exist in conceptual knowledge even in adulthood, mostly suggesting that early educational experiences have long term implications for how well the concepts of inversion and associativity are understood.

In summary, the growing body of research on both additive and multiplicative concepts demonstrates the complexity of factors relating to not only conceptual knowledge of arithmetic but also to how conceptual knowledge is assessed. As Robinson and LeFevre (2012) pointed out, current research, and theories of how children's concepts develop are based primarily on additive concepts, typically rely on studies which focus on only one concept at a time, and do not include a broad range of measures to investigate the many factors that are involved in children's developing conceptual knowledge of arithmetic.

CONCLUSIONS AND FUTURE DIRECTIONS

The NMAP (2008) proposed that knowledge of mathematical concepts is an important predictor of success in later mathematics learning. They took specific note of the inverse relation between addition and subtraction as well as the inverse relation between multiplication and division but they concluded that American children's weak understanding of these and other core arithmetic concepts was not acceptable and highlighted the need for curricular change.

The NMAP's (2008) singling out of a multiplicative concept as a specific area of concern highlights the need for researchers to increase their focus on children's understanding of multiplication and division. These two operations and the relations between them are more difficult for children to understand and may require explicit instruction. In relation to the operations of addition and subtraction, Nunes et al. (2009) proposed that explicit teaching of concepts, such as inversion and their associated conceptually-based shortcuts is important. Although teaching children about concepts, such as inversion or associativity in the classroom might present a challenge to researchers who rely on the novelty of the problems to assess use of the conceptually-based shortcuts, the benefits to children clearly far outweigh the costs to researchers. By explicitly drawing children's attention to how operations are related to one another and how conceptual knowledge can be applied via problem solving procedures are likely important steps toward increasing children's knowledge of mathematics (Schneider & Stern, 2009). It can also help reinforce the need to look for patterns and relations—a central aspect of mathematics (Lai et al., 2008). For children to gain a deep understanding of mathematics, children may need to be taught not only about facts and procedures, but also about concepts as outlined by the U.S. Common Core State Standards (National Governors Association for Best Practices and Council of Chief School Officers, 2010).

The studies discussed in this chapter not only highlight the importance of arithmetic concepts but also reveals some important research and educational issues. From a research perspective, it should be clear that more attention should be paid to multiplicative concepts. They are more challenging for children than additive concepts and take much longer to develop. Further, to gain a more complete understanding of how concepts develop, whether they are additive or multiplicative, different patterns of development may be identified depending on the concept being investigated. This expanded research agenda should also include a focus on developmental differences in the emergence of additive and multiplicative concepts. Finally, given the striking individual variability from the early elementary years to the university years in how well arithmetic concepts are understood and applied during problem solving, a more thorough examination is needed of the factors or, better yet, the combination of factors relating to the understanding of arithmetic concepts and the application of conceptually-based shortcuts.

From an educational perspective, teachers and parents need to be more aware that what may seem obvious to them (e.g., that multiplication and division are the inverse operations of each other), is not necessarily obvious to children. Just as teachers overestimate their students' understanding of the equal sign (Sherman, 2007), teachers may overestimate their students' understanding of simple concepts including inversion and associativity. By teaching children about the relations between operations and by directly comparing and contrasting additive and multiplicative concepts, children will have the opportunity to become deeper and more flexible mathematical thinkers—qualities that will be invaluable as they move from dealing with simple arithmetic problems and concepts to more complex mathematical concepts and ideas.

REFERENCES

Ashcraft, M. H. (1982). The development of mental arithmetic: A chronometric approach. *Developmental Review*, *2*, 213–236.

Ashcraft, M. H., & Rudig, N. O. (2012). Higher cognition is altered by noncognitive factors: How affect enhances and disrupts mathematics performance in adolescence and young adulthood. In V. F. Reyna (Ed.), *The adolescent brain: Learning, reasoning, and decision making* (pp. 243–263). Washington, DC: American Psychological Association.

Baroody, A. J. (2003). The development of adaptive expertise and flexibility: The integration of conceptual and procedural knowledge. In A. J. Baroody, & A. Dowker (Eds.), *The development of arithmetic concepts and skills* (pp. 1–33). Mahwah, NJ: Erlbaum.

Baroody, A. J., & Ginsburg, H. P. (1986). The relationship between initial meaningful and mechanical knowledge of arithmetic. In J. Hiebert (Ed.), *Conceptual and procedural knowledge: The case of mathematics* (pp. 75–112). Hillsdale, NJ: Lawrence Erlbaum Associates.

Baroody, A. J., Ginsburg, H. P., & Waxman, B. (1983). Children's use of mathematical structure. *Journal for Research in Mathematics Education*, *14*, 156–168.

Bisanz, J. (1999). The development of mathematical cognition: Arithmetic. *Journal of Experimental Child Psychology, 74*, 153–156.

Bisanz, J., & LeFevre, J. (1990). Strategic and nonstrategic processing in the development of mathematical cognition. In D. F. Bjorklund (Ed.), *Children's strategies: Contemporary views of cognitive development* (pp. 213–244). Hillsdale, NJ: Erlbaum.

Bisanz, J., & LeFevre, J. (1992). Understanding elementary mathematics. In J. I. D. Campbell (Ed.), *The nature and origins of mathematical skills* (pp. 113–136). Amsterdam: North Holland, Elsevier Science.

Bisanz, J., Watchorn, R. P. D., Piatt, C., & Sherman, J. (2009). On "understanding" children's developing use of inversion. *Mathematical Thinking and Learning, 11*, 10–24.

Canobi, K. H. (2005). Children's profile of addition and subtraction understanding. *Journal of Experimental Child Psychology, 92*, 220–246.

Canobi, K. H. (2009). Concept-procedure interactions in children's addition and subtraction. *Journal of Experimental Child Psychology, 102*, 131–149.

Canobi, K. H., & Bethune, N. E. (2008). Number words in young children's conceptual and procedural knowledge of addition, subtraction, and inversion. *Cognition, 108*, 675–686.

Carpenter, T. P. (1986). Conceptual knowledge as a foundation for procedural knowledge: Implications from research on the initial learning of arithmetic. In J. Hiebert (Ed.), *Conceptual and procedural knowledge: The case of mathematics* (pp. 113–132). Hillsdale, NJ: Lawrence Erlbaum Associates.

Cowan, R. (2003). Does it all add up? Changes in children's knowledge of addition combinations, strategies, and principles. In A. J. Baroody, & A. Dowker (Eds.), *The development of arithmetic concepts and skills: Constructing adaptive expertise* (pp. 35–74). Mahwah, NJ: Lawrence Erlbaum Associates.

Cowan, R., & Renton, M. (1996). Do they know what they are doing? Children's use of economical addition strategies and knowledge of commutativity. *Educational Psychology, 16*, 407–420.

Cragg, L., & Gilmore, C. (2014). Skills underlying mathematics: The role of executive function in the development of mathematics proficiency. *Trends in Neuroscience and Education, 3*, 63–68.

Crooks, N. M., & Alibali, M. W. (2014). Defining and measuring conceptual knowledge in mathematics. *Developmental Review, 34*, 344–377.

Dixon, J. A., & Boncoddo, R. (2009). Strategies and problem representations: Implications for models of changing cognitive structure. Commentary on "Learning new problem solving strategies leads to changes in problem representation" by M. W. Alibali, K.M. Ockuly, & A. D. Fischer. *Cognitive Development, 24*, 102–105.

Dixon, J. A., & Moore, C. F. (1996). The developmental role of intuitive principles in choosing mathematical strategies. *Developmental Psychology, 32*, 241–253, doi: 10.1037/0012-1649.32.2.241.

Dixon, J. A., Deets, J. K., & Bangert, A. (2001). The representations of the arithmetic operations including functional relationships. *Memory & Cognition, 29*, 462–477.

Dubé, A. K. (2014). Adolescents' understanding of inversion and associativity. *Learning and Individual Differences, 36*, 49–59.

Dubé, A. K., & Robinson, K. M. (2010a). The relationship between adults' conceptual understanding of inversion and associativity. *Canadian Journal of Experimental Psychology, 64*, 60–66.

Dubé, A. K., & Robinson, K. M. (2010b). Accounting for individual variability in inversion shortcut use. *Learning and Individual Differences, 20*, 687–693.

Ellis, S. (1997). Strategy choice in sociocultural context. *Developmental Review, 17*, 490–524.

Fuson, K. C. (1988). *Children's counting and concepts of number*. New York: Springer-Verlag.

Fuson, K. C. (2009). Avoiding misinterpretations of Piaget and Vygotsky: Mathematical teaching without learning, learning without teaching, or helpful learning-path teaching? *Cognitive Development, 24*, 343–361.

Gilmore, C. K., & Bryant, P. (2006). Individual differences in children's understanding of inversion and arithmetical skill. *British Journal of Educational Psychology, 76*, 309–331.

Gilmore, C. K., & Papadatou-Pastou, M. (2009). Patterns of individual differences in conceptual understanding and arithmetical skill: A meta-analysis. *Mathematical Thinking and Learning, 11*, 25–40.

Hallett, D., Nunes, T., Bryant, P., & Thorpe, C. M. (2012). Individual differences in conceptual and procedural fraction understanding: The role of abilities and school experience. *Journal of Experimental Child Psychology, 113*, 469–486.

Hiebert, J., & Lefevre, P. (1986). Conceptual and procedural knowledge in mathematics: An introductory analysis. In J. Hiebert (Ed.), *Conceptual and procedural knowledge: The case of mathematics* (pp. 199–223). Hillsdale, NJ: Lawrence Erlbaum Associates.

Ho, C. S. -H., & Fuson, K. C. (1998). Children's knowledge of teen quantities as tens and ones: Comparisons of Chinese, British, and American kindergartners. *Journal of Educational Psychology, 90*, 536–544.

Jordan, N. C., Hanich, L. B., & Uberti, H. Z. (2003). Mathematical thinking and learning difficulties. In A. J. Baroody, & A. Dowker (Eds.), *The development of arithmetic concepts and skills* (pp. 351–383). Mahwah, NJ: Erlbaum.

Kilpatrick, J., Swafford, J., & Findell, B. (2001). *Adding it up: Helping children learn mathematics*. Washington, DC: National Academy Press.

Klein, J. S., & Bisanz, J. (2000). Preschoolers doing arithmetic: The concepts are willing but the working memory is weak. *Canadian Journal of Experimental Psychology, 54*, 105–116.

Lai, M. -L., Baroody, A. J., & Johnson, A. R. (2008). Fostering Taiwanese preschoolers' understanding of the addition-subtraction inverse principle. *Cognitive Development, 23*, 216–235, http://dx.doi.org/10.1016/j.cogdev.2007.06.002.

Laski, E. V., & Yu, Q. (2014). Number line estimation and mental addition: Examining the potential roles of language and education. *Journal of Experimental Child Psychology, 117*, 29–44.

Lemaire, P., & Lecacheur, M. (2011). Age-related changes in children's executive functions and strategy selection: A study in computational estimation. *Cognitive Development, 26*, 282–294.

Ma, L. (1999). *Knowing and teaching elementary mathematics: Mathematics in China and the United States*. Mahwah, NJ: Erlbaum.

Martin, A. J., Anderson, J., Bobis, J., Vellar, R., & Way, J. (2012). Switching on and switching off in mathematics: An ecological study of future intent and disengagement among middle school students. *Journal of Educational Psychology, 104*, 1–18.

McNeil, N. M., Rittle-Johnson, B., Hattikudur, S., & Petersen, L. A. (2010). Continuity in representation between children and adults: Arithmetic knowledge hinders undergraduates' algebraic problem solving. *Journal of Cognition and Development, 11*, 437–457.

Miller, K., Major, S. M., Shu, H., & Zhang, H. (2000). Ordinal knowledge: Number names and number concepts in Chinese and English. *Canadian Journal of Psychology, 54*, 129–140.

National Council of Teachers in Mathematics. (2014). *Principles to actions: Ensuring mathematical success for all*. Reston, VA: NCTM.

National Council of Teachers of Mathematics. (2000). *Principles and standards for school mathematics*. Reston, VA: NCTM.

National Governors' Association Center for Best Practices, & Council of Chief State School Officers. (2010). *Common core state standards for mathematics*. Available from: www.corestandards.org/wp-content/uploads/Math_Standards1.pdf

National Mathematics Advisory Panel. (2008). *Foundations for success: The final report of the National Mathematics Advisory Panel.* Washington, D.C: U.S. Department of Education.

Nunes, T., Bryant, P., Burman, D., Bell, D., Evans, D., Hallett, D., & Montgomery, L. (2008). Deaf children's understanding of inverse relationships. In M. Marschark, & P. C. Hauser (Eds.), *Deaf cognition: Foundations and outcomes* (pp. 201–225). New York, NY: Oxford.

Nunes, T., Bryant, P., Hallett, D., Bell, D., & Evans, D. (2009). Teaching children about the inverse relation between addition and subtraction. *Mathematical Thinking and Learning, 11*, 61–78.

Piaget, J. (1952). *The child's conception of number.* London, UK: Routledge & Kegan Paul.

Prather, R. W., & Alibali, M. W. (2009). The development of arithmetic principle knowledge: How do we know what learners know? *Developmental Review, 29*, 221–248.

Ramirez, G., Chang, H., Maloney, E. A., Levine, S. C., & Beilock, S. L. (2016). On the relationship between math anxiety and math achievement in early elementary school: The role of problem solving strategies. *Journal of Experimental Child Psychology, 141*, 83–100.

Rasmussen, C., Ho, E., & Bisanz, J. (2003). Use of the mathematical principle of inversion in young children. *Journal of Experimental Child Psychology, 85*, 89–102.

Rittle-Johnson, B., & Alibali, M. W. (1999). Conceptual and procedural knowledge of mathematics: Does one lead to the other? *Journal of Educational Psychology, 91*, 175–189.

Rittle-Johnson, B., Schneider, M., & Star, J.R. (2015). Not a one-way street: Bidirectional relations between procedural and conceptual knowledge of mathematics. *Educational Psychology Review. 27*, 587–597.

Robinson, K.M., & Beatch, J.-A. (2015). Conceptual knowledge of arithmetic for Asian- and Canadian-educated adults. *Canadian Journal of Experimental Psychology.* Advance online publication.

Robinson, K. M., & Dubé, A. K. (2009a). Children's understanding of addition and subtraction concepts. *Journal of Experimental Child Psychology, 103*, 532–545.

Robinson, K. M., & Dubé, A. K. (2009b). Children's understanding of the inverse relation between multiplication and division. *Cognitive Development, 24*, 310–321.

Robinson, K. M., & Dubé, A. K. (2009c). A microgenetic study of the multiplication and division inversion concept. *Canadian Journal of Experimental Psychology, 63*, 193–200.

Robinson, K. M., & Dubé, A. K. (2012). Children's use of arithmetic shortcuts: The role of attitudes in strategy choice. *Child Development Research, 2012*, 10.

Robinson, K. M., & Dubé, A. K. (2013). Children's additive concepts: Promoting understanding and the role of inhibition. *Learning and Individual Differences, 23*, 101–107.

Robinson, K.M., & Dubé, A.K. (2015). *Children's understanding of multiplication and division: Novel effects identified through a meta-analysis of 7 studies.* Philadelphia, PA: Poster presented at the biennial meeting of the Society for Research in Child Development.

Robinson, K. M., & LeFevre, J. (2012). The inverse relation between multiplication and division: Concepts, procedures, and a cognitive framework. *Educational Studies in Mathematics, 79*, 409–428.

Robinson, K. M., & Ninowski, J. E. (2003). Adults' understanding of inversion concepts: How does performance on addition and subtraction inversion problems compare to performance on multiplication and division inversion problems? *Canadian Journal of Experimental Psychology, 57*, 321–330.

Robinson, K. M., Arbuthnott, K. D., Rose, D., McCarron, M. C., Globa, C. A., & Phonexay, S. D. (2006). Children's strategies for solving simple division problems. *Journal of Experimental Child Psychology, 93*, 224–238.

Robinson, K. M., Ninowski, J. E., & Gray, M. L. (2006). Children's understanding of the arithmetic concepts of inversion and associativity. *Journal of Experimental Child Psychology, 94*, 349–362.

Robinson, K.M., Dubé, A.K., & Beatch, J.-A. (2016). Children's multiplication and division shortcuts: Increasing shortcut use depends on how the shortcuts are evaluated. Learning and Individual Differences. Advance online publication.

Schneider, M., & Stern, E. (2009). The inverse relation of addition and subtraction: A knowledge integration perspective. *Mathematical Thinking and Learning*, *11*(1), 92–101.

Schneider, M., Rittle-Johnson, B., & Star, J. R. (2011). Relations among conceptual knowledge, procedural knowledge, and procedural flexibility in two samples differing in prior knowledge. *Developmental Psychology*, *47*, 1525–1538.

Sherman, J. (2007, April). *From failure to success on equivalence problems and how teachers perceive the process.* Paper presented at the biennial conference for the Society for Research in Child Development, Boston, MA.

Sherman, J., & Bisanz, J. (2007). Evidence for use of mathematical inversion by three-year-old children. *Journal of Cognition and Development*, *8*, 334–344.

Shumway, R. J. (1974). Negative instances in mathematical concept acquisition: Transfer effects between the concepts of commutativity and associativity. *Journal for Research in Mathematics Education*, *5*, 197–211.

Siegler, R. S. (1988). Strategy choice procedures and the development of multiplication skill. *Journal of Experimental Psychology: General*, *117*, 258–275.

Siegler, R. S., & Araya, R. (2005). A computational model of conscious and unconscious strategy discovery. In R. V. Kail (Ed.), *Advances in child development and behaviour* (pp. 1–42). (Vol. 33). New York, NY: Elsevier.

Siegler, R. S., & Lortie-Forgues, H. (2015). Conceptual knowledge of fraction arithmetic. *Journal of Educational Psychology*, *107*, 909–918.

Siegler, R. S., & Stern, E. (1998). Conscious and unconscious strategy discoveries: A microgenetic analysis. *Journal of Experimental Psychology: General*, *127*, 377–397.

Starkey, P., & Gelman, R. (1982). The development of addition and subtraction abilities prior to formal schooling in arithmetic. In T. P. Carpenter, J. M. Moser, & T. A. Romberg (Eds.), *Addition and subtraction: A cognitive perspective* (pp. 99–116). Hillsdale, NJ: Erlbaum.

Stevenson, H. W., Lee, S., Chen, C., Stigler, J. W., Hsu, C. -C., Kitamura, S., & Hatano, G. (1990). Contexts of achievement: A study of American, Chinese, and Japanese children. *Monographs of the Society for Research in Child Development*, *55*(1–2, 221), 123.

Torbeyns, J., Schneider, M., Xin, Z., & Siegler, R. S. (2015). Bridging the gap: Fraction understanding is central to mathematics achievement in students from three different continents. *Learning and Instruction*, *37*, 5–13.

Vilette, B. (2002). Do young children grasp the inversion relationship between addition and subtraction? Evidence against early arithmetic. *Cognitive Development*, *17*, 1365–1383.

Watchorn, R. P. D., Bisanz, J., Fast, L., LeFevre, J., Skwarchuk, S. -L., & Smith-Chant, B. L. (2014). Development of mathematical knowledge in young children: Attentional skill and the use of inversion. *Journal of Cognition and Development*, *15*, 161–180.

Young, L. K., & Booth, J. L. (2015). Student magnitude knowledge of negative numbers. *Journal of Numerical Cognition*, *1*, 38–55.

Chapter 3

Arithmetic Word Problem Solving: The Role of Prior Knowledge

Catherine Thevenot
University of Lausanne, Institute of Psychology, Lausanne, Switzerland

INTRODUCTION

Solving a word problem, such as "Nat has 36 candies to share among himself and 3 friends. How many candies will each of them receive?" requires the correct use of arithmetic procedures but, beforehand and crucially, the identification of the right procedure (division by 4 in the example). To do so, the situation described in the text of the problem has to be understood and the mental representation constructed from the wording of the problem needs to be translated into a mathematical model (Verschaffel, Greer, & De Corte, 2000). As a result, word problem solving is a complex activity that requires reading skills, text comprehension aptitudes, along with arithmetical, and sometimes more complex mathematical skills. As we will see in this chapter, the comprehension demands of word problem has extensively been studied within different theoretical frameworks, which are not mutually exclusive. Despite some debate, these theories share the assumption that solvers' prior knowledge determines the way in which the problem is understood and solved.

Prior knowledge can either correspond to concrete daily life experiences that sometimes trigger naïve and simplistic conceptions about mathematical concepts, or to rigid problem schemata stored in long-term memory (see also DeWolf and Melissa, this volume). As discussed in this chapter, prior knowledge can sometimes facilitate the construction of mental representations of the problems but, unfortunately, at other times, it can be quite detrimental to problem solving. Although daily life experience can strongly influence the way word problems are understood, it turns out that, conversely, individuals sometimes disregard concrete knowledge about the real world. This is true even when using this knowledge is necessary to solve the problem or, less dramatically, when

Acquisition of Complex Arithmetic Skills and Higher-Order Mathematics Concepts
http://dx.doi.org/10.1016/B978-0-12-805086-6.00003-5

using this knowledge would have been helpful in the construction of accurate mental representations of the problems. In this chapter, I therefore discuss the disconnection between word problem solving and real world knowledge and the fact that it is often due to general erroneous conceptions that children and adults have about how to solve word problems. On a more positive note, it is possible to help students in overcoming these difficulties and, in the last part of this chapter, I will discuss programs that have been developed to this aim.

To sum up, I present in this chapter a review of the beneficial and detrimental roles of prior knowledge on arithmetical word problem solving. Special attention will be given to adults and children's naïve and simplistic conception of word problems.

THE ROLE OF DAILY LIFE EXPERIENCE IN SOLVING ARITHMETIC WORD PROBLEMS

One goal of teaching children to solve arithmetic word problems is to help them in concretizing abstract mathematical concepts by relating these to their daily lives. For example, the problem "Susan sold two-thirds of her jam jars for 44 cents each. She has 12 left for sale. How much money does she have from the jars already sold?" should allow children to apply their knowledge about fractions in a plausible and concrete situation. However, the relation can work both ways and our daily life experience can also impact the way arithmetic word problems are mentally represented and solved.

As an example of how the framing of a word problem can sometimes dramatically influence performance, consider the study by Fischbein, Deri, Nello, and Marino (1985) where 5th, 7th, and 9th graders were asked to choose the correct operations for two word problems. The first problem was: "From 1 quintal of wheat you get 0.75 quintals of flour. How much flour do you get from 15 quintals of wheat?" And the second problem was: "1 kilo of detergent is used in making 15 kilos of soap. How much soap can be made from 0.75 kilos of detergent?" For the first problem, the percentages of correct answer were 79% for 5th graders, 74% for 7th graders, and 76% for 9th graders. However, for the second problem, the percentages of correct answer were only of 27, 18, and 35% in 5th, 7th, and 9th graders, respectively. Despite the fact that both problems require the same multiplication (15×0.75), the second problem was far more difficult to solve than the first one.

According to Fischbein et al. (1985), this is due to the fact that in the first case the operator is a whole number but in the second case, the operator is a decimal (see also DeWolf and Melissa, this volume). In this second case, the problem contradicts the tacit mental representation that students have about multiplication, which is often understood as repeated addition. Within this naïve conception, the operator must be a whole number. For example, it is easy to understand that three quarters (0.75) of a liter of milk are multiplied by 3 to feed 3 cats ($0.75 + 0.75 + 0.75$), but a problem in which 3 L of milk is multiplied by

0.75 has no intuitive meaning. Similarly, the conceptualization of multiplication as repeated addition makes it difficult for students to understand that the multiplicand (i.e., first number in the example just above) can be a negative number. Another constraint of the repeated addition model for multiplication is the property that multiplication "makes bigger." This sometimes faulty conception is reinforced by everyday experience because when one thing is multiplied by something else, the result is always a larger number (see also Van Hoof and Jo, this volume). For example, if we multiply the price of a glass of wine by the number of guests at a party, the result is unfortunately higher than the price of a single glass. More generally, Fischbein et al.'s experiment shows that the inferences arising from people's daily experiences can be inappropriately applied during the solving of arithmetic word problems and other mathematical problems.

At other times, these everyday inferences can help students mentally represent word problems. Coquin-Viennot and Moreau (2003) showed that a problem such as "For a prize, the florist prepares 5 roses and 7 tulips for each of the 14 candidates. How many flowers does the florist use in total?" is likely to be solved by children using a distributed strategy, that is, (14 x/5) +/(14 x/7). However, a different strategy is used when the same problem is framed "For a prize, the florist prepares a bouquet composed of 5 roses and 7 tulips for each of the 14 candidates. How many flowers does the florist use in total?" With this wording, the problem is more likely to be solved with a factorizing strategy, by performing 14 x/(5 + /7). In the second formulation of the problem, the term "bouquet," and more precisely the knowledge that it corresponds to a bunch of flowers, structures the organization of the elements contained in the mental representation constructed to solve the problem.

Similar results were obtained by Bassok, Chase, and Martin (1998). They showed that the semantic relations between the everyday objects represented in word problems can influence how college students construct mathematical representations. More concretely, division involves an asymmetric mathematical relation between dividend and divisor ($a/b \neq b/a$). When Bassok et al.'s participants were asked to create word problems, they applied this asymmetric relation to object sets that readily evoked functionally asymmetric semantic relations (e.g., apples and baskets), but refrained from dividing object sets that did not evoke functionally asymmetric relations (e.g., apples and oranges). Conversely, when apples and oranges had to be integrated in the text of the problem, college students usually related such functionally symmetric sets with the mathematically symmetric operation of addition ($a + b = b + a$). As in Coquin-Viennot and Moreau's (2003) study, students' knowledge of everyday relationships shaped the mental representations they constructed about mathematical relations.

These illustrations suggest that students' mental representations constructed to solve arithmetic word problems are strongly influenced by their everyday knowledge. In other words, individuals process the concrete situations described within the text of the problem and do not focus directly on the mathematical relations between the different elements of the problem. According to Brissiaud and Sander (2010), the initial mental representation constructed from a problem text always activates a situation-based strategy first, even after instruction (see also Brissiaud, 1994). In other words, students will always first attempt to solve word problems using their knowledge about concrete daily life situations. This approach is modified only when the strategy based on the simulation of the situation described in the text does not produce an answer. For example, the situation-based strategy for a problem such as "Luc is playing with 42 marbles. During the game, he loses 39 marbles. How many marbles does Luc have now?" would consist of counting backward from 42, which is cognitively demanding. A less demanding and thus more efficient approach would be to "rerepresent" the problem and count forward from 39 to 42. In accordance with their Situation Strategy First model, the authors show that such problems are indeed more difficult for 7- and 8-year-old children than problems where the situation-based strategy maps directly onto the most efficient strategy (i.e., "Luc is playing with 42 marbles. During the game he loses 3 marbles. How many marbles does Luc have now?").

We have seen so far that real world knowledge can sometimes prevent the construction of mathematically accurate mental representations of the situations described in word problems. We have also seen that everyday knowledge can shape the mental representations of word problems and even support the mathematical representation of the information presented in them. Nevertheless and conversely, as we will see in the next section, real word knowledge is sometimes ignored when it should not be because individuals wrongly assume that everyday experience has nothing to do with mathematical problem solving.

THE ROLE OF STEREOTYPIC REPRESENTATIONS ABOUT PROBLEM SOLVING IN SCHOOL

Verschaffel and coworkers (Verschaffel, De Corte, & Lasure, 1994; Verschaffel et al., 2000) studied what they call P-items. These are items in which individuals fail to make realistic considerations or, following Verschaffel's wording, items that are problematic for a realistic modeling. For example, consider the problem: "450 soldiers must be bussed to their training site. Each army bus can hold 36 soldiers. How many buses are needed?" Here the student has to realize that the mathematical answer 12.5 does not make sense in real life and that the answer of 13 buses is more realistic. As another example, consider the problem "Ann has bought 4 2.5 meter planks. How many 1 meter planks can she saw from these planks." In the real world, the correct answer is 8, although many students answer 10(4×2.5) because they ignore the everyday context. Verschaffel et al. (1994) showed that most students process these problems as if they were completely disconnected from reality and simply apply the arithmetic operations suggested in the described situation. Strikingly, even after encouraging students to use real-life considerations either by highlighting the nonroutine aspects of the problems (Yoshida, Verschaffel, & De Corte, 1997) or adding some details in the texts to make them more authentic (Palm, 2008), many students still fail to integrate realistic considerations to their mental representations of the problems and give unreasonable answers from a practical point of view.

Therefore, these studies show that students are sometimes reluctant to use their everyday knowledge to solve arithmetic word problems. This is probably due to the fact that they possess restricted assumptions about the nature of these problems. In other words, they create constraints that do not exist. For example, children have a strong tendency to believe that every word problem has a solution (Reusser & Stebler, 1997). More concretely, De Corte and Verschaffel (1985) presented 1st graders with the problem "Pete had some apples. He gave 4 apples to Ann. How many apples does Pete have now?" The majority of children did not realize that there was a missing value and, therefore, that the problem was unsolvable. Likewise, children have great difficulties in conceiving that several different answers can be provided for a specific problem. For them, one problem has one unique solution. To illustrate, consider the problem "Sven and Karen are nature lovers. They counted the number of ducks in a lake. Sven has counted 12 ducks and Karen 18. How many ducks are there in the lake?" Many children state there are 30 ducks without considering that Sven and Karen might have counted some of the same ducks. Similarly, for the problem "Bruce and Alice go to the same school. Bruce lives at a distance of 17 kilometers from the school and Alice at 8 kilometers. How far do Bruce and Alice live from each other?", only 5% of upper elementary school children realized that several answers can be given (Verschaffel et al., 1994).

Another strong belief that children have about word problem is that all of the provided values need to be used in order to reach the answer. To illustrate, consider the problem "On the first day of the big race, Anthony ran 12 miles. He ran 9 miles the second day. Shelley sold 15 quarts of lemonade during the race. She gave away 8 quarts of ice water. How far did Anthony run?". Among low achieving 5th graders, 65% of them added all four numbers provided in the problem (Littlefield & Rieser, 1993; see Jimenez & Verschaffel, 2014 for a review). Finally, many students falsely believe that everyday knowledge cannot interfere with the solving of word problems. Then, as an additional example, problems such as "The distance between 2 poles is 12 meters. How many pieces rope of 1.5 meters would Ann need to tie together to stretch between the poles?" will classically lead to the answer 8. Pupils know that an extra but undetermined length of rope will be needed to tie the ropes together and to tie the two end pieces to the poles. Nevertheless, they probably ignore these concrete considerations because they think that they have nothing to do with mathematical reasoning.

In fact, it is not so surprising that children do not always process arithmetic word problem solving as real complex situations because the problems studied in school generate a disconnect. Indeed, the vast majority of problems in textbooks are solvable, require the use of all the provided numbers, and have only a single solution that is directly given by the calculation to be performed. This is probably why 30% of German 1st graders provide an answer to the problem "There are 26 sheep and 10 goats on a ship. How old is the captain?" (Radatz, 1983, cited in Verschaffel et al., 2000). Even more striking, only 10% of kindergarten children provided an answer to this problem but this percentage reached 60% in 3rd grade. Based on these patterns, Reusser (1988) came to the alarming conclusion that such absurdities are created in school and sustained by school practice. Whereas young children try to process the problem sensibly, older students blindly apply ready-made strategies as they would use kitchen recipes without careful examination.

We have seen so far that naïve representations about mathematical concepts and narrow representations of how word problems can be solved can impact students' understanding of what is being asked in these problems (see Table 3.1 for a summary). In the next part of this chapter, we will see that these two factors cannot explain all solvers' behaviors and misconceptions and that specific problem schemata can also shape individuals' word problem mental representations.

THE ROLE OF PROBLEM SCHEMATA STORED IN LONG-TERM MEMORY

As noted, problem solvers have great difficulties in mentally representing multiplication problems with a decimal number as the operator (Fischbein et al., 1985). As already explained, this is most probably because in daily life situations, multiplication is conceived as repeated addition. Within this limited conception, the repetition of a collection by a decimal number does not make intuitive sense (see DeWolf and Melissa, this volume). However, incorrect inferences based on everyday experience are not the only source of misconceptions about word problems. Adults also have great difficulty in imagining a division problem wherein the answer is larger than the numbers used in the problem (Fischbein et al., 1985), whereas these quotative situations in which counting or measuring the number of times you can subtract the divisor from the dividend often occur in everyday life. Imagine that you have 6 L of coffee and you want to determine how many cups of 25 cL you will be able to serve. You will need to divide 6 by 0.25 to determine that 24 people can be served. Therefore, the difficulty that adults have in constructing nontypical word problems not only stems from their naïve conception of mathematical concepts, which is strongly based on their everyday experience, but also from prototypic mental schemata for solving specific types of problems. In the case of division, the mental predominance of the partitive (e.g., equal sharing) over the quotative schema is probably due to the fact that division is mainly taught in school through the use of partitive examples.

More generally, mental schemata are abstract problem frames stored in long-term memory. They take the form of semantic structures representing typical relations between elements described in arithmetic word problems (Kintsch

TABLE 3.1 Example of Factors That can Shape the Mental Representations of Arithmetic Word Problems

Daily life considerations	Examples
Naïve conceptions about arithmetic operations (Fischbein et al., 1985)	Multiplication by a decimal number has no intuitive meaning: from 1 quintal of wheat you get 0.75 quintals of flour. How much flour do you get from 15 quintals of wheat?" is easier than "1 kilo of detergent is used in making 15 kilos of soap. How much soap can be made from 0.75 kilos of detergent?"
Structuring terms (Coquin-Viennot & Moreau, 2003; Bassok et al., 1998)	(14x/5) +/(14x/7) for "For a prize, the florist prepares 5 roses and 7 tulips for each of the 14 candidates. How many flowers does the florist use in total?" but 14 x/(5 +/7) for "For a prize, the florist prepares a bouquet composed of 5 roses and 7 tulips for each of the 14 candidates. How many flowers does the florist use in total?"
Mental simulation (Brissiaud & Sander, 2010)	42, 41, 40, 39 for "Luc is playing with 42 marbles. During the game he loses 3 marbles. How many marbles does Luc have now?" but 39, 40, 41, 42 for "Luc is playing with 42 marbles. During the game, he loses 39 marbles. How many marbles does Luc have now?"
Stereotypic representations of word problems	
All problems have a solution (Reusser & Stebler, 1997; De Corte & Verschaffel, 1985)	Children try to solve: "Pete had some apples. He gave 4 apples to Ann. How many apples does Pete have now?"
A problem does not have several answers (Verschaffel et al., 1994)	"Sven and Karen are nature lovers. They counted the number of ducks in a lake. Sven has counted 12 ducks and Karen 18. How many ducks are there in the lake?" Many children answer 30.
All values must be used (Littlefield & Rieser, 1993)	"On the first day of the big race, Anthony ran 12 miles. He ran 9 miles the second day. Shelley sold 15 quarts of lemonade during the race. She gave away 8 quarts of ice water. How far did Anthony run?" Some children add up the 4 numbers.
Everyday experience has nothing to do with problem solving (Verschaffel et al., 1994)	"Ann has bought 4 planks each 2.5 meters long. How many planks 1 meter long can she saw from these planks". Students can give the answer 10.

& Greeno, 1985). More concretely, a schema is a "construct that allows problem solvers to group problems into categories in which the problems in each category require similar solutions" (Cooper & Sweller, 1987, p. 348). This type of schematic knowledge is only found in proficient word problem solvers because schemata are constructed through practice and abstracted from numerous encounters with the same exemplar of a problem category. Schemata are abstract

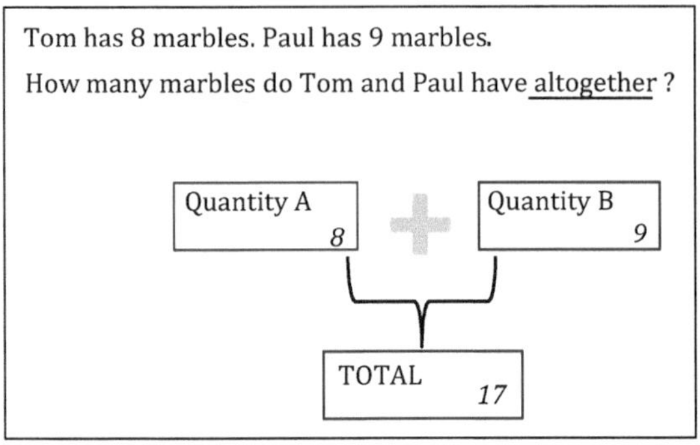

FIGURE 3.1 **Example of an activation and instantiation of a Combine schema.**

in the sense that they only contain the general frame related to a problem category and contain empty slots that are filled by the specific information described in the text (see Fig. 3.1 for an example). Indeed, the specific values or the name of protagonists described in the problem are integrated within the problem schema as soon as they are read. Another characteristic of schemata is that they are activated by specific expressions used in the text of the problem. For example, if a problem contains the word "altogether," an addition is often needed to solve the problem and problem solvers thus activate a Combine schema.

More precisely, Riley, Greeno, and Heller (1983) have classified typical addition problems as a function of their semantic structure (see Table 3.2 for examples of the problems in the different categories). Change problems describe actions that cause increases or decreases in quantities and the unknown value is either the resulting amount, the amount of the change, or the starting amount. In contrast to such dynamic problems, Combine and Comparison problems represent static situations. Combine problems describe two quantities that have to be considered in combination to produce the answer. The question can be relative to the total amount or the amount in one of the subsets. Finally, in Comparison problems, a comparison is made between two quantities using the expressions "more than" and "less than." The unknown is either the difference between the two quantities, the compared quantity, or the referent.

As already mentioned, specific expressions, such as "more than," "altogether," "give," or "loose" in the text of the problem activate the corresponding schema. These schemata contain the procedure required to solve the problem and they are triggered as soon as all empty slots (i.e., number to be operated on) are filled in (Kintsch & Greeno, 1985). Lewis and Mayer (1987) provided good experimental evidence for the use of word problem schemata by studying two different forms of comparison problems, namely consistent and inconsistent problems. In consistent problems, the relational term maps directly onto the arithmetic operation to be performed. For example, the expression "has more" is consistent with the fact that an addition is required to solve the problem "John has 3 marbles. Tom has 5 more marbles than Joe. How many marbles does Tom have." In contrast, in

TABLE 3.2 **Examples of Word Problems From the Classification of Riley et al. (1983)**

Type of problems	Unknown variable	Examples
Change problems	Resulting amount	John has 6 marbles. Tom gives him 3 more marbles. How many marbles does John have now?
	Amount of the change	John had 4 marbles. Tom gave him some more marbles. Now, John has 6 marbles. How many marbles did Tom give him?
	Starting amount	John had some marbles. Tom gave him 5 marbles. Now, John has 7 marbles. How many marbles did John have at the beginning?
Combine problems	Total amount	John has 2 marbles. Peter has 4 marbles. How many marbles do John and Tom have altogether?
	Amount in one of the subset	Joe and Tom have 8 marbles altogether. Joe has 3 marbles. How many marbles does Tom have?
Comparison problems	Difference between the two quantities	John has 3 marbles. Tom has 5 marbles. How many marbles does John have less than Tom?
	Compared quantity	John has 3 marbles. Tom has 5 more marbles than John. How many marbles does Tom have?
	Referent	John has 3 marbles. He has 5 less marbles than Tom. How many marbles does Tom have?

inconsistent problems, the relational term conflicts with the necessary arithmetic operation. As an example, the term "less" is inconsistent with the addition that has to be performed to solve the problem "Joe has 3 marbles. He has 5 marbles less than Tom. How many marbles does Tom have?" It turns out that inconsistent problems are more error prone and are solved more slowly than consistent problems. By default, an addition procedure is activated as soon as the word "have more" is read and, conversely, a subtraction procedure is activated as soon as the expression "have less" is read. If the default procedure is incorrect for the problem and not inhibited, the problem will not be solved accurately. Even if the default procedure is inhibited, more time will be needed for this inhibition and for mentally rearranging the problem information so that it can be correctly solved.

High mental availability of schemata constructed after numerous encounters of typical word problems explains why, when asked to imagine an arithmetic word problem that can be solved by a subtraction between two numbers (e.g., 8 – 3), the vast majority of 8- and 9-year-olds produce a text such: "[Tom, John…] had 8 [marbles, candies…]. He [lost, gave…] 3 of them. How many [marbles, candies…] does he have left?" (Torres-Alcedo, Nogry, Bosc-Miné, & Sander, in preparation). However, as described by Riley et al. (1983), a subtraction is not needed only in situations where something is lost and, as a matter of fact, some subtraction problems involve gains. This is the case for Change problems when the amount of change is unknown (e.g., John had three marbles. Tom gave him some marbles. Now, John has eight marbles. How many marbles did Tom give him?), or when the starting amount is unknown (John had some marbles. Tom gave him three marbles. Now, John has eight marbles. How many marbles did John have at the beginning?).

A subtraction is also required in Combine problems when one of the subset is unknown (e.g., Joe and Tom have eight marbles altogether. Joe has three marbles. How many marbles does Tom have?), or in simple Comparison problems wherein a comparison is made between two quantities using the expressions "more than" and "less than" (e.g., John has three marbles. Tom has eight marbles. How many marbles does John have less than Tom?). The reason why adults consistently imagine subtraction problems as stories where some objects are lost or given is due to the fact that the word "subtraction" convokes its most typical definition in mathematics according to which a number is taken away from another (Fischbein et al., 1985). Another common conception of subtraction is "building up" as in the problem "John has 6 marbles. How many more does he need in order to have 10 marbles" (Fischbein et al., 1985). When this conception is activated, children have a strong tendency to count from 6 to 10 in order to solve the problem rather than to subtract 6 from 10 (see Brissiaud, 1994). In relation with the first part of this chapter, note that many readily available schemata are not only constructed through school practice but also through numerous everyday experiences.

A different illustration of the obstacles caused by the activation of highly typical word problem schemata can be provided within the domain of multiple step problems. Thevenot and Oakhill (2005) presented adults with problems, such as "How many more marbles do John and Tom have altogether than Paul? John has 44 marbles. Tom has 24 marbles. Paul has 41 marbles." The strategy preferred by participants to solve this problem was to calculate: (44 + 24) – 41. The word "altogether" strongly activated a Combine problem schema and the expression "have more" easily activated a Comparison schema (Riley et al., 1983). However, a more efficient strategy to solve the problem would be to consider that John has 3 more marbles than Paul (44 – 41) and that, as a consequence, the quantity of Tom's marbles (24) has simply to be added to 3. Nevertheless, this strategy supposes that the solver managed to inhibit the activation of typical problem schemata in order to construct an alternative mental representation of the problem text. Interestingly, when three-digit numbers were used in the text of the problem, (e.g., How many more marbles do John and

Tom have altogether than Paul? John has 749 marbles. Tom has 323 marbles. Paul has 746 marbles), college students were more likely to use the more efficient strategy. With three-digit numbers, the calculations required through the mobilization of the classic Combine and Comparison schemata were probably too demanding for participants, who then searched for an alternative representation of the problem in order to reduce the cognitive load associated with solving processes.

In other words, individuals seem to be able to weigh the costs and benefits of either blindly applying straightforward schemata or engaging in a cognitively more demanding activity to construct an alternative mental representation (see Thevenot & Oakhill, 2008 for a more precise evaluation of these costs and benefits). A noticeable result in Thevenot and Oakhill's study (2005) was that, despite the objectively higher difficulty of three-digit than two-digit word problem, three-digit problems were solved as accurately as two-digit problems. Therefore, the alternative representation constructed and used by participants to solve the problems allowed them to overcome the complexity of three-digit additions and subtractions. In a similar study, Thevenot and Oakhill (2006) showed that adults with lower working memory capacities were more likely to construct alternative representations than adults with higher capacities, even when two-digit numbers were used in the text of the problems. The interpretation was that solving the problems with classic schemata was considered as too demanding for these individuals and thus they benefited from constructing alternative representations to the ones directly induced by the wording of the problem.

In summary, schemata stored in long-term memory and activated as soon as specific expressions are read in the text of problem ("altogether," "have more," "have less"...) can help identify which procedures are required to solve a problem. Nevertheless, an automatic application of these schemata can undermine performance and can sometimes prevent individuals from setting up more efficient and less cognitively demanding strategies.

Proportional problems, which are related with linear functions, are also strongly associated with highly mentally available schemata and automatic procedures. This type of problem (e.g., Sam needs four tablespoons of sugar for two cakes, how many would he need for three cakes?) is usually learned during 5th or 6th grades and the way they are formulated is standardized. As a consequence, when nonproportional problems share some wording characteristics with proportional problems, they can create an illusion of linearity. Indeed, according to Freudenthal (1983, p. 267), "linearity is such a suggestive property of relations that one readily yields to the seduction to deal with each numerical relation as though it were linear." As an example, 32 of 33 preservice teachers incorrectly answered 45 (3×15) for the problem "Sue and Julie were running equally fast around a track. Sue started first. When she had run 9 laps, Julie had run 3 laps. When Julie completed 15 laps, how many laps had Sue run?" In other words, almost all apprentice teachers processed the problem as if it were a proportional rather than an additive problem. Indeed, the statement explicitly mentions that Sue and Julie are equally fast so 6 laps separate them. When Julie has completed 15 laps, Sue has therefore completed 15 + 6 = 21 laps (Cramer, Post, & Currier, 1993).

Also in line with the illusion of linearity, Van Dooren, De Bock, Depaepe, Jansen, and Verschaffel (2003) observed that many secondary school and university students believe that the probability of getting a 6 when a dice is rolled twice is 2/6 (i.e., 1 chance out of 6 of getting a 6 on a single roll ×2, whereas the answer is 11/36: 6 × 6 possibilities when 2 dice are rolled and 11 combinations with at least a 6). Even when students are given strong suggestions (e.g., "keep in mind that all problems are not proportional") for solving this kind of problem, they persist with their first assumption and keep considering them as proportional problems (De Bock, Verschaffel, & Janssens, 2002; De Bock, Verschaffel, Janssens, Van Dooren, & Claes, 2003). Strikingly, students who eventually admit that the problems are not proportional then have difficulties solving typical proportional problems because they are now unsure whether the problem is proportional or not (Van Dooren, De Bock, Hessels, Janssens, & Verschaffel, 2004; De Bock, Van Dooren, Janssens, & Verschaffel, 2007 for a review). As in the age of the captain problem described earlier, the fact that the linearity illusion increases from 2nd grade to 5th or 6th grade provides evidence that, again, schooling is partly responsible for the illusion (Van Dooren, De Bock, Gillard, & Verschaffel, 2009; Van Dooren, De Bock, Hessels, Janssens, & Verschaffel, 2005).

In all, individuals easily activate mental schemata to solve highly typical arithmetic word problems. Nevertheless, as I describe more directly in the next part of this chapter and as already indicated earlier, the schema theory cannot by itself explain all aspects of problem solving. Specifically, schemata are rigid mental structures that cannot account for contextual or subtle wording effects.

THE USE OF SCHEMATA VERSUS SITUATION-BASED MODELS

According to Staub and Reusser (1995) as well as Nathan, Kintsch, and Young (1992), the schema theory cannot account for all of the effects observed in the domain of arithmetic word problem solving. Without being explicit, I have already provided examples supporting the idea that rigid schemata stored in long-term memory cannot explain all children's and adults' problem solving performance and strategies. Indeed, the schema theory, which postulates that specific expressions will trigger a particular procedure, cannot account for results, such as those obtained by Coquin-Viennot and Moreau (2003) showing that the integration of the term "bouquet" within the text of a problem can modify students' problem solving strategy. Therefore, "nonmathematical" representations are sometimes necessarily constructed by individuals to solve arithmetic word problems. The Situation Model was the first to integrate these nonmathematical representations (Van Dijk & Kintsch, 1983). It corresponds to a level of representation that specifies the agents, the actions and the relationships between the events in everyday contexts. Within this framework, the difficulty of the problem is determined by the difficulty in understanding the described situation and not by the difficulty in activating the right mental schema. The same type of representation, namely a Mental Model, is described by Johnson-Laird (1983) in the domain of text comprehension and reasoning. For Johnson-Laird, solving a problem requires the construction of a mental representation of the situation described by the problem and not a mental representation of the problem itself (Zwaan & Radvansky, 1998).

The relevance of these two types of theories, namely the schema theory on the one hand and the mental or situation model on the other hand, has been directly compared by manipulating the position of the questions asked in the problem (Thevenot, Devidal, Barrouillet, & Fayol, 2007). It is known that placing the question at the beginning rather than at the end of an arithmetic word problem facilitates problem solving (e.g., How many marbles do Tom, John, and Paul have altogether? Tom has 31 marbles, John has 22 marbles, and Paul has 17 marbles). This was first interpreted by the early activation of the adequate problem schema, triggered as soon as a specific expression used in the question is read (e.g., "more than" or "altogether") (Devidal, Fayol, & Barrouillet, 1997; Fayol, Abdi, & Gombert, 1987). This early activation would enable the integration of the numerical data to the schema as soon as they are encountered, which, in turn, would allow the problem solver to execute the calculation procedures while reading the problem. This on-line processing necessarily reduces working memory demands, which could explain the improvement in performance.

However, Thevenot, Barrouillet, and Fayol (2004) showed that the positive effects associated with the position of the question before a text problem is also observed when the calculations cannot be executed while reading. The authors proposed that the question before the text operates as a problem solving guide and help individuals in constructing their mental representation of the problem. As with a title before a narrative text (Rawson & Kintsch, 2002), the question before an arithmetic problem would facilitate the subsequent encoding and integration of the information in the representation. Thevenot et al. (2007) confirmed this interpretation by showing that the facilitating effect of the question before the text is more pronounced for children with poor abilities in mathematics. This last result is not compatible with the schema theory. If the facilitating effect of the question before the text is due to the early activation of the appropriate schema, high-skilled children would benefit the most from this effect. Indeed, because schemata are abstracted from numerous encounters of a specific problem, high-skilled children are, by definition, those children who have more often successfully solved word problems in the past. As a consequence, they are also the children who are the most likely to have abstracted and stored problem frames in long-term memory. Therefore, within a schema-based approach, high-skilled children should have benefited the most from an early schema activation whereas it was low-skilled children who benefitted the most. It has been shown that low-skilled children experience great difficulties in the construction of coherent and accurate mental models of situations (Cain & Oakhill, 1999; Oakhill, 1996; Oakhill, Cain, & Yuill, 1998; Yuill, Oakhill, & Parkin, 1989). Thus, within the mental model framework, it is easy to understand why lower-skilled benefit more than high-skilled individuals from an aid to the construction of the representation.

Thevenot (2010) provided another direct test of the relevance of the mental model and schema theories in the domain of arithmetic word problem solving. In her experiment, college students were asked to solve Comparison problems wherein the question took a form such as "How many marbles does Louis have more than Jean?" After solving a series of such problems, participants' ability to recognize previously solved problems was assessed. They were presented either with the original problems, with inconsistent problems that had never been solved, or with paraphrases that respected the relational structure of the original problems, but not their exact wording. These paraphrases were constructed by inversing the terms and the linguistic expressions in the original problems (my example became "How many marbles does Jean have less than Louis?). Inconsistent problems were constructed by keeping the same relational terms as in the original problem, but by inversing the names of the protagonists (i.e., "How many marbles does Jean have more than Louis"). Whereas the literal form of the paraphrased problems bore the least resemblance to original problems, these problems were nevertheless associated to higher recognition rates than inconsistent problems. These results provided strong evidence that it is the structure of the arithmetic word problem and not its exact wording or its propositional level of representation that is mentally represented by individuals engaged in problem solving task. This result is obviously at odds with the schema theory postulating that the relational expression ("have more" in our example) plays a crucial role in triggering the appropriate mental schema.

In conclusion, the schema theory alone cannot account for all the findings in the domain of arithmetic problem solving. Of course, as described in the preceding part of this chapter, this is not to say that schemata are never evoked in word problem solving activities. In fact, we saw that they can sometimes obstruct problem-solving processes when they are automatically activated and should have been inhibited (e.g., Lewis & Mayer, 1987; Thevenot & Oakhill, 2005, 2006).

HOW CAN WE HELP STUDENTS

The different studies described in this chapter show that individuals sometimes rely excessively on their everyday experience when solving solve arithmetic word problems. At other times, they reflexively apply mental schemata derived from these experiences or from highly typical problem situations encountered in textbooks. At still other times, students fail to correctly solve word problems because they disconnect arithmetic problem solving activities from real world considerations that can be used to solve the problem. Therefore, helping students to correctly mentally represent the situations described in the texts of arithmetic world problems can be quite challenging.

Indeed, we have already mentioned that providing explicit hints about the need to take the realistic context into account has only minimal effects on children's performance (Yoshida et al., 1997). Moreover, providing drawings or figures highlighting the realistic dimensions that should be considered is not more efficient than not providing them (Weyns, Van Dooren, & Verschaffel, submitted). More generally, alerting pupils on the problematic nature of problems that require realistic considerations has only small effects on the number of realistic answers students provide (Verschaffel, De Corte, & Lasure, 1999). In contrast, an effective way of improving children's performance is to set up more authentic situations. More precisely, DeFranco and Curcio (1997, described in Verschaffel et al., 2000) modified the bus problem described earlier (450 soldiers must be bussed to their training site. Each army bus can hold 36 soldiers. How many buses are needed?) and asked 6th graders to make a telephone call to order minivans for a class party. More precisely, 32 children were supposed to go to the party and participants had to order a certain number of minivans, knowing that each minivan can seat 5 children. In this realistic setting, the number of realistic answers increased significantly (i.e., from 10% to 95%). According to Verschaffel et al. (2000), the positive effects of removing arithmetic word problems from the routine context of typical school arithmetic lessons were obtained because students did not engage in the strong schema-based constraints related to word problem solving that they have created over instruction years.

Fischbein (1989) suggested a three-step intervention program concerning students' tendency to rely on their naïve and intuitive conception of mathematical concepts when they solve arithmetic word problem. First, students' production (i.e., the answer and all work leading to it) should be carefully analyzed in order to understand which tacit mathematical model is responsible for the errors. Second, the identified tacit models should be explained to make students aware of the conflict between their models and more formal models. Third, the same problem should be presented in various contexts to assess the transfer of the new acquired formal mathematical concepts. However, according to Fischbein, students should not be taught that intuitive models are always wrong but rather that their use is not always appropriate.

The conceptual change theory can be used to develop is another pedagogical approach to reduce students overuse of mental schemata and failure to construct more flexible and adaptive mental representations (e.g., Vosniadou, 2003). The key idea is that students sometimes need to change their mental representations of problems because their initial representations are contaminated by prior too simplistic or even flawed knowledge. Within the conceptual change theory, the first step is to determine students' prior knowledge that might be used to solve the presented problems. The second step is to explicitly address their misconceptions about how this prior knowledge is relevant to the current problem. The third step is to promote students' metaconceptual awareness and metacognition about their misconceptions, mainly through group discussions. Conceptual change programs rely on external representations, such as tables, diagrams, or graphs and the level of students' motivation is kept high through the use of attractive and meaningful problems. Van Dooren et al. (2004) nicely applied this framework to proportional reasoning and showed that conceptual change was indeed possible in this domain. Indeed, the authors reduced students' linearity illusions but noticed that this was not accompanied by a deep conceptual understanding of the newly acquired notions.

Finally and quite differently, students can also be helped when they have not yet stored the adequate problem schemata in long-term memory. Jitendra, Griffin, Deatline-Buchman, and Sczesniak (2007) designed an intervention tool called the schema-based instruction (SBI). SBI promotes the explicit analysis of the problem schema and the relationships between its different elements. According to the author, "SBI allows students to approach the problem by focusing on the underlying problem structure, thus facilitating conceptual understanding and adequate word-problem-solving skills" (Jitendra et al., 2007, p. 286). The instructional program consists of asking children to read and retell the problem to discover its type (e.g., Change, Combine, and Comparison problems described earlier). Then, the use of a schematic diagram is provided in order to decide which operation needs to be performed. Once the answer is reached, the last important step is checking the answer. SBI applications improved student word problem performance in regular and special education classrooms (Jitendra & Hoff, 1996; Jitendra & Star, 2011).

CONCLUSIONS AND FUTURE DIRECTIONS

Prior knowledge based on everyday experience to schema-based representations of problem solving and to prototypic problems have strong influences on the way adults and children mentally represent arithmetic word problems. When a prototypic problem is encountered (e.g., a Change problem), a schema is activated from long-term memory and the problem can be solved quickly. Unfortunately, because of superficial analyses of the problem text, the wrong or, at least, not the optimum schema is often activated, resulting in problem solving difficulties. Schemata are often constructed through school practice but can also be abstracted from numerous everyday experiences of situations superficially similar to the word problem. Therefore and according to Fischbein et al. (1985), "enactive prototype of an arithmetical operation may remain rigidly attached to the concept long after the concept has acquired a formal status" (p. 6). The third type of prior knowledge strongly influencing arithmetic word problem solving is stereotypical beliefs about problem solving activities, such as "all numbers have to be used" or "world knowledge has nothing to do with word problems." Such stereotypes create a disconnection between word problem solving and real word knowledge, which can be quite detrimental to solving processes.

Nevertheless, despite possible interferences between prior knowledge and arithmetic word problem solving, it is important to note that successful solving cannot occur without reliance on knowledge stored in long-term memory. Each problem-solving step, such as reading, comprehension, translation of the situation in a mathematic sentence or arithmetic operations, requires the mobilization of basic (e.g., letter recognition) or more complex (e.g., word or sentence meaning) knowledge. The key is to utilize the most relevant knowledge for the problem at hand. Explicit interventions are needed to facilitate students' ability to identify and use the most appropriate schema knowledge or to promote conceptual change. More generally, mental flexibility seems to constitute a crucial skill that children must apply in the domain of arithmetic problem solving. Indeed, a large part of children's and adults' errors, flawed mental representations, and incorrect answers described in this chapter are the consequence of rigid problem processing. Therefore, promoting mental flexibility should be one of the teachers' priorities in arithmetic word problem solving classrooms. To this aim, the use of multiple-step problems, such as the ones described earlier and studied by Thevenot and Oakhill (2005, 2006) could be very beneficial to children (e.g., how many marbles do Tom and Paul have more than John? Tom has 44 marbles, Paul has 27 marbles, and John has 29 marbles). As seen before, these problems can be solved through different strategies and, importantly, the best strategy depends on the specific values of the numbers in the text.

In this way and through extensive practice, children could eventually understand that there is no such thing as "magic recipes" to solve arithmetic word problems and that a deep conscious analysis of the problem to be solved is a necessary step during problem solving. Superficial considerations of the problems could be minimized and this could prevent children from committing the errors I have described throughout this chapter. Additionally, introducing P-items in school curriculum could also be of great help for children who blindly apply problem-solving routines without considering the more practical aspects of the problem. If children were sometimes presented with unsolvable problems, problems with several possible answers, unrealistic problems, or problems needing real world considerations in order to be solved sensibly, then, a deep understanding of each problem encountered in classrooms would be required. As with multiple-step problems, P-items could therefore improve children's sense of analysis and could help them in developing their ability to construct flexible and accurate mental representations of the situations described in arithmetic word problems.

REFERENCES

Bassok, M., Chase, V. M., & Martin, S. A. (1998). Adding apples and oranges: alignment of semantic and formal knowledge. *Cognitive Psychology*, *35*, 99–134.

Brissiaud, R. (1994). Teaching and development: solving "missing addend" problems using subtraction. *European Journal of Psychology of Education*, *4*, 343–365.

Brissiaud, R., & Sander, E. (2010). Arithmetic word problem solving: a Situation Strategy First Framework. *Developmental Science, 13*, 92–107.

Cain, K., & Oakhill, J. (1999). Inference making ability and its relation to comprehension failure in young children. *Reading and Writing, 11*, 489–503.

Coquin-Viennot, D., & Moreau, S. (2003). Highlighting the role of the episodic situation model in the solving of arithmetical problems. *European Journal of Psychology of Education, 18*, 267–279.

Cooper, G., & Sweller, J. (1987). Effects of schema acquisition and rule automation on mathematical problem-solving transfer. *Journal of Educational Psychology, 79*, 347–362.

Cramer, K., Post, T., & Currier, S. (1993). Learning and teaching ratio and proportion: research implications. In D. T. Owens (Ed.), *Research ideas for the classroom: Middle grades mathematics* (pp. 159–178). New York: Macmillan.

De Bock, D., Verschaffel, L., & Janssens, D. (2002). The effects of different problem presentations and formulations on the illusion of linearity in secondary school students. *Mathematical Thinking and Learning, 4*, 65–89.

De Bock, D., Verschaffel, L., Janssens, D., Van Dooren, W., & Claes, K. (2003). Do realistic contexts and graphical representations always have a beneficial impact on students' performance? Negative evidence from a study on modelling non-linear geometry problems. *Learning and Instruction, 13*, 441–463.

De Bock, D., Van Dooren, W., Janssens, D., & Verschaffel, L. (2007). *The illusion of linearity: From analysis to improvement*. New York: Springer.

De Corte, E., & Verschaffel, L. (1985). Beginning first graders' initial representation of arithmetic word problems. *Journal of Mathematical Behavior, 4*, 3–21.

DeFranco, T. C., & Curcio, F. R. (1997). A division problem with a remainder embedded across two contexts: children's solutions in restrictive vs. real-world Settings. *Focus on Learning Problems in Mathematics, 19*, 58–72.

Devidal, M., Fayol, M., & Barrouillet, P. (1997). Stratégies de lecture et résolution de problèmes arithmétiques. *L'Année Psychologique, 97*, 9–31.

Fayol, M., Abdi, H., & Gombert, J. -E. (1987). Arithmetic problem formulation and working memory load. *Cognition and Instruction, 4*, 183–202.

Fischbein, E. (1989). Tacit models and mathematical reasoning. *For the Learning of Mathematics, 9*, 9–14.

Fischbein, E., Deri, M., Nello, M. S., & Marino, M. S. (1985). The role of implicit models in solving verbal problems in multiplication and division. *Journal for Research in Mathematics Education, 16*, 3–17.

Freudenthal, H. (1983). *Didactical phenomenology of mathematical structures*. Dordrecht: Reidel.

Jiménez, L., & Verschaffel, L. (2014). Development of children's solutions of non-standard arithmetic word problem solving. *Journal of Psychodidactics, 19*, 93–123.

Jitendra, A. K., Griffin, C. C., Deatline-Buchman, A., & Sczesniak, E. (2007). Mathematical word problem solving in third-grade classrooms. *The Journal of Educational Research, 100*, 283–302.

Jitendra, A. K., & Hoff, K. (1996). The effects of schema-based instruction on the mathematical word-problem solving performance of students with learning disabilities. *Journal of Learning Disabilities, 29*, 422–431.

Jitendra, A. K., & Star, J. R. (2011). Meeting the needs of students with learning disabilities in inclusive mathematics classrooms: The role of schema-based instruction on mathematical problem-solving. *Theory into Practice, 50*, 12–19.

Johnson-Laird, P. N. (1983). *Mental models: Towards a cognitive science of language, inference, and consciousness*. Cambridge, MA: Harvard University Press.

Kintsch, W., & Greeno, J. G. (1985). Understanding and solving word arithmetic problems. *Psychological Review*, *92*, 109–129.

Lewis, A., & Mayer, R. E. (1987). Students' miscomprehension of relational statements in arithmetic word problems. *Journal of Educational Psychology*, *79*, 363–371.

Littlefield, J., & Rieser, J. J. (1993). Semantic features of similarity and children's strategies for identifying relevant information in mathematical story problems. *Cognition and Instruction*, *11*, 133–188.

Nathan, M. J., Kintsch, W., & Young, E. (1992). A theory of algebra word problem comprehension and its implication for the design of learning environments. *Cognition and Instruction*, *4*, 329–390.

Oakhill, J. (1996). Mental models in children's text comprehension. In J. Oakhill, & A. Garnham (Eds.), *Mental models in cognitive science* (pp. 77–94). Hove: Psychology Press.

Oakhill, J., Cain, K., & Yuill, N. (1998). Individual differences in children's comprehension skill: Toward an integrated model. In C. Hulme, & R. M. Joshi (Eds.), *Reading and spelling: Development and disorders* (pp. 343–367). Mahwah, NJ, US: Lawrence Erlbaum Associates, Publishers.

Palm, T. (2008). Impact of authenticity on sense making in word problem solving. *Educational Studies in Mathematics*, *67*, 37–58.

Radatz, H. (1983). Untersuchungen zum Lösen eingekleideter Aufgaben. *Zeitschrift fur Mathematik-Didaktik*, *4*, 205–217.

Rawson, K. A., & Kintsch, W. (2002). How does background information improve memory for test content? *Memory & Cognition*, *30*, 768–778.

Reusser, K. (1988). Problem solving beyond the logic of things: contextual effects on understanding and solving word problems. *Instructional Science*, *17*, 309–338.

Reusser, K., & Stebler, R. (1997). Every word problem has a solution—the social rationality of mathematical modeling in schools. *Learning and Instruction*, *7*(4), 309–327.

Riley, M. S., Greeno, J. G., & Heller, J. I. (1983). Development of children's problem solving ability in arithmetic. In H. P. Ginsburg (Ed.), *The development of mathematical thinking* (pp. 153–196). New York: Academic Press.

Staub, F. C., & Reusser, K. (1995). The role of presentational structures in understanding and solving mathematical word problems. In C. A. Weaver, S. Mannes, & C. R. Fletcher (Eds.), *Discourse comprehension. Essays in honor of Walter Kintsch* (pp. 285–305). Hillsdale, NJ: Lawrence Erlbaum.

Thevenot, C. (2010). Arithmetic word problem solving: evidence for the construction of a mental model. *Acta Psychologica*, *133*, 90–95.

Thevenot, C., Barrouillet, P., & Fayol, M. (2004). Mental representation and procedures in arithmetic word problems: The effect of the position of the question. *L'Année Psychologique*, *104*, 683–699.

Thevenot, C., Devidal, M., Barrouillet, P., & Fayol, M. (2007). Why does placing the question before an arithmetic word problem improve performance? A situation model account. *Quarterly Journal of Experimental Psychology*, *60*, 43–56.

Thevenot, C., & Oakhill, J. (2005). The strategic use of alternative representation in arithmetic word problem solving. *Quarterly Journal of Experimental Psychology*, *58*, 1311–1323.

Thevenot, C., & Oakhill, J. (2006). Representations and strategies for solving dynamic and static arithmetic word problems: the role of working memory capacities. *European Journal of Cognitive Psychology*, *18*, 756–775.

Thevenot, C., & Oakhill, J. (2008). A generalization of the representational change theory from insight to non-insight problems: the case of arithmetic word problems. *Acta Psychologica*, *129*, 315–324.

Torres-Alcedo, C., Nogry, S., Bosc-Miné, C., & Sander, E. (In preparation). Typicality effects in problem posing.

Van Dijk, T., & Kintsch, W. (1983). *Strategies of discourse comprehension*. New York: Academic Press.

Van Dooren, W., De Bock, D., Depaepe, F., Janssens, D., & Verschaffel, L. (2003). The illusion of linearity: expanding the evidence towards probabilistic reasoning. *Educational Studies in Mathematics*, *53*, 113–138.

Van Dooren, W., De Bock, D., Gillard, E., & Verschaffel, L. (2009). Add or multiply? A study on the development of primary school students' proportional reasoning skills. Proceedings of the 33 Annual Conference of the International Group for the Psychology of Mathematics Education; Thessaloniki, Greece.

Van Dooren, W., De Bock, D., Hessels, A., Janssens, D., & Verschaffel, L. (2004). Remedying secondary school students' illusion of linearity: a teaching experiment aiming at conceptual change. *Learning and Instruction*, *14*, 485–501.

Van Dooren, W., De Bock, D., Hessels, A., Janssens, D., & Verschaffel, L. (2005). Not everything is proportional: effect of age and problem type on propensities for overgeneralization. *Cognition and Instruction*, *23*, 57–86.

Verschaffel, L., De Corte, E., & Lasure, S. (1994). Realistic considerations in mathematical modeling of school arithmetic word problems. *Learning and Instruction*, *4*, 273–294.

Verschaffel, L., De Corte, E., & Lasure, S. (1999). Children's conceptions about the role of real-world knowledge in mathematical modeling of school word problems. In W. Schnotz, S. Vosniadou, & M. Carretero (Eds.), *New perspectives on conceptual change* (pp. 175–189). Oxford: Elsevier.

Verschaffel, L., Greer, B., & De Corte, E. (2000). *Making sense of word problems*. Netherlands: Swets & Zeitlinger Publishers.

Vosniadou, S. (2003). Exploring the relationships between conceptual change and intentional learning. In G. M. Sinatra, & P. R. Pintrich (Eds.), *Intentional conceptual change* (pp. 377–406). London: LawrenceErlbaum Associates.

Weyns, A., Van Dooren, W., & Verschaffel, L. (submitted). The effect of emphasizing the realistic modeling complexity in the text and/or picture on pupils realistic solutions of P-items. *Educational Psychology*

Yoshida, H., Verschaffel, L., & De Corte, E. (1997). Realistic considerations in solving problematic word problems: do Japanese and Belgian children have the same difficulties? *Learning and Instruction*, *7*, 329–338.

Yuill, N., Oakhill, J., & Parkin, A. (1989). Working memory, comprehension ability and the resolution of text anomaly. *British Journal of Psychology*, *80*, 351–361.

Zwaan, R., & Radvansky, G. (1998). Situation models in language comprehension and memory. *Psychological Bulletin*, *123*, 162–185.

Chapter 4

Neurodevelopmental Disorders as Model Systems for Understanding Typical and Atypical Mathematical Development

Marcia A. Barnes*, Kimberly P. Raghubar**

*Department of Special Education & Meadows Center for Preventing Educational Risk, University of Texas, Austin, TX, United States; **Department of Pediatrics, Baylor College of Medicine and Psychology Service, Texas Children's Hospital & Department of Psychology, University of Houston, Houston, TX, United States

INTRODUCTION

Understanding the factors that are associated with the development of mathematical ability and disability is essential for developmental models of mathematical cognition and may be useful for informing identification and intervention strategies for at risk children. Many correlational studies, including those that compare children with and without mathematical difficulties, have been used to hypothesize about the core abilities and deficits that lead to typical or atypical development of mathematical competence. As has been argued for reading, academic outcomes are correlated with a large number of variables, but it is important to determine which of these are critical "pressure points" for academic skill development and disability (Perfetti & Adlof, 2012). Experimental studies in which hypothesized mathematics-related abilities are manipulated are the gold standard in that they provide "low inference" information relevant to this question; that is, manipulations of putative mathematics-related factors that lead to changes in some aspect of mathematics performance provide the strongest evidence for the importance of those factors for mathematical processing.

In contrast, experimental studies may be less relevant for developmental models of mathematical cognition in that they are not designed to determine

Acquisition of Complex Arithmetic Skills and Higher-Order Mathematics Concepts
http://dx.doi.org/10.1016/B978-0-12-805086-6.00004-7

whether the pressure points for mathematics change across development and for different types of mathematical tasks (Ansari, 2010; LeFevre et al., 2010). Another approach is to use longitudinal designs to determine the factors that distinguish typical from atypical mathematical development. Although longitudinal studies cannot lead to causal statements and are subject to third-variable confounds (Shadish, Cook, & Campbell, 2002), *temporal precedence* in such studies allows stronger statements to be made about the potential link between earlier developing skills and later developing academic outcomes. In this chapter we describe a series of studies we have conducted with individuals with SBM, a neurodevelopmental disorder associated with high rates of mathematics disability. We discuss findings from our descriptive studies of mathematical processing in this population and then focus on findings from our longitudinal studies comparing children with SBM to their typically developing peers.

SPINA BIFIDA AS A MODEL SYSTEM FOR UNDERSTANDING MATHEMATICAL LEARNING DISABILITIES

Dissociations in skill development that are produced by brain injury can be useful for understanding the development of academic abilities and disabilities more generally (Mazzocco, Murphy, & McCloskey, 2007). SBM is one such neurodevelopmental disorder. SBM is a disabling birth defect that is diagnosed before or at birth, which affects the development of the spine and brain. The modal cognitive and academic phenotype of SBM at school age includes word reading that is well-developed in contrast to difficulties in mathematics (Ayr, Yeates, & Enrile, 2005; Dennis & Barnes, 2010; Fletcher et al., 2005). About 50% of children with SBM who are not intellectually disabled have mathematical learning disabilities or MLD (Fletcher et al., 2005). Even when children do not meet criteria for MLD, mathematics achievement is most often significantly lower than reading achievement (Barnes et al., 2002, 2006). Although this pattern of discrepancies in academic skills may also be found in children without brain injuries, the prevalence of this pattern is much greater in SBM than it is for individuals without neurological impairment where there tends to be greater overlap in mathematics and reading achievement (Barbaresi, Katusic, Colligan, Weaver, & Jacobsen, 2005; Shalev, Auerbach, Manor, & Gross-Tsur, 2000).

In combination with the known high prevalence of MLD in SBM, the fact that SBM is diagnosed before or at birth makes this condition of particular interest to understanding mathematical development. A condition, such as SBM provides the opportunity to use powerful longitudinal designs to examine early developmental precursors of later learning abilities and disabilities. In other words, because we know that SBM is associated with high rates of MLD and given that SBM is diagnosed before or at birth, this neurodevelopmental disorder provides a *model system* for studying the development of mathematical abilities, possibly from birth. The term *model system* is taken from biology. It refers to the study of an organism (in biology) or to a particular group of children (in our studies) that allows one to study particular phenomena in ways that are not as feasible (i.e., mathematical development from early in life) or as easy to study in other groups of interest. It is assumed, however, that the findings based on the model system will apply to these other groups.

In order to determine whether SBM might be a good candidate for a model system for understanding the development of MLD, we first need to determine whether the MLD in these children looks similar to that in children with MLD and no neurological disorder. In a series of studies, we found evidence for striking similarities in the behavioral phenotype of mathematical difficulties of children with SBM compared to children with MLD and no neurological disorder. First, children with SBM are significantly lower than their age peers on a number of standardized measures of mathematics achievement and these effects are moderate to large depending on the particular mathematics achievement outcome (Barnes et al., 2002, 2006; Raghubar et al., 2015).

Second, core mathematical processing deficits of children with SBM and MLD look very similar to those in children with MLD and no neurological disorder. Children with SBM and MLD show one of the signature deficits of MLD (e.g., Jordan, Hanich, & Kaplan, 2003), specifically, slow and less accurate math fact retrieval than age peers, less use of direct retrieval, and greater use of finger counting strategies (Barnes et al., 2006). We also found remarkably similar patterns of errors on a written subtraction task for subgroups of children with SBM who had MLD or reading disability (RD) or MLD + RD (Barnes et al., 2006; also see Ayr et al., 2005) and for analogous subgroups of children with learning disabilities without neurodevelopmental disorder (Raghubar et al., 2009).

In these latter studies we used Van Lehn's (1990) method of coding of errors in a multidigit subtraction task. Errors were coded for: math facts, which are errors in single digit subtraction within the multidigit subtraction problem; procedural bugs, which are errors that occur more than once and are assumed to reflect lack of procedural knowledge (e.g., consistently subtracting the smaller from larger number regardless of whether the larger number is in the subtrahend or minuend; failing to borrow; ability to borrow, but not from zero); and procedural slips, which are procedural errors of the type just mentioned, but which occur only once and may reflect lapses in attention or procedural knowledge that is not fully consolidated in long term memory. We additionally coded for errors that appeared to be related to difficulties with visual–spatial processing (error due to misreading of numbers, crowding and/or misalignment of work, seeming neglect of one side or part of a problem), which is a type of error seen in adult patients with dyscalculia and hypothesized to be related to lesions to parts of the brain involved in spatial aspects of number processing (reviewed in Rourke & Conway, 1997). This type of error is thought to be related to mathematical performance in children with MLD that is not accompanied by RD (Rourke, 1993).

As shown in Table 4.1, there were no important distinctions to emerge from these studies in the performance of children with MLD with or without SBM.

TABLE 4.1 Types of Errors in MultiDigit Subtraction Problems

	Studies of Children with SBM *N* = 192 (Barnes et al., 2002, 2006)	Study of Children with MLD and/or RD *N* = 291 (Raghubar et al., 2009)
Math fact errors—error in subtraction at single digit level	Associated with severity of math disability not MLD + RD	Associated with severity of math disability not MLD + RD
Procedural bugs—error that occurs more than once (e.g., subtract smaller from larger; no borrow; problems subtracting from zero)	Common error of MLD + RD and MLD Only	Common error of MLD + RD and MLD Only
Procedural slips—like procedural bugs but occur only once–could reflect attention difficulties	Not differentially associated with learning disability group	Not differentially associated with learning disability group
"Visual–spatial" errors—error due to crowding of work, misalignment	Low rates, not associated with learning disability group	Low rates; RD Only had highest rates

Furthermore, problems in visual–spatial processing, a distinct area of weakness in SBM (Dennis & Barnes, 2010), did not translate in any transparent way into mathematical performance on a complex calculation task. Such findings are in keeping with the idea that neuropsychological studies of mathematical cognition in adults with brain lesions may not provide suitable model systems for understanding mathematical disorders in children with or without brain injury (Ansari, 2010). In sum, these cognitive studies of math fact retrieval and subtraction errors suggest that some of the core mathematical processing deficits are very similar in children with MLD with and without neurodevelopmental disorder.

A third line of evidence for comparability of mathematical difficulties in children with SBM and those with MLD but no neurodevelopmental disorder comes from our studies of early mathematical cognition. Children with SBM who are at high risk for MLD differ from their age peers on measures of informal number knowledge that have been shown to also characterize young children at risk for MLD without neurodevelopmental disorder (e.g., Jordan, Kaplan, Ramineni, & Locuniak, 2009). The performance of preschoolers with SBM was significantly lower than peers on oral counting (how high a child can count orally) and on a task assessing knowledge of counting principles as measured by an error detection task in which a puppet counts correctly or incorrectly, in the latter case, violating counting principles (Gelman & Meck, 1983). These errors included difficulties with one-to-one correspondence (e.g., puppet skip counted), stable order (reversed numbers in a count, such as counting "one" "three" two"), and cardinality (e.g., puppet counts to six and when asked how many objects there are, says "five"). Preschoolers with SBM were also less accurate on a nonverbal arithmetic task that involves addition and subtraction on small sets of objects (Jordan, Huttenlocher, & Levine, 1992). Here, children watch the examiner place poker chips on a mat (e.g., 3 chips placed), followed by the insertion of a screen that occludes what is on the mat. The examiner then inserts in full view of the child additional poker chips behind the screen (e.g., inserting 2 more chips would make a 3 + 2 problem) or removes poker chips from the display behind the screen (e.g., removing 2 chips would make a 3 – 2 problem). The child then shows her ability to perform the nonverbal calculation by placing on her mat what is on the examiner's mat behind the screen.

Children with SBM were also less accurate on a standardized measure of early mathematics at 60 months that assessed matching on the basis of number using blocks, block counting, and solving simple addition problems using blocks and pictures (Stanford-Binet-IV Quantitative subtest, Thorndike et al., 1989), and on a standardized measure of early counting and number knowledge at 36 months (Test of Early Mathematics Ability-2, Ginsburg & Baroody, 1990). The only aspect of early math for which the preschoolers with SBM did not differ from their peers was on "small" problems on the nonverbal arithmetic task. "Small" problems involved a sum or a minuend of 3 or less (e.g., 2 + 1; 3 – 2). As small problems are within the subitizable range they are thought to be solved differently from large problems, possibly by *nonnumerical* means using object files (Trick & Plyshyn, 1994), episodic visual representations involved in the storage and updating of information about small sets of objects as they move in time and space (Noles, Scholl, & Mitroff, 2005). The effect sizes on all of these preschool mathematics measures were in the moderate to large range ($d = 0.53 - 0.87$) with the exception of the small nonverbal arithmetic problems ($d = 0.10$) (Barnes et al., 2011).

In sum, there is considerable evidence from studies of school-age mathematics achievement outcomes, of mathematical processing, and of preschool mathematical knowledge, that children with SBM have difficulties in mathematics that appear remarkably similar to those in children at risk for MLD or with MLD. These findings suggest the appropriateness of using SBM as a model system for the longitudinal study of the development of mathematical ability and disability. However, we are also cognizant of the fact that findings from model systems research need to be interpreted with caution, particularly given that similarities in behavioral performance need not always mean that underlying neurocognitive processes are equivalent between the model system and other populations.

LONGITUDINAL APPROACHES TO THE STUDY OF MATHEMATICAL DEVELOPMENT AND DISABILITY

Studies that follow children from infancy to the elementary school grades or beyond are the most informative. Although there are some studies of mathematical cognition employing longitudinal designs starting in infancy (e.g., Starr, Libertus, & Brannon, 2013) or in the preschool years or early school years (e.g., Blair & Razza, 2007; Bull et al., 2008; Fuhs et al., 2014; LeFevre et al., 2010; Mussolin, Nys, Content, & Leybaert, 2014), these studies are often conducted over relatively short developmental time windows (but see Geary, 2011) that either do not extend into school age or that extend from school-entry into the early elementary grades. Most of these studies do not focus on MLD per se.

In contrast, some of the longitudinal studies of David Geary and Nancy Jordan and their coworkers have an explicit focus on children at high risk for MLD and/or RD. In these studies, children are followed from school entry or over the first few years in the primary grades (e.g., Geary, Hamson, & Hoard, 2000; Geary, Hoard, Nugent, & Bailey, 2012; Jordan, Kaplan, & Hanich, 2002). Major goals of such studies have been to determine: (1) what domain-specific mathematics and domain-general cognitive abilities account for growth in mathematics and (2) whether the association between these abilities and specific aspects of mathematical cognition differs for children with MLD only, RD, or RD + MLD. Some of these studies have also tested whether the factors associated with persistently low mathematical performance differ from those associated with below average mathematical achievement.

Studies employing large national and international longitudinal databases (e.g., Early Childhood Longitudinal Study-K; NICHD-Study of Early Childhood and Youth Development; British Cohort Study) have been focused on identifying the earlier developing mathematical abilities that account for unique variance in specific later mathematical outcomes, controlling for domain-general factors, such as working memory (e.g., Siegler et al., 2012). Another recent approach has been to investigate the relative contribution of "traits" (e.g., working memory, motivation, etc.) and "states" (e.g., prior mathematical knowledge) to individual differences in mathematics achievement and to across-grade stability in these differences (e.g., Bailey, Watts, Littlefield, & Geary, 2014). These studies take a regression-based approach and as such test hypotheses using school-age children with a range of abilities in math.

These different types of longitudinal studies have advantages and disadvantages for understanding the development of mathematical ability and disability. The studies vary in terms of the time windows across which mathematical and mathematics-related factors are measured; the range and types of mathematics-related abilities and mathematical outcomes that are measured; whether children with MLD or at risk for MLD are a focus of the study; and the nature and breadth of control variables that can be included in such studies (see discussions in Bailey et al., 2014; LeFevre et al., 2010; Watts et al., 2015). These differences across studies make it difficult to draw firm conclusions as to which domain-general (e.g., working memory) and early domain-specific mathematics abilities are the most critical for long-term competence with mathematics.

Our longitudinal approach has several strengths that solidify conclusions that can be drawn from more typical longitudinal studies. These strengths include: (1) the unusually long time window across which we assessed the development of mathematical abilities; (2) the fact that we were able to follow both a group at high risk for MLD across this developmental time window as well as their typically developing or TD peers; (3) the measurement of many domain-general cognitive abilities that have been shown to predict mathematics achievement, as well as several early numeracy skills prior to school entry; and (4) testing of the specificity or discriminant validity of longitudinal predictors by comparing outcomes for mathematics and reading. Weaknesses include the loss of information over time because not all children were assessed at all time points, the lack of experimental number sense measures (e.g., approximate number system acuity) prior to 60 months of age, and a constrained set of school-age mathematics outcomes focusing mostly on numbers and operations rather than on other complex outcomes, such as geometry, word problem solving, fractions, and the like (e.g., see Jordan et al., Chapter 6; Robinson, Chapter 2; Thevenot, Chapter 3). We first discuss the proposed sources of mathematical learning disabilities followed by how we used this information in our longitudinal studies.

Sources of Mathematical Disability

There are three general approaches for thinking about the sources of mathematics difficulty. First, the domain-specific view is that difficulties in math at school age have their origins in early core deficits in number-specific skills. One version of this view is that there are individual differences, perhaps from birth, in nonsymbolic representations of magnitude, which are thought to be part of the infant's "starter kit" for math (Butterworth, 2005; Feigenson, Dehaene, & Spelke, 2004; Libertus, Feigenson, & Halberda, 2011), although as described later the extent to which this early number sense contributes to later mathematics achievement is debated (De Smedt, Noël, Gilmore, & Ansari, 2013; Fazio, Bailey, Thompson, & Siegler, 2014; Piazza, Pica, Izzard, Spelke, & Dehaene, 2013; vanMarle, Chu, Li, & Geary, 2014).

Another version of the domain specific view is that early competence with number (e.g., knowledge of number symbols, such as number words, counting, and the ability to perform arithmetical manipulations on nonsymbolic quantities) that is acquired prior to schooling is important for individual differences in mathematics during elementary school (Jordan, Kaplan, Olah, & Locuniak, 2006). This version of the domain-specific view neither presupposes nor precludes the possibility that individual differences in core number-specific abilities are present from birth.

Second, the domain-general view is that limitations in general cognitive abilities that are not number-specific, such as language, working memory, and the like, affect the child's ability to learn mathematical concepts and procedures across both the preschool and school years. Third, an interesting but understudied view is that MLD arises from difficulties in the mapping of nonsymbolic representations of quantities to their language-based symbols (i.e., oral and written Arabic numerals). These three approaches to individual differences in mathematics, discussed later, are not necessarily mutually exclusive (Geary, 2013). We also report the strength of the relations of domain-specific and domain-general abilities to mathematics where such information exists.

Domain-Specific Approaches

The idea that core deficits in representations of magnitude explain MLDs is based on studies of typically developing infants (e.g., Xu & Spelke, 2000; Wynn, 1992) and cross-species studies showing both human and nonhuman organisms can discriminate differences in large quantities and precisely enumerate and manipulate very small quantities in the subitizing range (Feigenson et al., 2004; Geary, Berch, & Mann Koepke, 2015). Whether this inherent sense of number depends on one or two representational systems (i.e., one for precisely representing small quantities, another for approximately representing larger quantities) is a matter of debate (Geary, 2013; Mou & vanMarle, 2014), as is whether and how symbolic representations (i.e., Arabic numerals) come to be mapped from these nonsymbolic number systems (Lyons, Ansari, & Beilock, 2015). Individual differences in these nonsymbolic representational system(s) have been proposed to be an important contributor to MLD (e.g., Mazzocco, Feigenson, & Halberda, 2011). Approximate number system (ANS) acuity tasks where participants indicate the larger of two sets of dots after very brief exposures are thought to tap this core number ability. Reflecting the recent intense interest in the possibility that deficits in ANS acuity (smallest ratio between two large quantities at which a reliable difference can be detected) might account for at least some MLDs, 82% of all studies relating individual differences in ANS acuity to individual differences in different aspects of mathematical cognition were published between 2011 and 2014 (Schneider et al., 2016).

Despite the appeal of this strong domain-specific hypothesis about the origin of MLDs, the relation of ANS acuity to individual differences in mathematics achievement is not straightforward: effect sizes across three meta-analyses on this topic are all in the small range (about $r = 0.20$ from Chen & Li, 2014; Fazio et al., 2014; Schneider et al., 2016); the direction of the relation of ANS acuity and mathematics achievement is inconsistent across studies (e.g., Mussolin et al., 2014; Starr et al., 2013) and may even be reciprocal (Piazza et al., 2013); latent profiles analyses do not show that low ANS acuity is reliably associated with MLD (e.g., Hart et al., 2016); and although MLD is highly heritable, heritability of ANS acuity is relatively low (Tosto et al., 2014). A more nuanced view of the relation of ANS acuity to mathematics achievement is beginning to emerge. For example, this relationship might be developmentally limited (e.g., vanMarle et al., 2014) and nonlinear across the distribution of mathematical ability such that it might be a better predictor of risk for MLD very early in development and particularly for young children with very low mathematical ability (Purpura & Logan, 2015). Similarly, heritability estimates might also vary as a function of age and/or ability (Tosto et al., 2014).

It is interesting to note that rapid advances in theory and measurement, as discussed earlier for ANS acuity, may not be reflected in longitudinal studies. Our longitudinal studies were designed and carried out over the past 15 years or more and so reflect, to some extent, what was known (and unknown) about mathematical learning disabilities at the time. Although being able to study the development of mathematics and mathematics-related abilities is a strength of studies that employ long developmental time windows, these strengths may be offset, to some extent, when there are rapid changes in theory and associated measures that are impossible or difficult to incorporate into an ongoing or completed study. Data collection in our study, for example, ended before researchers began to measure ANS acuity as the marker of individual differences in nonsymbolic representations of quantity. We did, however, measure several aspects of early symbolic and nonsymbolic quantitative abilities in preschoolers (e.g., counting procedures and counting knowledge, nonverbal arithmetical transformations on number) that are acquired prior to formal instruction and that are good predictors of later mathematical performance in young school-age children (e.g., Jordan et al., 2006; Jordan et al., 2009).

Mapping of Nonsymbolic to Symbolic Quantity

One proposal is that a difficulty for some children with MLD lies in the mapping of nonsymbolic quantities to symbolic quantities (Rousselle & Noel, 2007). In keeping with this proposal is the consistent finding that the relation of magnitude comparison and mathematics achievement is greater when magnitude comparison is measured using tasks that involve symbols (i.e., Arabic numerals) compared to typical ANS acuity tasks using nonsymbolic materials (e.g. Lyons et al., 2014; Rousselle & Noel, 2007; Schneider et al., 2016). Geary (2013) has suggested that difficulty in mapping nonsymbolic information to their culturally determined symbols may reflect a more general learning difficulty that also applies to other academic domains, such as reading. The mechanism contributing to these mapping difficulties may be weaknesses in domain-general attentional control (Geary, 2013).

Domain-General Approach

From the discussion of more domain-specific approaches to MLD earlier, it seems that an adequate explanation of long-term mathematical development and disability may require a broader view that encompasses how general cognitive systems that may be involved in learning are also implicated in the development of specific mathematical abilities. Whether these domain-general factors are considered to be causal or facilitative mechanisms in mathematical development is beyond the scope of this chapter (see Carey, 2004; Gelman & Butterworth, 2005; Spelke, 2003 for differing views on the role of language in math). However, there is a great deal of evidence showing relations between difficulties in several aspects of mathematical cognition and deficits in these more general cognitive systems.

Language

Language abilities, such as those tapped by vocabulary and phonological awareness tasks are related to a variety of mathematical outcomes in longitudinal studies of both younger and older children (e.g., LeFevre et al., 2010; Purpura & Logan, 2015; Vukovic & Lesaux, 2013). Language abilities are required to map quantities to number words and their symbols, and to represent exact number (Dehaene, Piazza, Pinel, & Cohen, 2005). Language is understandably implicated in mathematical tasks, such as word problem solving (Fuchs, Fuchs, Compton, Hamlett, & Wang, 2015), but also for those, such as geometry that, on the surface, appear less language-based (Spelke & Tsivkin, 2001; Vukovic & Lesaux, 2013). General vocabulary knowledge, which is correlated with mathematical achievement in some studies may be a proxy for mathematics-related vocabulary (e.g., "more," "first," "equals," "expression"). In recent studies, mathematics-related vocabulary is related to mathematical performance in both younger (e.g. Purpura & Logan, 2015) and older (e.g., Fuchs, Geary, Fuchs, Compton, & Hamlett, 2016) children.

The quality of phonological representations as assessed by phonological awareness tasks is related to learning the number word sequence in counting (Krajewski & Schneider, 2009) and to the use of particular verbal strategies, such as direct retrieval of answers to small addition problems (De Smedt, Taylor, Archibald, & Ansari, 2010; Jordan et al., 2006). The contribution of phonological processing to addition and multiplication (but not subtraction and division) has also been demonstrated in a brain imaging study of children. Performance on these aspects of arithmetic was related to the arcuate fasciculus, an area associated with language and reading, but this correlation was no longer significant when phonological processing was covaried (Van Beek, Ghesquiere, Lagae, & De Smedt, 2014).

In both younger and older children, measures of language are significantly related to all or almost all mathematics outcomes (LeFevre et al., 2010; Purpura & Logan, 2015). In one study, effect sizes ranged from small ($r = 0.24$) for magnitude comparison to moderate ($r = 0.32$–0.54) on a range of math tasks, such as measurement, geometry, calculation, and number line estimation. These effect sizes for language and mathematics were as large as those for language and reading (LeFevre et al., 2010). Such findings are consistent with Spelke's (2003) proposal that language is involved in the construction of more powerful and flexible mathematical representations than those that emerge solely from nonsymbolic representations of magnitude. As such, language would be expected to contribute to the development of new systems of mathematical knowledge across different mathematical domains. It is also of interest that language abilities have been implicated in the development of domain-general abilities, such as executive processes including working memory, which are themselves related to mathematics achievement (e.g., Fuhs & Day, 2011; Kuhn, Willoughby, Wilbourn, Vernon-Feagans, & Blair, 2014; Schneider, Lockl, & Fernandez, 2005).

Working Memory

In keeping with dual-process theories of cognition, working memory is particularly important on tasks requiring less automatized, more effortful problem solving (Evans & Stanovich, 2013; Lee, Ng, and Bull, Chapter 9), which applies to a great deal of mathematical learning and performance, particularly for young children for whom most mathematical concepts and procedures are novel. Working memory, whether using verbal, visual–spatial, or numerical stimuli, is consistently related to mathematics achievement and performance (Peng, Namkung, Barnes, & Sun, 2015), in keeping with the idea that performance on working memory tasks draws on central executive resources needed to maintain the focus on task-relevant information and inhibit task-irrelevant information (Engle & Kane, 2004). Children, including preschoolers, may construct visual or visual–spatial mental models, which helps them to solve some mathematical problems without having to rely solely on verbal codes (e.g., Bisanz, Sherman, Rasmussen, & Ho, 2005). Children also use verbal codes to store partial answers while continuing to process successive parts of the mathematics task (reviewed in Raghubar, Barnes, & Hecht, 2010). And, during task performance, children and adults likely use a combination of visual–spatial and verbal codes in mathematics problem solving (McKenzie, Bull, & Gray, 2003; Raghubar et al., 2010) depending on the strategies they bring to bear to solve particular mathematical problems.

A recent meta-analysis of the relation between working memory and mathematics yielded a medium effect size ($r = 0.35$), which was not moderated by type of working memory task (i.e., visual–spatial vs. verbal), but which was moderated by type of mathematics outcome, with whole number calculations and word problem solving showing the highest relations to working memory (Peng et al., 2015). The size of the relation of working memory to mathematics achievement and performance varied across samples of typically developing children, children with MLD, and children with MLD accompanied by other developmental disorders (e.g., AD/HD, neurodevelopmental disorders), such that it was significantly higher for this latter group.

Finger Skills

The universal use by young children of their fingers to count and calculate is proposed to result in common neural representations of fingers and numbers (e.g., Butterworth, 1999) and finger skills have been linked to early numeracy and particularly to arithmetic (Gracie-Bafalluy and Noel, 2008). In typically developing children, for example, finger localization/gnosis, but also other skills, such as finger tapping speed predict performance on several different types of mathematical tests, such as calculation (Noël, 2005; Penner-Wilger et al., 2007) and number-line estimation and magnitude comparison (Penner-Wilger et al., 2007; Penner-Wilger et al., 2008). Variability in finger localization abilities has also been specifically tied to the use of fingers during counting and calculation in 5- to 7-year olds (Reeve & Humberstone, 2011). Finger skills involved in finger counting and in accurate pointing and differentiation of objects during counting may help children compensate for limited working memory capacity by reducing the need to internally store a mental representation of each counted object (Alibali & DiRusso, 1999). Problems in differentiating fingers or coordinating counting across hands might be expected to produce errors in simple arithmetic when fingers are used to count and calculate. In a study of first grade children (Wasner, Nuerk, Martignon, Roesch, & Moeller, 2016), the relation of finger localization abilities to performance on timed single digit addition and subtraction problems was significant, but small in the presence of other cognitive predictors ($r \sim 0.23$).

It is worth noting that domain-general cognitive abilities, such as working memory, inhibitory control, and language have been found to mediate or account for the relation of nonsymbolic magnitude processing to mathematics achievement (Bugden & Ansari, 2015; Fuhs & McNeil, 2013). And, as discussed earlier, domain-general cognitive abilities may also be implicated in learning the mapping of quantities to symbols (Geary, 2013), and to young children's performance on ANS acuity tasks (e.g., Bugden & Ansari, 2015). Therefore, domain-general abilities might affect mathematical cognition through several routes including their influence on: the development of domain-specific representations; the mapping of nonsymbolic to symbolic representations; the learning of new mathematical concepts and procedures; and supporting performance in multistep mathematical problem-solving situations.

The domain-specific abilities (i.e., nonsymbolic and symbolic quantitative abilities including counting knowledge and procedures, and nonverbal arithmetic) and domain-general abilities (i.e., language, working memory, and finger skills) that we assessed in our longitudinal study have all been related to various aspects of mathematics and mathematical disabilities as reviewed earlier, with effect sizes in the small to medium range depending on the specific mathematics-related ability and the type of mathematical task. It is worth noting that the use of different control variables across studies make relative comparisons of the *unique* effects of these mathematics-related correlates difficult. We have handled this issue in our studies by simultaneously testing multiple domain-general and domain-specific developmental predictors of later mathematical outcomes.

LONGITUDINAL STUDIES OF MATHEMATICAL COGNITION IN CHILDREN WITH SBM AND THEIR TYPICALLY DEVELOPING PEERS

We next discuss longitudinal studies of mathematical development in children with SBM and their typically developing peers (Barnes et al., 2011; Barnes et al., 2014; Barnes & Raghubar, 2016). The general methods we used, information about our sample, and our choice of predictor and outcome variables are presented briefly. This is followed by findings from the longitudinal studies including: (1) prediction of 60 month number knowledge by concurrent domain-general abilities and 36-month mathematical knowledge; (2) 36- and 60-month domain-general predictors of 9–10 year old mathematics achievement and comparison to reading achievement to address the specificity of mathematics-related developmental precursors; and (3) prediction of 9–10 year mathematics achievement combining 36- and 60-month domain-specific and domain-general abilities in the same longitudinal models.

Our longitudinal project was designed to provide information about the motor, cognitive, language, social, and academic development of children with SBM. Children born with SBM in Texas, and in Ontario were recruited over about a 3-year period to participate in this study. A cohort of typically developing (TD) children born at the same time as the children with SBM was also recruited. Most children began the study at 6 months of age, but we accepted children with SBM up to 22 months of age at time of first assessment. Children with SBM and their TD peers were assessed several times prior to 60 months of age (6, 12, 18, 24, 36, and 60 months) and then again at ages 7.5, 8.5, and 9.5 years. Detailed information on the study can be obtained from Lomax-Bream et al. (2007a,b). The study was not explicitly designed as an investigation of the development of mathematical cognition, but the collection of several mathematics-related cognitive correlates at very young ages, number knowledge at 60 months, and standardized and experimental measures of mathematics at school age (7–10 years of age) allowed us to test longitudinal models of mathematical development. We first present a study on the prediction of early number knowledge followed by longitudinal mediation studies in which early domain-general abilities and early number knowledge are used to predict school-age mathematical outcomes.

Are Domain-General Cognitive Abilities Related to Number Knowledge?

Based on hypotheses about some of the cognitive correlates that have been related to mathematics, we (Barnes et al., 2011) investigated whether 60-month domain-general cognitive abilities including vocabulary, phonological awareness, visual–spatial ability, and finger skills (accuracy and speed to place grooved pegs in a pegboard) predicted performance on number knowledge tasks described earlier (oral counting, detection of counting errors, nonverbal arithmetic, standardized measure of mixed math problem solving) and also assessed at 60 months. The longitudinal aspect of this study comes from using a number-specific autoregressor (performance on the Test of Early Mathematics-2 or TEMA-2) (Ginsburg & Baroody, 1990) administered 2 years earlier, at 36 months of age. At 36 months the TEMA-2 taps the ability to count smaller and larger arrays of objects, show number using fingers, compare quantities, such as "which has more," and knowledge of the cardinal value of number words and numerals. We then tested whether the domain-general cognitive abilities predicted unique variance on any of the early number knowledge tasks after accounting for mathematical knowledge at 3 years of age. The logic behind using an autoregressor is that if a domain-general cognitive predictor remains significant when autoregressor effects are controlled, stronger evidence is provided for the importance of that cognitive predictor for mathematical development and performance.

Early mathematical knowledge at 36-months accounted for significant variance on most of the 60-month number knowledge tasks confirming the importance of a foundation in early mathematical knowledge for later mathematical development (Jordan et al., 2009). However, in the presence of domain-general cognitive factors, which explained considerable additional variance in the 60-month number knowledge outcomes, early mathematical knowledge did not always remain significant. Across tasks, domain-general abilities accounted for up to 25% of the variance in early number knowledge with greater contributions of domain-general abilities on mixed quantitative problem solving and nonverbal arithmetic measures compared to counting. The findings also show different patterns of relationship of domain-general cognitive abilities to different number knowledge skills, with counting related to phonological ability, and visual–spatial ability related to mixed problem solving and nonverbal arithmetic. In keeping with the hypothesized specific relation of finger skills and arithmetic, the former were related only to nonverbal calculation. In the presence of phonological awareness, vocabulary did not uniquely predict performance on any measured aspect of number knowledge. Importantly, this pattern of predictors of performance on the early number knowledge tests did not differ for preschoolers with SBM and their typically developing peers (Barnes et al., 2011).

The findings suggest that domain-general cognitive abilities are uniquely related to number knowledge over and above the contribution of earlier number abilities. However, relations between concurrently measured abilities tend to be stronger than relations between variables measured at different times. A stronger test of the contribution of early domain-general and domain-specific factors in mathematical development requires that the assessment of predictors, both domain-specific and domain-general, precede the assessment of the mathematical outcomes. The next two studies use longitudinal mediation techniques to accomplish this goal.

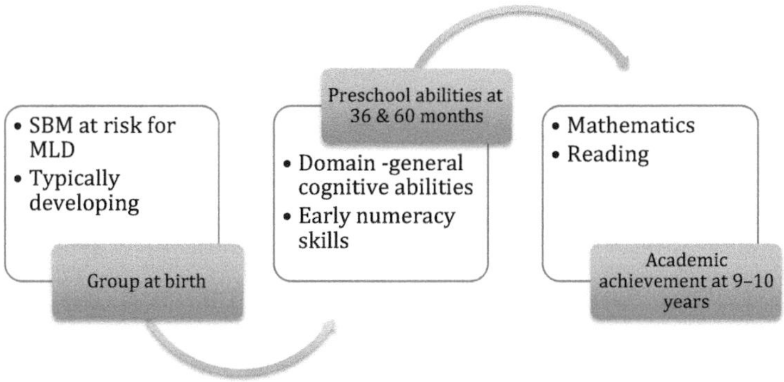

FIGURE 4.1 **Logic of longitudinal multiple mediation models.**

Longitudinal Mediation of School-Age Mathematics Achievement

The logic behind the multiple mediation longitudinal design that we use in our studies is illustrated in Fig. 4.1. We start with two groups at birth—one that is known to have a high risk for later MLD (SBM) and one that is typically developing. Based on both the epidemiological studies of SBM reviewed earlier as well as in the current data set, we know that *Group* is associated with large direct effects on mathematics achievement at 9–10 years of age; that is, there are large differences between children with SBM and their TD peers in mathematics achievement at school age. The question of interest is whether particular domain-general cognitive abilities and domain-specific mathematical abilities prior to school age mediate the relation of group and mathematics achievement. That is, if we know the levels of children's domain-general cognitive abilities and/or their domain-specific mathematics abilities when those children are only 36 or 60 months of age, can we explain or account for the differences between groups in mathematics achievement 4–6 years later? Of interest from a developmental perspective, then, is whether and to what extent, cognitive and mathematical abilities from a much earlier point in development and prior to schooling, predict mathematics achievement in the later elementary grades. A number of research questions can be addressed with this type of design:

1. Are both early domain-general and domain-specific abilities good predictors of later mathematical outcomes?
2. Do the early domain-general and domain-specific predictors vary for different types of school-age mathematical outcomes?
3. To what extent are these early domain-general predictors specific to mathematics? Which of these are also implicated in learning how to read words and understand text?

The answers to these questions are important for theoretical reasons, but they also have practical relevance for early assessment and intervention as addressed later.

Three domain-general cognitive abilities were used in these models. Language was assessed at 60 months of age using phonological awareness tasks (sound analysis and elision from the Comprehensive Test of Phonological Processing). Finger skills were assessed at 36 months of age using the pegboard task described earlier. We did not use a finger gnosis or finger localization test because the measurement of finger gnosis at the age at which we wished to measure finger skills is unreliable. Working memory was also assessed at 36 months using a preschool version of a self-ordered pointing task (Diamond, Prevor, Callender, & Druin, 1997; Petrides & Milner, 1982). A set of 6 Cheerios was shown to the child and then a box with a picture of an animal was placed over each Cheerio. The child lifted a box and was rewarded with a Cheerio. Then a screen was placed between the child and the display of boxes to introduce a memory delay. Cheerios were not replaced under previously searched boxes. The screen was then removed and the child was allowed to search for another Cheerio. Whether the child turned over the previously searched box and there was no Cheerio or she turned over a new box and retrieved a Cheerio, the next trial started with the placement of the screen. Trials stopped when the child had retrieved all of the Cheerios such that higher scores are associated with less efficient working memory. This task taps central executive components of working memory: there is a concurrent storage and processing requirement as well as an inhibitory processing component. As the working memory system in the first few years of life is not as well differentiated into visual–spatial and verbal components as it is a little later in development (reviewed in Garon, Bryson, & Smith, 2008), we do not claim that we are specifically measuring visual–spatial working memory.

At 9.5–10 years of age, the two math outcomes assessed were single-digit math fluency where children had a limited amount of time to write the answers to a running list of single-digit addition, subtraction, and multiplication problems, and a complex calculation test where children completed written single and multidigit calculations (Math Fluency and Calculation subtests from the Woodcock-Johnson Tests of Achievement—Revised or WJ-III) (Woodcock & Johnson, 1989). At 8.5–9 years of age, children were assessed on a measure of mixed mathematics problem solving that included numerical reasoning and knowledge and application of numerical concepts, such as division, fractions, decimals, time, and measurement, using picture supports (Stanford Binet IV Quantitative subtest) (Thorndike et al., 1989). Experimental tests of single-digit addition and estimation were also administered, but not reported here (see Raghubar et al., 2015).

TABLE 4.2 Domain-General Preschool Cognitive Mediators of School-Age Mathematics and Reading (Barnes et al., 2014)

School-age outcomes[a]	Phonological awareness at 60 months	Working memory at 36 months	Finger skills at 36 months
MultiDigit Calculations	Yes	Yes	
Math Fluency	Yes		
Mixed Quantitative Problem Solving		Yes	
Word Reading	Yes		
Passage Comprehension	Yes		
Inferential Comprehension	Yes	Yes	

[a]*School-age outcomes were assessed at 9.5 years, except for Mixed Quantitative Problem Solving, which was assessed at 8.5 years.*

What is the Relation of Different Preschool Domain-General Cognitive Abilities to Different Mathematics Outcomes at School Age?

In order to answer this question, we simultaneously tested whether phonological awareness, working memory, and finger skills at 36–60 months of age mediated the relationship between group (SBM vs. TD) and each mathematics achievement outcome. First, it should be noted that there are significant correlations between the domain-general abilities and the mathematical outcomes, but the question of interest in multiple mediation is whether a domain-general cognitive ability *uniquely* mediates particular mathematical outcomes in the presence of the other domain-general cognitive abilities. A summary of the findings is shown in Table 4.2 and detailed results can be found in Barnes et al. (2014).

First, note that finger skills did not mediate the relation of group to any of the mathematical outcomes. Based on hypotheses about the relation of finger skills to accuracy in the use of finger counting strategies, which are, in turn, hypothesized to be important for single-digit arithmetic, we might have expected early finger skills at 36 months to account for some of the group differences in later math fact fluency. It may be that with several other domain-general predictors in the model, finger skills do not contribute much to the prediction of later single-digit arithmetic (also see Wasner et al., 2016). On the other hand, it is also possible that the predictive value of finger skills for arithmetic is developmentally constrained. Consider that by 9–10 years of age many children will solve single-digit arithmetic problems through direct retrieval or counting verbally rather

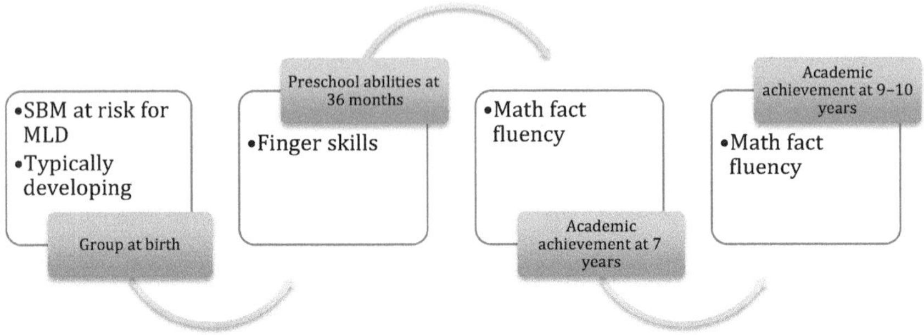

FIGURE 4.2 Hypothetical sequential longitudinal mediation model for finger skills and single digit arithmetic (math fact fluency).

than using finger-counting strategies. It is possible that finger skills mediate the effect of group on math fact fluency, but only at an earlier age where counting strategies are in more frequent use (reviewed in Geary, 1993).

We could extend the use of SBM as a model system to test such hypothesized developmental constraints on the cognitive abilities that are implicated in the acquisition and performance of mathematical skills. One way to do this is to employ *Sequential Longitudinal Mediation Models*. Using math fact fluency as an example we could test the following hypothetical developmental sequence: finger skills at 36 months mediate the relation of group to single-digit arithmetic at age 7 but not at age 9–10 and single-digit arithmetic at age 7 mediates the effect of group on single-digit arithmetic at age 9–10. In this case, finger skills at 36 months of age would be *directly* related to single-digit arithmetic at 7, but *indirectly* related to single-digit arithmetic at age 9 through their effects on single-digit arithmetic at age 7. This hypothetical sequential longitudinal mediation model is illustrated in Fig. 4.2. Knowing that a domain-general ability, such as early finger skill has a developmentally limited relationship with the acquisition and performance of particular aspects of math is highly relevant for developmental models of mathematical cognition. Such developmental modeling would complement findings from cross-sectional studies suggesting that behavioral and neural markers of mathematics vary with the shifting strategies that the child brings to bear for solving the same mathematical task at different ages (Ansari & Dhital, 2006; De Smedt et al., 2010; McKenzie et al., 2003; Meyer, Salimpoor, Wu, Geary, & Menon, 2010). As well, such findings might be useful for informing developmental timing of interventions to improve finger skills for transfer to counting and calculation in experimental studies (Gracie-Bafalluy & Noel, 2008; Moeller, Martignon, Wessolowski, Engel, & Nuerk, 2011).

Working memory proved to be a powerful early mediator of both complex calculation and mixed types of mathematical problems. *To our knowledge, this is the first study to show a strong and direct effect of very early working memory abilities on mathematics achievement 6–7 years later.* What are we to make of such findings? Consistent with dual process models of cognition (Evans & Stanovich, 2013), performance on higher-level mathematical skills as tapped by multidigit calculation tasks and mathematical problem solving tasks (i.e., vs. single digit arithmetic) require working memory resources. The relation of working memory and mathematics is consistently found in studies of individual differences and mathematics (reviewed in Peng et al., 2015; Raghubar et al., 2010), in longitudinal studies of typically developing children across short developmental time windows (e.g., Bull et al., 2008; Lee, Ng, and Bull, Chapter 9; LeFevre et al., 2010), and in concurrent mediation models of mathematics in both neurodevelopmental (SBM; Raghubar et al., 2015) and acquired child neurological disorders, such as traumatic brain injury (Raghubar et al., 2013). The current study adds an important developmental piece to this picture, namely that deficits in working memory in the very early preschool years partially account for the differences in mathematics achievement several years later between children with and without MLD. This finding is consistent with those from Bailey et al. (2014) in which longitudinal data provided evidence for the importance of "trait-like" domain-general cognitive abilities in mathematical development. In concurrent mediation models we have also found that working memory assessed at 9–10 years of age mediates the effect of group on a range of mathematical outcomes at 9–10 years of age including performance on standardized mathematics achievement tests as well as performance on an experimental measure of estimation (Raghubar et al., 2015).

Phonological awareness was a significant and unique longitudinal mediator of both multidigit calculation and single digit math fluency. When assessing broad academic outcomes, such as multidigit calculation on standardized tests, it is not possible to take on line measures or use dual-task procedures that might provide clues as to what representational systems are being used to solve problems. However, the longitudinal data suggest that the quality of language-based representations in the preschool years, in addition to working memory, is associated with calculation abilities several years later. That phonological awareness in the preschool years was the only significant mediator of single-digit arithmetic 4–5 years later is consistent with the idea that proficiency in single-digit arithmetic in middle childhood relies on verbal retrieval from long-term memory; the longitudinal mediation findings add a developmental angle in that children with higher quality lexical representations in the preschool years may be those that develop more efficient math fact retrieval by the late elementary grades.

Do the Longitudinal Mediators Differ for Math and Reading?

The findings suggest that the answer to this question has less to do with mathematics versus reading per se and more to do with the specific component of reading or mathematics that is under consideration. These findings are summarized in Table 4.2. Our interpretation of the results is that by 9–10 years of age, those aspects of both reading and mathematics that are more automatized (i.e., word decoding, single sentence reading fluency, single digit arithmetic) are predicted primarily by the quality of early lexical representations, while those that are less automatized (i.e., inferential comprehension of text passages, multidigit calculations and mathematical problem solving) are predicted by early working memory or by both working memory and language. Although the fact that working memory was not found to mediate performance on reading comprehension seems to violate our interpretation earlier, the comprehension measure in this study (Passage Comprehension from the WJ-III Tests of Achievement) has been shown to rely more on reading decoding and less on general comprehension abilities than other measures of passage-level reading comprehension (Keenan, Betjemann, & Olson, 2008). Note that working memory did mediate performance on the reading measure that required the integration of information between sentences in passages of text. Our interpretation of the longitudinal findings across mathematics and reading is consistent with behavior genetic studies, which show stronger shared heritability for math calculations and passage-level reading comprehension than for calculation and word recognition (Christopher et al., 2012). One possibility is that greater shared heritability could be related to shared cognitive processes, with one candidate being working memory. The findings are also generally consistent with those from a longitudinal study that assessed domain-general and domain-specific predictors of both calculation and word reading from the beginning of first grade to the end of third grade (Fuchs et al., 2016). In that study, first grade domain-general processes were more influential in the development of calculation than they were for the development of word reading.

What is the Relation of Early Domain-General Abilities and Domain-Specific Number Knowledge to Later Mathematical Achievement?

Although the longitudinal mediation findings suggest that early working memory and the quality of language representations are important developmental precursors of later mathematical achievement, it is important to know whether these domain-general abilities remain strong predictors across time when early numeracy skills are also taken into account. In some longitudinal studies using large national databases, for example, the predictive value of domain-general abilities is significantly diminished when prior mathematical abilities are also modeled (e.g., Siegler et al., 2012; but see Bailey et al., 2014). As we assessed number knowledge at 60 months of age, we are able to test both early domain-general cognitive abilities and domain-specific number knowledge in the same multiple mediation models. In our most recent analyses we have used this combined approach, testing whether domain-general abilities (phonological awareness, working memory, and finger skills) and domain-specific abilities (oral counting, counting knowledge, and nonverbal arithmetic) assessed in the preschool years mediate the relation of group to later mathematics achievement (Barnes & Raghubar, 2016).

For multidigit calculation, phonological awareness and working memory remained significant longitudinal mediators, and nonverbal arithmetic was also a significant unique mediator. These findings have two implications. First, because early working memory and language remain important developmental precursors of later mathematics calculations in the presence of early number knowledge, the level of domain-general abilities in the preschool years has important predictive value for the development of later complex calculation skills beyond that provided by early foundational number knowledge. Second, there is some developmental continuity in the early domain-specific predictors of later calculation in that the ability to perform transformations on nonsymbolic number (i.e., nonverbal arithmetic) predicts complex symbolic arithmetic abilities 4–5 years later. Group differences in mixed mathematical problem solving were mediated by working memory, but not by any of the early number knowledge skills, again underlining the importance of working memory abilities for solving novel quantitative problems. Fuchs et al. (2016) have argued that novel and more complex mathematical problems (i.e., in our study this would refer to both multidigit calculation and mixed types of mathematical problems) are likely to engage domain-general processes, such as working memory and reasoning in comparison to other less novel and less complex academic tasks, such as word reading or math fact retrieval, which may rely more on associative learning processes. Our findings are consistent with these ideas; in our studies, performance on novel and complex mathematics and reading tasks was mediated by the level of working memory abilities measured 6–7 years earlier.

For single digit math fluency, phonological awareness was no longer a significant mediator with counting knowledge in the model. These findings, in combination with those from the two previously discussed longitudinal studies, suggest that a domain-specific developmental precursor, counting knowledge, is an important determinant of math fact acquisition. The quality of linguistic representations may be important for the acquisition of counting, which is, in turn, important for the acquisition of math facts, but we have not directly tested this hypothetical developmental sequence with our sample (see theoretical model of this type of developmental sequence in Krajewski & Schneider, 2009; also see Libertus, Feigenson, & Halberda, 2013 for an example of developmentally constrained relations for counting).

What are the Implications of the Findings for Assessment and Intervention?

The findings have implications for the assessment of MLD and perhaps also for the science of interventions for MLD. We focus our discussion on working memory. Assessment of cognitive processes, such as working memory is considered to be *unnecessary* for diagnosing MLD in school-age children, and empirical tests and simulations of the inclusion of cognitive processes in diagnostic schemes for learning disabilities has been shown to lack reliability and validity (Miciak, Fletcher, Stuebing, Vaughn, & Tolar, 2014; Miciak, Taylor, Denton, & Fletcher, 2015). We suggest that one exception to this general rule has to do with early identification of risk for MLD; that is, the strong relationship of early deficits in working memory to later mathematics achievement suggests that early screening for learning disabilities, particularly in mathematics, include an assessment of working memory. This suggestion is also informed by the fact that in very young children the direct assessment of mathematical knowledge may not provide a complete estimate of mathematics learning potential. For example, for children who are just entering school, the lack of exposure to mathematical concepts in the child's early home and preschool environment may result in false positives for risk on tests of mathematical knowledge. Thus, the prediction of risk for difficulties in mathematics learning in preschool children might benefit from assessments of number-specific knowledge that are supplemented by assessments of working memory (and language).

With respect to intervention, there is little evidence at present for the transfer of training of working memory to mathematics learning and performance (Jacob & Parkinson, 2015; Melby-Lervag & Hulme, 2012). However, working memory training studies have rarely involved students with learning disabilities and there remain many unresolved issues regarding working memory training itself and its transfer to other cognitive domains (review in Shipstead, Redick, & Engle, 2012). These include: whether a younger age at training is better (e.g., Wass, Scerif, & Johnson, 2012); whether the sequence of working memory training in relation to academic skill learning matters (e.g., working memory first or integrated with math intervention); what constitutes adequate dosage (Schwaighofer, Fischer, & Bühner, 2015); whether massed versus distributed practice for cognitive training is more effective (Wang, Zhou, & Shah, 2014); and whether the content and procedures of existing working memory training programs map onto what is known about how working memory increases in capacity in typical development. Not enough is known about the effectiveness of working memory interventions at present for improving working memory or academic skill learning, and there is some skepticism about the value of such training and even about the value of further research on this topic.

However, recent findings on how working memory *moderates* the effects of different types of mathematics interventions suggests that a child's working memory abilities are important to consider in the instruction of higher level mathematics. Fuchs et al. (2014) have shown that children with MLD and very low working memory benefit from an explicit self-explanation component in a fractions intervention, whereas students with somewhat better developed working memory do not seem to gain as much from this intervention component. H.L. Swanson and his coworkers have shown that different types of strategies for word problem solving are effective for different subgroups of children with MLD—those with better developed working memory versus those with low working memory (Swanson, 2015; Swanson, Lussier, & Orosco, 2015). Although these findings are intriguing, we currently lack explanations for the pattern of findings; that is, what are the features of intervention strategy *A* that makes it more effective for students with MLD and low working memory compared to intervention strategy *B* that is effective for students with MLD with better working memory?

To conclude, the strength of the relationship between working memory and mathematics achievement across a long developmental time window suggests that working memory might be an appropriate assessment target of risk for later MLD, particularly for very young children. Although studies are finding that different levels of working memory moderate the effects of different mathematics intervention components, our understanding of "why" certain strategies are more or less effective for children with MLD with different levels of working memory is in its infancy. However, there is promise in this new intervention work for customizing domain-specific interventions that are based on the child's particular domain-general cognitive characteristics.

CONCLUSIONS AND FUTURE DIRECTIONS

In this chapter we proposed that neurodevelopmental disorders associated with high rates of mathematics disability at school age and that are diagnosed before or at birth provide the field with model systems for testing hypotheses that may be more difficult to test in populations without clear neurological disorders. The longitudinal studies of children with SBM and their typically developing peers provide support for the importance of domain-general abilities—working memory and language—not just in the *performance* of mathematical tasks, but in the *development* of mathematical competencies. They also provide support for the idea that early number knowledge is an important developmental precursor of later mathematical achievement. Furthermore, these number-specific precursors show some developmental continuity and differentiation; for example, the early ability to perform arithmetical manipulations on nonsymbolic quantities uniquely predicts the later ability to perform complex manipulations (calculations) on symbolic quantities.

Although these findings are intriguing and add to a rapidly changing knowledge base on the development of mathematical cognition and the factors associated with difficulties in acquiring mathematical abilities, there are a few caveats with our longitudinal approach that merit consideration. First, although our findings converge with many others in the literature, phenotypic similarities in performance on mathematical tasks between groups (SBM vs. TD; younger vs. older children) need not mean that underlying neurocognitive processes are identical (Ansari, 2010). For example, the fact that the relation of working memory to mathematics may be larger in children whose MLD is associated with other developmental disorders than for those with MLD only (Peng et al., 2015) could also mean that the large contributions of early working memory to later mathematics that we found in our studies may not be generalizable to all other groups of children with MLD. Second, our studies were neither initially nor primarily designed to test models of mathematical development and disability even though they yielded data amenable to investigating aspects of these models over long developmental time windows. This means there are several model-driven hypotheses about mathematical development that our studies were not equipped to test. Based on our findings, but also these caveats, we think there would be benefit in conducting studies that begin assessment in infancy and follow children into school age, oversampling for children at high risk for MLD. Furthermore, the field might particularly benefit from studies that strike a balance between the power that is provided by large numbers and breadth in measurement that characterize investigations using national longitudinal databases and the theoretical coherence and depth of measurement at finer grain sizes that characterize smaller more recent studies of children assessed in infancy and again in the preschool years (e.g., Starr et al., 2013).

In conclusion, the findings from our longitudinal studies in which children at high risk for MLD were followed across a long developmental time window are largely consistent with those from correlational studies of children with and without MLD and from longitudinal studies of school-age children. The findings point to the importance of domain-general and domain-specific abilities for *growth* in mathematics. We see this convergence of findings across different research designs and with different populations as a positive sign for the field of mathematical development and disability because unifying theories that can handle both typical and atypical development within the same framework are likely to lead to more focused and integrated research in mathematical cognition.

ACKNOWLEDGMENTS

This research was supported by grants from the National Institute of Child Health and Development, P01 HD35946, Spina Bifida: Cognitive and Neurobiological Variability and R01HD046609, Longitudinal Effects of Spina Bifida on Learning, and by funding from the Canadian Institutes of Health Research and from the CIHR Institute of Human Development, Child and Youth Health.

REFERENCES

Alibali, M. W., & DiRusso, A. A. (1999). The function of gesture in learning to count: more than keeping track. *Cognitive Development*, *14*(1), 37–56.

Ansari, D. (2010). Neurocognitive approaches to developmental disorders of numerical and mathematical cognition: the perils of neglecting the role of development. *Learning and Individual Differences*, *20*(2), 123–129.

Ansari, D., & Dhital, B. (2006). Age-related changes in the activation of the intraparietal sulcus during nonsymbolic magnitude processing: an event-related functional magnetic resonance imaging study. *Journal of Cognitive Neuroscience*, *18*(11), 1820–1828.

Ayr, L. K., Yeates, K. O., & Enrile, B. G. (2005). Arithmetic skills and their cognitive correlates in children with acquired and congenital brain disorder. *Journal of the International Neuropsychological Society*, *11*, 249–262.

Bailey, D. H., Watts, T. W., Littlefield, A. K., & Geary, D. C. (2014). State and trait effects on individual differences in children's mathematical development. *Psychological Science*, *25*(11), 2017–2026.

Barbaresi, W. J., Katusic, S. K., Colligan, R. C., Weaver, A. L., & Jacobsen, S. J. (2005). Math learning disorder: Incidence in a population-based birth cohort, 1976—1982, Rochester, Minn. *Ambulatory Pediatrics*, *5*, 281–289.

Barnes, M. A., Pengelly, S., Dennis, M., Wilkinson, M., Rogers, T., & Faulkner, H. (2002). Mathematics skills in good readers with hydrocephalus. *Journal of the International Neuropsychological Society*, *8*, 72–82.

Barnes, M.A., & Raghubar, K.P. (2016). Neurodevelopmental disorders as model systems for understanding math disabilities. Paper presented at the *Pacific Coast Research Conference*, San Diego, California, February 6, 2016.

Barnes, M. A., Raghubar, K. P., English, L., Williams, J. M., Taylor, H., & Landry, S. (2014). Longitudinal mediators of achievement in mathematics and reading in typical and atypical development. *Journal of Experimental Child Psychology*, *119*, 1–16.

Barnes, M. A., Stubbs, A., Raghubar, K. P., Agostino, A., Taylor, H., Landry, S., & Fletcher, J. M. (2011). Mathematical skills in 3- and 5-year olds with spina bifida and their typically developing peers: a longitudinal approach. *Journal of International Neuropsychological Society*, *17*, 1–14.

Barnes, M. A., Wilkinson, M., Khemani, E., Boudesquie, A., Dennis, M., & Fletcher, J. M. (2006). Arithmetic processing in children with spina bifida calculation accuracy, strategy use, and fact retrieval fluency. *Journal of Learning Disabilities*, *39*(2), 174–187.

Bisanz, J., Sherman, J. L., Rasmussen, C., & Ho, E. (2005). Development of arithmetic skills and knowledge in preschool children. In J. I. D. Campbell (Ed.), *Handbook of Mathematical Cognition* (pp. 143–162). Hove: Psychology Press.

Blair, C., & Razza, R. P. (2007). Relating effortful control, executive function, and false belief understanding to emerging math and literacy ability in kindergarten. *Child Development*, *78*, 647–663.

Bugden, S., & Ansari, D. (2015). Probing the nature of deficits in the 'Approximate Number System' in children with persistent Developmental Dyscalculia. *Developmental Science*, 1–17.

Bull, R., Espy, K. A., & Wiebe, S. A. (2008). Short-term memory, working memory, and executive functioning in preschoolers: longitudinal predictors of mathematical achievement at 7 years. *Developmental Neuropsychology*, *33*, 205–228.

Butterworth, B. (1999). *The mathematical brain*. London: Macmillan.

Butterworth, B. (2005). The development of arithmetical abilities. *Journal of Child Psychology and Psychiatry*, *46*(1), 3–18.

Carey, S. (2004). Bootstrapping & the origin of concepts. *Daedalus*, *133*(1), 59–68.

Chen, Q., & Li, J. (2014). Association between individual differences in non-symbolic number acuity and math performance: a meta-analysis. *Acta Psychologica*, *148*, 163–172.

Christopher, M. E., Miyake, A., Keenan, J. M., Pennington, B., DeFries, J. C., Wadsworth, S. J., …, & Olson, R. K. (2012). Predicting word reading and comprehension with executive function and speed measures across development: a latent variable analysis. *Journal of Experimental Psychology: General*, *141*(3), 470–488.

Dehaene, S., Piazza, M., Pinel, P., & Cohen, L. (2005). Three parietal circuits for number processing. In J. I. D. Campbell (Ed.), *Handbook of mathematical cognition* (pp. 433–453). New York: Psychology Press.

Dennis, M., & Barnes, M. A. (2010). The cognitive phenotype of spina bifida meningomyelocele. *Developmental Disabilities Research Reviews*, *16*(1), 31–39.

De Smedt, B., Noël, M. P., Gilmore, C., & Ansari, D. (2013). How do symbolic and non-symbolic numerical magnitude processing skills relate to individual differences in children's mathematical skills? A review of evidence from brain and behavior. *Trends in Neuroscience and Education*, *2*(2), 48–55.

De Smedt, B., Taylor, J., Archibald, L., & Ansari, D. (2010). How is phonological processing related to individual differences in children's arithmetic skills? *Developmental Science*, *13*, 508–520.

Diamond, A., Prevor, M. B., Callender, G., & Druin, D. P. (1997). Prefrontal cortex cognitive deficits in children treated early and continuously for PKU. *Monographs of the Society for Research in Child Development*, *62*(4), i–i10.

Engle, R. W., & Kane, M. J. (2004). Executive attention, working memory capacity, and a two-factor theory of cognitive control. *Psychology of Learning and Motivation*, *44*, 145–200.

Evans, J. S. B., & Stanovich, K. E. (2013). Dual-process theories of higher cognition: advancing the debate. *Perspectives on Psychological Science*, *8*(3), 223–241.

Fazio, L. K., Bailey, D. H., Thompson, C. A., & Siegler, R. S. (2014). Relations of different types of numerical magnitude representations to each other and to mathematics achievement. *Journal of Experimental Child Psychology*, *123*, 53–72.

Feigenson, L., Dehaene, S., & Spelke, E. (2004). Core systems of number. *Trends in Cognitive Sciences*, *8*(7), 307–314.

Fletcher, J. M., Copeland, K., Fredrick, J. A., Blaser, S. E., Kramer, L. A., Northrup, H., … Dennis, M. (2005). Spinal lesion level in spina bifida: A source of neural and cognitive heterogeneity. *Journal of Neurosurgery, 102* (Suppl. 3), 268–279.

Fuchs, L. S., Fuchs, D., Compton, D. L., Hamlett, C. L., & Wang, A. Y. (2015a). Is word-problem solving a form of text comprehension? *Scientific Studies of Reading*, *19*(3), 204–223.

Fuchs, L. S., Geary, D. C., Fuchs, D., Compton, D. L., & Hamlett, C. L. (2016b). Pathways to third-grade calculation versus word-reading competence: are they more alike or different? *Child Development*, *87*, 558–567.

Fuchs, L. S., Schumacher, R. F., Sterba, S. K., Long, J., Namkung, J., Malone, A., … & Changas, P. (2014). Does working memory moderate the effects of fraction intervention? An aptitude–treatment interaction. *Journal of Educational Psychology*, *106*(2), 499–514.

Fuhs, M. W., & Day, J. D. (2011). Verbal ability and executive functioning development in preschoolers at head start. *Developmental Psychology*, *47*(2), 404–416.

Fuhs, M. W., & McNeil, N. M. (2013). ANS acuity and mathematics ability in preschoolers from low income homes: contributions of inhibitory control. *Developmental Science*, *16*(1), 136–148.

Fuhs, M. W., Nesbitt, K. T., Farran, D. C., & Dong, N. (2014). Longitudinal associations between executive functioning and academic skills across content areas. *Developmental Psychology*, *50*(6), 1698–1709.

Garon, N., Bryson, S. E., & Smith, I. M. (2008). Executive function in preschoolers: a review using an integrative framework. *Psychological Bulletin*, *134*(1), 31–60.

Geary, D. C. (1993). Mathematical disabilities: cognitive, neuropsychological, and genetic components. *Psychological Bulletin*, *114*(2), 345–362.

Geary, D. C. (2011). Cognitive predictors of achievement growth in mathematics: a 5-year longitudinal study. *Developmental Psychology*, *47*(6), 1539–1552.

Geary, D. C. (2013). Early foundations for mathematics learning and their relations to learning disabilities. *Current Directions in Psychological Science*, *22*(1), 23–27.

Geary, D. C., Berch, D. B., & Mann Koepke, K. (Eds.). (2015). *Evolutionary origins and early development of number processing (Vol. 1 Mathematical Cognition and Learning)*. San Diego, CA: Elsevier Academic Press.

Geary, D. C., Hamson, C. O., & Hoard, M. K. (2000). Numerical and arithmetical cognition: a longitudinal study of process and concept deficits in children with learning disability. *Journal of Experimental Child Psychology, 77*(3), 236–263.

Geary, D. C., Hoard, M. K., Nugent, L., & Bailey, D. H. (2012). Mathematical cognition deficits in children with learning disabilities and persistent low achievement: a five-year prospective study. *Journal of Educational Psychology, 104*(1), 206–223.

Gelman, R., & Butterworth, B. (2005). Number and language: how are they related? *Trends in Cognitive Sciences, 9*(1), 6–10.

Gelman, R., & Meck, E. (1983). Preschoolers' counting: principles before skill. *Cognition, 13*(3), 343–359.

Ginsburg, H. P., & Baroody, A. J. (1990). *Test of early mathematics ability-2*. Austin, TX: Pro-Ed.

Gracie-Bafalluy, M., & Noel, M. -P. (2008). Does finger training increase young children's numerical performance? *Cortex, 44*, 368–375.

Hart, S. A., Logan, J. A. R., Thompson, L., Kovas, Y., McLoughlin, G., & Petrill, S. A. (2016). A latent profile analysis of math achievement, numerosity, and math anxiety in twins. *Journal of Educational Psychology, 108*(2), 181–193.

Jacob, R., & Parkinson, J. (2015). The potential for school-based interventions that target executive function to improve academic achievement: a review. *Review of Educational Research, 85*(4), 512–552.

Jordan, N. C., Hanich, L. B., & Kaplan, D. (2003). A longitudinal study of mathematical competencies in children with specific mathematics difficulties versus children with comorbid mathematics and reading difficulties. *Child Development, 74*(3), 834–850.

Jordan, N. C., Huttenlocher, J., & Levine, S. C. (1992). Differential calculation abilities in young children from middle-and low-income families. *Developmental Psychology, 28*(4), 644–653.

Jordan, N. C., Kaplan, D., & Hanich, L. B. (2002). Achievement growth in children with learning difficulties in mathematics: findings of a two-year longitudinal study. *Journal of Educational Psychology, 94*(3), 586–597.

Jordan, N. C., Kaplan, D., Olah, L. N., & Locuniak, M. N. (2006). Number sense growth in kindergarten: a longitudinal investigation of children at risk for mathematics difficulties. *Child Development, 77*(1), 153–175.

Jordan, N. C., Kaplan, D., Ramineni, C., & Locuniak, M. N. (2009). Early math matters: kindergarten number competence and later mathematics outcomes. *Developmental Psychology, 45*(3), 850–867.

Keenan, J. M., Betjemann, R. S., & Olson, R. K. (2008). Reading comprehension tests vary in the skills they assess: differential dependence on decoding and oral comprehension. *Scientific Studies of Reading, 12*(3), 281–300.

Krajewski, K., & Schneider, W. (2009). Exploring the impact of phonological awareness, visual–spatial working memory, and preschool quantity-number competencies on mathematics achievement in elementary school: findings from a 3-year longitudinal study. *Journal of Experimental Child Psychology, 103*, 516–531.

Kuhn, L. J., Willoughby, M. T., Wilbourn, M. P., Vernon-Feagans, L., & Blair, C. B. (2014). Early communicative gestures prospectively predict language development and executive function in early childhood. *Child Ddevelopment, 85*(5), 1898–1914.

LeFevre, J. -A., Fast, L., Skwarchuk, S. -L., Smith-Chant, B. L., Bisanz, J., Kamawar, D., & Penner-Wilger, M. (2010). Pathways to mathematics: longitudinal predictors of performance. *Child Development, 81*, 1753–1767.

Libertus, M. E., Feigenson, L., & Halberda, J. (2011). Preschool acuity of the approximate number system correlates with school math ability. *Developmental Science*, *14*(6), 1292–1300.

Libertus, M. E., Feigenson, L., & Halberda, J. (2013). Is approximate number precision a stable predictor of math ability? *Learning and Individual Differences*, *25*, 126–133.

Lomax-Bream, L. E., Barnes, M. A., Copeland, K., Taylor, H. B., & Landry, S. H. (2007a). The impact of spina bifida on development across the first three years. *Developmental Neuropsychology*, *31*, 1–20.

Lomax-Bream, L. E., Taylor, H. B., Landry, S. H., Barnes, M. A., Fletcher, J., & Swank, P. (2007b). Role of early parenting and motor skills in development of children with spina bifida. *Journal of Applied Developmental Psychology*, *28*, 250–263.

Lyons, I. M., Price, G. R., Vaessen, A., Blomert, L., & Ansari, D. (2014). Numerical predictors of arithmetic success in grades 1–6. *Developmental Science*, *17*(5), 714–726.

Lyons, I. M., Ansari, D., & Beilock, S. L. (2015). Qualitatively different coding of symbolic and nonsymbolic numbers in the human brain. *Human Brain Mapping*, *36*(2), 475–488.

Mazzocco, M. M. M., Feigenson, L., & Halberda, J. (2011). Impaired acuity of the approximate number system underlies mathematical learning disability (dyscalculia). *Child Development*, *82*, 1224–1237.

Mazzocco, M. M., Murphy, M. M., & McCloskey, M. (2007). The contribution of syndrome research to understanding mathematical learning disability: the case of fragile X and Turner syndromes. In D. Berch, & M. M. Mazzocco (Eds.), *Why is math so hard for some children? The nature and origins of mathematical learning difficulties and disabilities* (pp. 173–193). Baltimore, MD: Brookes.

McKenzie, B., Bull, R., & Gray, C. (2003). The effects of phonological and visual–spatial interference on children's arithmetical performance. *Educational and Child Psychology*, *20*, 93–108.

Melby-Lervag, M., & Hulme, C. (2012). Is working memory training effective? A meta-analytic review. *Developmental Psychology*, *49*(2), 270–291.

Meyer, M. L., Salimpoor, V. N., Wu, S. S., Geary, D. C., & Menon, V. (2010). Differential contribution of specific working memory components to mathematics achievement in 2nd and 3rd graders. *Learning and Individual Differences*, *20*(2), 101–109.

Miciak, J., Fletcher, J. M., Stuebing, K. K., Vaughn, S., & Tolar, T. D. (2014). Patterns of cognitive strengths and weaknesses: Identification rates, agreement, and validity for learning disabilities identification. *School Psychology Quarterly*, *29*(1), 21.

Miciak, J., Taylor, W. P., Denton, C. A., & Fletcher, J. M. (2015). The effect of achievement test selection on identification of learning disabilities within a patterns of strengths and weaknesses framework. *School Psychology Quarterly*, *30*(3), 321–334.

Moeller, K., Martignon, L., Wessolowski, S., Engel, J., & Nuerk, H. C. (2011). Effects of finger counting on numerical development–the opposing views of neurocognition and mathematics education. *Frontiers in psychology*, *2*, 75–79.

Mou, Y., & vanMarle, K. (2014). Two core systems of numerical representation in infants. *Developmental Review*, *34*(1), 1–25.

Mussolin, C., Nys, J., Content, A., & Leybaert, J. (2014). Symbolic number abilities predict later approximate number system acuity in preschool children. *PLoS ONE*, *9*(3), e91839.

Noël, M. P. (2005). Finger gnosia: a predictor of numerical abilities in children? *Child Neuropsychology*, *11*(5), 413–430.

Noles, N. S., Scholl, B. J., & Mitroff, S. R. (2005). The persistence of object file representations. *Perception & Psychophysics*, *67*(2), 324–334.

Peng, P., Namkung, J., Barnes, M., & Sun, C. (2015). A meta-analysis of mathematics and working memory: moderating effects of working memory domain, type of mathematics skill, and sample characteristics. *Journal of Educational Psychology.*, Advance online publication. doi: http://dx.doi.org/10.1037/edu0000079.

Penner-Wilger, M., Fast, L., LeFevre, J., Smith-Chant, B. L., Skwarchuk, S., Kamawar, D., & Bisanz, J. (2007). The foundations of numeracy: subitizing, finger gnosia, and fine motor ability. In D. S. McNamara, & J. G. Trafton (Eds.), *Proceedings of the 29th Annual Cognitive Science Society* (pp. 1385–1390). Austin, TX: Cognitive Science Society.

Penner-Wilger, M., Fast, L., Lefevre, J. A., Smith-Chant, B. L., Skwarchuk, S. L., Kamawar, D., … & Deslauriers, W. A. (2008). Investigating the building blocks of numerical representations: Subitizing and finger gnosis. In *Proceedings of the 30th Annual Conference of the Cognitive Science Society.* Austin, TX: Cognitive Science Society.

Perfetti, C. A., & Adlof, S. M. (2012). Reading comprehension: a conceptual framework from word meaning to text meaning. In J. P. Sabatini, E. R. Albro, & T. O'Reilly (Eds.), *Measuring up: advances in how we assess reading ability* (pp. 3–20). Lanham, MD: Rowman & Littlefield Education.

Petrides, M., & Milner, B. (1982). Deficits on subject-ordered tasks after frontal- and temporal-lobe lesions in man. *Neuropsychologia, 20*(3), 249–262.

Piazza, M., Pica, P., Izard, V., Spelke, E. S., & Dehaene, S. (2013). Education enhances the acuity of the nonverbal approximate number system. *Psychological Science, 24*(6), 1037–1043.

Purpura, D. J., & Logan, J. A. (2015). The nonlinear relations of the approximate number system and mathematical language to early mathematics development. *Developmental Psychology, 51*(12), 1717–1724.

Raghubar, K. P., Barnes, M. A., Dennis, M., Cirino, P. T., Taylor, H., & Landry, S. (2015). Neurocognitive predictors of mathematical processing in school-aged children with spina bifida and their typically developing peers: attention, working memory, and fine motor skills. *Neuropsychology, 29*(6), 861–873.

Raghubar, K. P., Barnes, M. A., & Hecht, S. A. (2010). Working memory and mathematics: a review of developmental, individual difference, and cognitive approaches. *Learning and Individual Differences, 20*(2), 110–122.

Raghubar, K. P., Barnes, M. A., Prasad, M., Johnson, C. P., & Ewing-Cobbs, L. (2013). Mathematical outcomes and working memory in children with TBI and orthopedic injury. *Journal of the International Neuropsychological Society, 19*(03), 254–263.

Raghubar, K., Cirino, P., Barnes, M., Ewing-Cobbs, L., Fletcher, J., & Fuchs, L. (2009). Errors in multi-digit arithmetic and behavioral inattention in children with math difficulties. *Journal of Learning Disabilities, 42*(4), 356–371.

Reeve, R., & Humberstone, J. (2011). Five- to 7-year-olds' finger gnosia and calculation abilities. *Frontiers in Psychology, 2*(359), 1–10.

Rourke, B. P. (1993). Arithmetic disabilities, specific and otherwise: a neuropsychological perspective. *Journal of Learning Disabilities, 26*, 214–226.

Rourke, B. P., & Conway, J. A. (1997). Disabilities of arithmetic and mathematical reasoning perspectives from neurology and neuropsychology. *Journal of Learning disabilities, 30*(1), 34–46.

Rousselle, L., & Noel, M. P. (2007). Basic numerical skills in children with mathematics learning disabilities: a comparison of symbolic vs. non-symbolic number magnitude processing. *Cognition, 102*(3), 361–395.

Schneider, M., Beeres, K., Coban, L., Merz, S., Schmidt, S. S., Stricker, J., & De Smedt, B. (2016). Associations of non-symbolic and symbolic numerical magnitude processing with mathematical competence: a meta-analysis. *Developmental Science.*

Schneider, W., Lockl, K., & Fernandez, O. (2005). Interrelationships among theory of mind, executive control, language development, and working memory in young children: a longitudinal analysis. In W. Schneider, R. Schumann-Hensteler, & B. Sodian (Eds.), *Young Children's cognitive development: Interrelationships among executive functioning, working memory, verbal ability, and theory of mind* (pp. 259–284). New York: Psychology Press.

Schwaighofer, M., Fischer, F., & Bühner, M. (2015). Does working memory training transfer? A meta-analysis including training conditions as moderators. *Educational Psychologist, 50*(2), 138–166.

Shadish, W. R., Cook, T. D., & Campbell, D. T. (2002). Statistical conclusion validity and internal validity. *Experimental and quasi-experimental designs for generalized causal inference*. Boston: Houghton-Mifflin, pp. 45-48.

Shalev, R. S., Auerbach, J., Manor, O., & Gross-Tsur, V. (2000). Developmental dyscalculia: prevalence and prognosis. *European Child & Adolescent Psychiatry, 9*(2), 58–64.

Shipstead, Z., Redick, T. S., & Engle, R. W. (2012). Is working memory training effective? *Psychological Bulletin, 138*(4), 628–651.

Siegler, R. S., Duncan, G. J., Davis-Kean, P. E., Duckworth, K., Claessens, A., Engel, M., ... & Chen, M. (2012). Early predictors of high school mathematics achievement. *Psychological Science, 23*(7), 691–697.

Spelke, E. S. (2003). What makes us smart? Core knowledge and natural language. In D. Gentner, & S. Goldin-Meadow (Eds.), *Language in mind: Advances in the study of language and thought,* (pp. 277–311). Cambridge, MA: MIT Press.

Spelke, E. S., & Tsivkin, S. (2001). Language and number: a bilingual training study. *Cognition, 78*(1), 45–88.

Starr, A., Libertus, M. E., & Brannon, E. M. (2013). Number sense in infancy predicts mathematical abilities in childhood. *Proceedings of the National Academy of Sciences, 110*(45), 18116–18120.

Swanson, H. L. (2015). Cognitive strategy interventions improve word problem solving and working memory in children with math disabilities. *Frontiers in Psychology, 6*(1099), 1–13.

Swanson, H. L., Lussier, C. M., & Orosco, M. J. (2015). Cognitive strategies, working memory, and growth in word problem solving in children with math difficulties. *Journal of Learning Disabilities, 48*(4), 339–358.

Thorndike, R. L., Hagen, E. P., & Sattler, J. M. (1989). *Stanford-Binet intelligence scale* (4th Edition). Chicago: Riverside.

Tosto, M. G., Petrill, S. A., Halberda, J., Trzaskowski, M., Tikhomirova, T. N., Bogdanova, O. Y., ... & Plomin, R. (2014). Why do we differ in number sense? Evidence from a genetically sensitive investigation. *Intelligence, 43*, 35–46.

Trick, L. M., & Pylyshyn, Z. W. (1994). Why are small and large numbers enumerated differently? A limited-capacity preattentive stage in vision. *Psychological Review, 101*(1), 80–102.

Van Beek, L., Ghesquière, P., Lagae, L., & De Smedt, B. (2014). Left fronto-parietal white matter correlates with individual differences in children's ability to solve additions and multiplications: a tractography study. *Neuroimage, 90*, 117–127.

Van Lehn, K. (1990). *Mind bugs: the origin of arithmetical misconceptions*. Cambridge, MA: MIT Press.

VanMarle, K., Chu, F. W., Li, Y., & Geary, D. C. (2014). Acuity of the approximate number system and preschoolers' quantitative development. *Developmental Science, 17*(4), 492–505.

Vukovic, R. K., & Lesaux, N. K. (2013). The language of mathematics: investigating the ways language counts for children's mathematical development. *Journal of Experimental Child Psychology, 115*(2), 227–244.

Wang, Z., Zhou, R., & Shah, P. (2014). Spaced cognitive training promotes training transfer. *Frontiers in Human Neuroscience, 8*(217), 1–8.

Watts, T. W., Duncan, G. J., Chen, M., Claessens, A., Davis-Kean, P. E., Duckworth, K., ... & Susperreguy, M. I. (2015). The role of mediators in the development of longitudinal mathematics achievement associations. *Child Development, 86*(6), 1892–1907.

Wasner, M., Nuerk, H. C., Martignon, L., Roesch, S., & Moeller, K. (2016). Finger gnosis predicts a unique but small part of variance in initial arithmetic performance. *Journal of Experimental Child Psychology*, *146*, 1–16.

Wass, S. V., Scerif, G., & Johnson, M. H. (2012). Training attentional control and working memory: is younger, better? *Developmental Review*, *32*(4), 360–387.

Woodcock, R. W., & Johnson, M. B. (1989). *Woodcock-Johnson psycho-educational battery-revised*. Chicago: Riverside Publishing.

Wynn, K. (1992). Addition and subtraction by human infants. *Nature*, *358*(6389), 749–750.

Xu, F., & Spelke, E. S. (2000). Large number discrimination in 6-month-old infants. *Cognition*, *74*(1), 1–11.

Part II

Rational Number Processing

Chapter 5

The Transition from Natural to Rational Number Knowledge

Jo Van Hoof*, Xenia Vamvakoussi**, Wim Van Dooren*, Lieven Verschaffel*
**KU Leuven, Leuven, Flanders, Belgium; **University of Ioannina, Ioannina, Epirus, Greece*

From a mathematics point of view, a rational number is commonly defined as a number that is or can be expressed in the form *a*/*b*, where *a* and *b* are integers, and *b* is not zero. As integers, terminating decimals, and also repeating decimals can be expressed in this form, they are rational numbers. For the purposes of this chapter, we will refer to numbers that are actually in the *a*/*b* form as fractions, and to rational numbers written in base-10 notation as decimal numbers.

Considerable research has indicated that a strong understanding of rational numbers is of critical importance for mathematics achievement in general and for performance in important content domains of the mathematics curriculum in particular, such as algebra (e.g., Siegler et al., 2012). Given that students' understanding of rational numbers has been shown to relate to their later mathematics achievement, it is problematic that many learners have trouble understanding several aspects of rational numbers, even through the end of secondary school and into adulthood (Vamvakoussi, Van Dooren, & Verschaffel, 2012). In this chapter we elaborate on learners' difficulties with rational numbers. First, we discuss the importance of robust rational number understanding for learners' general mathematics achievement. Second, we highlight the challenges that many learners face with respect to various aspects of rational numbers. Third, we elaborate on previous research that focused on a major source of learners' difficulty, namely the natural number bias. Fourth, we introduce two complementary theoretical perspectives that have been employed in the past decade to study the natural number bias, namely the framework theory approach to conceptual change and the dual process perspective of reasoning, and then we review some studies that we conducted from these perspectives. Finally, we discuss theoretical and educational implications of these studies and provide suggestions for further research.

Acquisition of Complex Arithmetic Skills and Higher-Order Mathematics Concepts
http://dx.doi.org/10.1016/B978-0-12-805086-6.00005-9

THE IMPORTANCE OF RATIONAL NUMBERS

A strong understanding of rational numbers is of critical importance for mathematics achievement in general and for particular domains of the mathematics curriculum. For example, in a study by Booth and Newton (2012), learners' fraction knowledge was a better predictor for their algebra readiness than their whole number magnitude knowledge. Further, Siegler, Thompson, and Schneider (2011) found high correlations between upper elementary school children's fraction magnitude knowledge and their general mathematics achievement. Siegler et al. (2012) extended these findings with longitudinal data from the United States and United Kingdom showing that 5th graders' fraction knowledge significantly predicted their general mathematics and algebra achievement scores in high school, even after controlling for reading achievement, IQ, working memory, whole number knowledge, family income, and family education.

Bailey, Hoard, Nugent, and Geary (2012) found that the predictive relation between learners' rational number knowledge and their later mathematics achievement could not be explained by their general mathematical ability. As a possible explanation for this predictive relation, they mentioned that learners with a high general mathematical ability might understand rational numbers more easily than those with a low general mathematical ability. Their results showed that this alternative explanation was not the sole explanation for the predictive relation. After controlling for general mathematics achievement in 6th grade and rational number knowledge in 7th grade, as well as controlling for working memory and intelligence, they still found that 6th graders' rational number knowledge significantly predicted mathematics achievement in grade 7.

RATIONAL NUMBERS: A CHALLENGE FOR LEARNERS AND FOR MATHEMATICS EDUCATION

Although the importance of rational number learning is widely acknowledged and considerable time and effort are invested in rational number teaching, numerous studies have documented that learners in elementary and secondary school, educated adults, and even (prospective) teachers face great difficulties with understanding various features of rational numbers (see, for example, Cramer, Post, & delMas, 2002; Depaepe et al., 2015; Hasemann, 1981; Janssen, Verschaffel, Tuerlinckx, Van den Noortgate, & De Fraine, 2010; Lortie-Forgues, Tian, & Siegler, 2015; Mazzocco & Devlin, 2008; McMullen, Laakkonen, Hannula-Sormunen, & Lehtinen, 2015; Meert, Grégoire, & Noël, 2010; Obersteiner, Van Dooren, Van Hoof, & Verschaffel, 2013, 2016; Siegler et al., 2012; Vamvakoussi, Christou, Mertens, & Van Dooren, 2011; Vamvakoussi et al., 2012; Vamvakoussi, Van Dooren, & Verschaffel, 2013; Vamvakoussi & Vosniadou, 2004, 2010; Van Hoof, Janssen, Verschaffel, & Van Dooren, 2015; Van Hoof, Lijnen, Verschaffel, & Van Dooren, 2013; Van Hoof, Vandewalle, Verschaffel, & Van Dooren, 2015; Van Hoof, Verschaffel,

& Van Dooren, 2015). To elaborate on one example, a recent study that included a nationally representative sample of Flemish 8th graders revealed that only about one fourth of them succeeded in achieving the minimum standards for operations in arithmetic, including operations with rational numbers (Vlaamse Overheid, 2010). An example item used in this assessment is: "in the bookstore, customers get a 5% reduction, which makes that the book I bought is 1€ cheaper. What was the price without the reduction?"

There are many reasons why learners have difficulty with rational numbers, some of which are summarized by Moss (2005): (1) there are several conceptually distinct meanings attached to rational numbers that need to be understood and coordinated (e.g., fractions as parts of a whole, fractions as ratios, fractions as numbers); (2) new symbols and representations are introduced that, again, need to be understood and coordinated (e.g., 1/2, 3/60.5, 0.50, 50%); (3) learners need to construct a complex knowledge network for number based on multiplicative rather than on additive relations; (4) the understanding of the unit and the arithmetical operations need to be reconceptualized. All in all, there is considerable new material to be learned and the content is highly complex. Moreover, learners' prior knowledge and experience in the domain of number is not always compatible with the new content on rational numbers, as already indicated by the aforementioned reasons (3) and (4). As we discuss in the following paragraph, this turns out to be a major source of difficulties with rational numbers.

THE INTERFERENCE OF NATURAL NUMBER KNOWLEDGE IN RATIONAL NUMBER TASKS

A much noticed, studied, and reported phenomenon is the interference of natural number knowledge in learners' attempts to make sense of rational numbers and deal with rational number tasks (see, for example, Kilpatrick, Swafford, & Findell, 2001; Moss, 2005; Ni & Zhou, 2005; Van Dooren, Lehtinen, & Verschaffel, 2015). It appears that learners transfer their natural number knowledge to the domain of rational numbers, using it to interpret information coming from instruction, and to solve rational number problems. The most salient indication is learners' systematic errors that occur when their prior knowledge and experiences with natural numbers are not appropriate for the task at hand. Such tasks are commonly called incongruent. For example, if learners are asked to compare the numerical sizes of 0.32 and 0.316, they may be inclined to rely on the knowledge that 316 is larger than 32, as well as on the knowledge that 0.316 has more digits than 0.32, leading them to the wrong conclusion that 0.316 is larger than 0.32. In contrast, items where reasoning purely in terms of knowledge of natural numbers leads to a correct answer are called congruent items. For instance, if learners are asked to compare the numerical sizes of 0.32 and 0.3, relying on their natural number knowledge that 32 is larger than 3, and/or that 0.32 has more digits will lead them to the correct conclusion that 0.32 is larger than 0.3. So, incongruency emerges when there is a difference between the solutions that would emerge with the use of natural number knowledge as contrasted with rational number knowledge. Rational numbers differ from natural numbers in many and important ways, each of which may be overlooked by learners (see Stafylidou & Vosniadou, 2004; Vamvakoussi & Vosniadou, 2010, for a detailed account). In what follows, we focus on three main differences, relating to size, operations, and density, which are reported to lead to those systematic errors.

The Size of Rational Numbers

Research indicates that errors in judging or comparing the size (i.e., magnitude) of decimal numbers are frequently made because of learners' misconception that, just as it is the case with natural numbers, the size of decimal numbers can be determined by considering differences in their number of digits ("longer decimals are larger" and "shorter decimals are smaller") (see, for example, Resnick et al., 1989). This leads to systematic mistakes such as concluding that 0.32 is larger than 0.5 because 32 is larger than 5. Similarly, learners' errors in understanding the magnitudes of fractions are largely based on the numerical value of each of the components of the fractions ("a fraction's numerical value increases when its denominator, numerator, or both increase") (see, for example, Clarke & Roche, 2009; Stafylidou & Vosniadou, 2004). As with decimal numbers, this leads to systematic errors such as concluding that 5/9 is larger than 3/4.

The Effect of Arithmetic Operations

When first learning arithmetic with natural numbers, many learners implicitly deduce that multiplication and addition will always result in a larger outcome whereas division and subtraction will always result in a smaller outcome. In the domain of rational numbers these rules no longer necessarily apply. However, learners may still rely on them, leading to mistakes, such as thinking that "0.99 * 5" leads to an outcome larger than 5 or accepting nonwhole numbers as multiplicands but not as multipliers (De Corte, Verschaffel, & Van Coillie, 1988; Fischbein, Deri, Nello, & Marino, 1985; Hasemann, 1981).

The Dense Structure of Rational Numbers

While natural numbers are discrete (one can always point out which number comes next), rational numbers are dense; rational numbers do not have a successor number, because between any two rational numbers are always infinitely many other numbers. This difference between natural and rational numbers leads to the common assumption that there are no (or finitely many) numbers between two pseudo-successive numbers (for example, between 1.2 and 1.3). Numerous studies have indicated that the dense ordering of rational numbers is difficult to grasp both for elementary and secondary school learners (Hannula,

Pehkonen, Maijala, & Soro, 2006; Merenluoto & Lehtinen, 2002, 2004; Smith, Solomon, & Carey, 2005).

Representation of Numbers as an Intersecting Difficulty

A fourth difficulty that intersects with these three aforementioned aspects is the representation of numbers. In this chapter we focus on two difficulties learners encounter concerning the representation of numbers and more specifically in the domain of rational numbers (see also DeWolf, Bassok, & Holyoak, Chapter 7).

First, rational numbers can be represented by fractions and decimals, and within each of these two representational types by an infinite number of possible representations. For example, "one half" can be represented as 0.5, but also as 0.50, 0.500, 1/2, 8/16, 50%, ... While making meaningful connections between different representational forms is highly important in mathematics, many learners struggle with it (Jordan et al., Chapter 6). As stated by Ni and Zhou (2005), this is especially the case in the domain of rational numbers. Learners further tend to think of a fraction as two (natural) numbers rather than as a number in its own right (see, for example, Smith et al., 2005; Stafylidou & Vosniadou, 2004).

Second, several studies indicate that students have difficulty interpreting literal symbols in algebra (e.g., Kieran, 2006). When first learning algebra, many students have difficulties accepting that a literal symbol can stand for a number and is not, for example, merely an abbreviation of an object's name (e.g., "h" for "height") (e.g., Booth, 1984). When learners do start to make the association between literal symbols and numbers, they tend to think that a literal symbol always stands for one single number (Asquith, Stephens, Knuth, & Alibali, 2007). Moreover, research indicates that many learners have a tendency to substitute these literal symbols in algebra only with natural numbers (Christou & Vosniadou, 2005, 2012; Van Dooren & Vamvakoussi, 2010; Van Hoof et al., 2015b) and not with other types of numbers, such as rational numbers.

THE NATURAL NUMBER BIAS

The examples presented previously are instances of natural number knowledge interference with rational number learning. Ni and Zhou (2005) coined the term "whole number bias" to refer to this phenomenon. We opt for the term "natural number bias" because unlike rational numbers, natural numbers are always positive; this specific feature may also be inappropriately assumed when dealing with rational number tasks, particularly when number are represented by literal symbols. The origin of the natural number bias—and, more specifically, whether there is an innate aspect or not—is still a matter of debate (Ni & Zhou, 2005). There is, however, substantial consensus in the literature that natural numbers are culturally privileged (Dantzig, 1954;

Greer, 2004; Ni & Zhou, 2005; Vamvakoussi et al., 2012). Indeed, natural number knowledge is supported from early on by cultural representational tools and practices, such as finger counting as well as by experiences, such as counting songs, rhymes, and board games. Furthermore, instruction on numbers in the first few elementary-school years is almost exclusively devoted to natural numbers. As a result, natural number ideas—validated and strengthened in and out of school settings—are firmly established and dominate people's numerical reasoning, which justifies the term natural number bias.

It is important to note that the natural number bias has various manifestations, other than systematic errors in incongruent tasks (see Vamvakoussi, 2015, for a concise account). For instance, much higher accuracy levels are found in congruent rational number tasks where the correct reasoning is in line with reasoning about natural numbers (Nunes & Bryant, 2008); consider also that the part-whole interpretation of fractions—which is compatible with natural number reasoning—is typically used for the introduction of fractions because it is deemed more accessible than other aspects, such as ratio (Moss, 2005). That is, the natural number bias acts as a facilitator in contexts that are compatible (congruent) with natural number knowledge. Further, it has been noticed that biased responses are commonly generated quickly, are deemed self-evident, and may be temporarily revised if one is explicitly guided to reconsider; however, they are not necessarily eradicated and may reappear at a different moment or in a different context (Fischbein, 1987; Merenluoto & Lehtinen, 2004).

THEORETICAL FRAMEWORKS FOR STUDYING THE NATURAL NUMBER BIAS

The interference of prior knowledge in mathematics learning in general, and in rational number learning in particular, has been studied through various theoretical lenses (Vamvakoussi, Vosniadou, & Van Dooren, 2013). We focus on two theoretical perspectives that capture different aspects of the natural number bias. The first is the framework theory approach to conceptual change that can account for the origin and the development of the bias (e.g., Vosniadou, 2013; Vosniadou, Vamvakoussi, & Skopeliti, 2008). The second is the dual process perspective of reasoning that can capture the processes underlying people's responses to rational number tasks, accounting for the immediacy and the perseverance of biased responses, even when one possesses the necessary knowledge to deal with the task at hand (Evans & Over, 1996).

The Conceptual Change Perspective

Thus far, most of the research on the natural number bias has interpreted this phenomenon from a conceptual change perspective, and more specifically in terms of the framework theory approach to conceptual change (Vosniadou, 1994; Vosniadou et al., 2008). The origins of conceptual change research are situated in the field of physics and physics education. More recently, however, it has been shown that the framework theory approach to conceptual change can also be successfully applied in research concerning the learning of mathematics (see, for example, Vosniadou & Verschaffel, 2004; Vamvakoussi et al., 2013b). Applied to the domain of number (Vamvakoussi & Vosniadou, 2010), the key assumption of this approach is that, beginning early in life, people organize their daily informal as well as formal experiences with numbers in a relatively coherent framework theory that relies on implicit assumptions about what numbers are and how they behave. As discussed previously, natural numbers are culturally privileged in and out of school settings. Thus, before rational numbers are introduced in instruction, a framework theory of numbers is assumed to be in place. Within this framework theory, numbers behave essentially as natural numbers. For instance, they are discrete, they can be ordered based on their place in the count list or on the number of digits ("longer numbers are bigger"), and they "get bigger" with addition or multiplication, and "smaller" with subtraction or division.

Learners' initial framework theory of number acts as an interpretive filter for new information about nonnatural numbers. We note that this is the defining feature of the natural number bias as defined by Ni and Zhou (2005). This presents no problems when new information is compatible with what is already known. However, incompatible information may be neglected or distorted in the process. It is assumed that learners typically employ additive mechanisms of learning to gradually incorporate new information into their initial framework theory of number. Adding incompatible information, however, destroys the coherence of the initial framework theory and results in inconsistencies and misconceptions. A central element of the framework theory approach is that conceptual change is a slow and gradual process. This is because people do not revise all background assumptions of the initial framework theory at once, largely because they are typically not aware of their initial implicit assumptions (e.g., Vosniadou, 1994).

In the transition from the initial framework theory of number to the scientific perspective, a special class of misconceptions appears, namely synthetic conceptions (or synthetic models). Synthetic conceptions combine elements of the initial framework theory of number with elements of the new information assimilated in the knowledge structure. An example is the synthetic conception of rational numbers as a collection of distinct sets of numbers, namely natural numbers, decimals, and fractions. One consequence is the assumption that these different types of numbers have different properties. For instance, decimals may be deemed densely ordered, whereas fractions are viewed as discrete (Vamvakoussi & Vosniadou, 2010). Another example of a synthetic conception appears in the comparison of fractions, when learners assume that the fraction with the smallest denominator is larger than a fraction with a larger denominator (e.g., 4/6 vs. 1/2), neglecting the relation between the terms of each fraction (e.g., Gómez, Jiménez, Bobadilla, Reyes, & Dartnell, 2014; Stafylidou & Vosniadou, 2004).

It is predicted that learners' initial framework theories are not completely abandoned as they develop a more sophisticated understanding of number

(DeWolf and Vosniadou, 2015). This is particularly relevant in the case of the initial framework theories of number, because natural numbers, although subsumed under the broader category of the rational numbers, preserve all their particular characteristics *within* the natural numbers set; thus initial framework theories of number remain useful as well as mathematically correct in this particular context (Vamvakoussi & Vosniadou, 2010).

The Dual Process Perspective on Reasoning

The dual process perspective originated as an attempt to explain errors committed by people on problems for which they in principle have the required knowledge to reach a correct solution (Gillard, Van Dooren, Schaeken, & Verschaffel, 2009). It postulates a distinction between intuitive or heuristic reasoning (fast, automatic, and with minimal demands of working memory) and analytical reasoning (slow, controlled, and demanding of working memory) (see, for example, Evans & Over, 1996). Intuitive reasoning is automatically engaged and will often lead to the correct answer, but sometimes more extensive, analytical thought processes are needed. In the latter case there are two possibilities. First, an incorrect intuitive thought process may occur without any intervention by the analytical processing system. In this case a wrong answer will be generated very quickly and without much effort. The second possibility is that the analytical processing system does intervene, evaluates the intuitively obtained answer, and attempts to inhibit it and generate an alternative answer. Depending on the success of this time-consuming and effortful analytic processing, a correct answer may or may not be obtained. Incorrect answers, in such a dual process view, can thus be the result of either an intuitive thought process that will result in the wrong answer and is not controlled by the analytical processing system or an unsuccessful intervention of the analytical processing system (Gillard et al., 2009).

The dual process perspective is a cognitive psychological one that so far has been primarily applied to logical reasoning tasks (Stanovich & West, 2000). However, several researchers in the field of mathematics education have put forward that this perspective may be used to operationalize the notion of intuition in mathematical reasoning (see, for example, Babai, Levyadun, Stavy, & Tirosh, 2006; Geary, 2008; Gillard et al., 2009).

Characterizing intuition in mathematical reasoning and learning is a challenging task that was undertaken in the seminal work by Fischbein (1987). Among the defining features of intuitions according to Fischbein, there are two particularly relevant for this discussion: Intuitive judgments are fast, grounded on an immediate grasp of the situation at hand; and intuitions, once established, are robust and not easily eradicated by instruction, often coexisting with mathematically correct ideas throughout a person's life.

Reaction times can be used to disentangle intuitive and analytical reasoning during mathematical problem solving (see, for example, De Neys, 2006; Gillard et al., 2009). Vamvakoussi et al. (2012) argued that when faced with rational number tasks, individuals' first, intuitive, response is grounded in natural number reasoning. Thus, errors in rational number tasks can result from a failure to inhibit intuitive knowledge and engage in analytical reasoning. This phenomenon emerges in reaction time studies, whereby it takes longer to correctly solve incongruent problems than congruent problems, due to the needed intervention by the analytical processing system. Even when people correctly solve incongruent problems, they still take longer than when solving congruent problems, reflecting the continued interference of natural number knowledge.

Combining the Conceptual Change Theory and Dual Process Perspective to Study Mathematical Thinking and Learning

We argue that the combination of the conceptual change theory and the dual process perspective is a promising way to study rational number understanding and the occurrence of the natural number bias. Conceptual change theory enables us to describe and explain the learner's evolution of the mathematical concept of rational number, starting from a very naïve, intuitive natural-number based idea to the scientifically correct idea. It explains why so many learners have difficulties with several aspects of rational number understanding; specifically, because they are biased by their prior knowledge of natural numbers and conceptual change is needed to fully understand the various aspects of rational numbers (see, for example, Vosniadou, 2013).

Moreover, it allows us to describe and predict the occurrence of specific synthetic conceptions (or synthetic models). The conceptual change theory is, however, less specific about the underlying ongoing reasoning processes that take place at the moment when one is solving a particular task once a more sophisticated understanding of rational numbers is acquired. The dual process perspective allows us to do so. This approach goes further than the conceptual change theory by explaining how it is possible that learners who already have the correct, scientific idea of what a number is (and have in other words succeeded in their conceptual change of a rational number) can still make mistakes (Evans & Over, 1996). As claimed previously, learners' correct scientific idea about number may come to coexist with their initial, naïve concept. So, although learners can correctly understand number from a scientific perspective, they can still make mistakes due to the influence of intuitive thought processes that are not or not sufficiently controlled by the analytical processing system.

OVERVIEW OF OUR STUDIES USING BOTH CONCEPTUAL CHANGE THEORY AND DUAL PROCESS PERSPECTIVE

In the past decade, we conducted several studies using both theoretical perspectives to increase our understanding of learners' transition from natural to rational number understanding, and the role of the natural number bias in this transition. Whereas paper-and-pencil tests and interviews are the two most common methods to investigate the natural number bias, we also applied a third, less common method, namely collecting reaction time data. As stated previously, learners can correctly understand a concept, but still make mistakes due to the intrusion of incorrect, intuitive knowledge. We investigated whether this is also applicable in the case of rational number understanding, by collecting reaction time data in several of our studies.

Size

In a cross-sectional study, Van Hoof et al. (2013) investigated Flemish 7th and 11th graders' natural number bias by measuring their accuracy levels and reaction times on fraction comparison tasks in which the fractions either had a common denominator (congruent items, for example, 11/26, vs. 17/26) or a common numerator (incongruent items, for example, 10/27 vs. 10/21). Accuracy was high for all problems, thus preventing the evaluation of a potential natural number bias, but this did emerge in problem solving reaction times. Incongruent problems took longer to solve correctly than congruent problems for students in both grades. Further, the natural number bias did not decrease (neither in terms of accuracy nor in terms of reaction time) between 7th and 11th grade. Similar results have been obtained with educated adults. Specifically, Vamvakoussi et al. (2013a) tested a group of Flemish university students in fraction comparison tasks. Both fractions with and without common components were included in the study (see Table 5.1 for examples). In the comparison tasks including fractions with common components, no effect of congruency was found in students' accuracy, due to ceiling levels. However, a significant effect of congruency was again found for reaction times: Similar to the secondary school students, the university students needed more time to respond correctly to incongruent items (i.e., fractions with common numerators). Concerning the fractions without common components, congruency had no significant effect either on accuracy or on reaction time.

Obersteiner et al. (2013) went further by assessing mathematics and computer science faculty members on the same type of tasks. They were all faculty members of a Flemish university. The results indicated that, similar to secondary school students and college students, congruency had a significant effect on

TABLE 5.1 Examples of Congruent and Incongruent Fraction Comparison Tasks, Both With and Without Common Components

	Congruent	Incongruent
Common components	2/5 versus 3/5	1/8 versus 1/3
No common components	1/4 versus 19/20	28/90 versus 1/3

reaction time when the fractions had common components (denominators or numerators). More specifically, experts needed more time to accurately solve incongruent tasks (e.g., 4/17 vs. 4/39) than congruent tasks (e.g., 22/49 vs. 18/49). In fraction comparison tasks without common components, congruency had no significant effect on accuracy or on reaction time data.

Operations

Based on the work of Christou and Vosniadou (2005), Van Hoof et al. (2015b) asked Flemish 8th graders to evaluate the accuracy of algebraic expressions, some of which were congruent (i.e., interpreting the letter as a natural number leads to a correct answer; for example, judging that "$x < x * 4$" can be true) and others incongruent (i.e., interpreting the letter as a natural number leads to an incorrect answer, for example, judging that "$x * 5 < x$" cannot be true). Higher accuracy rates for congruent items than for incongruent items were found. Additional interviews, in which the students were asked after each item to explain how they solved it, showed that these learners used similar strategies when they incorrectly solved incongruent items as when they correctly solved congruent items. In both cases, they almost always either referred to a natural number rule (for example: division always results in a smaller answer) or substituted the unknown term with one or more natural numbers. While the natural number bias was manifest in multiplication and division items, only a nonsignificant trend was found for addition and subtraction items. The same researchers administered the same algebraic expressions to a large group of Flemish 10th and 12th graders. Again, an overall natural number bias was evidenced by higher accuracy on congruent than on incongruent items. As with the 8th graders, traces of the natural number bias were found only in multiplication and division items. Based on the odds ratios for the difference in accuracy between congruent and incongruent items, results indicated that the strength of the natural number bias did not decrease from grade 8 to grade 12.

In a related study, Vamvakoussi et al. (2013a) asked university students to judge the correctness of algebraic statements that could be considered as either congruent (for example: "$5 + 2x$ can be bigger than 5") or incongruent items (for example: "$5 \div x$ can be bigger than 5"). The natural number bias was reflected in the significantly lower accuracy levels on incongruent than on congruent items; and also in the significantly longer reaction times needed to correctly answer incongruent than congruent items. Unlike the secondary school students studied by Van Hoof et al. (2015b), the college students who participated in this study showed similar results for addition and subtraction as those found for multiplication and division. Obersteiner et al. (2016) compared the performance of German secondary school students and Flemish expert mathematicians when they were asked to judge the correctness of algebraic expressions that were similar to the ones described in the studies previously (Vamvakoussi et al., 2013a; Van Hoof et al., 2015b). Unlike the studies described previously, the items were presented in two blocks that consisted of the same set of items, but in a first block, *x* was defined as a natural number, while in the second block, *x* was defined as a rational number. Results indicated that for secondary students accuracy was higher in the natural number block compared to the rational number block. Moreover, within the rational number block, they had higher accuracy levels on congruent items compared to incongruent items.

The experts, in contrast, did not show traces of the natural number bias. Within the rational number block, no differences could be found for either accuracy or reaction times across congruent and incongruent items. Moreover, the experts were more accurate in the rational number block compared to the natural number block. This finding appears to be in contrast with the findings of Obersteiner et al. (2013) that showed traces of the natural number bias in a population of experts. However, task differences need to be considered: it appears that in the context of algebraic expressions, the experts did not draw on their number knowledge; rather, they relied on their knowledge about the solvability of inequalities. As Obersteiner et al. (2016) concluded, it is possible for experts to circumvent the problem of the natural number bias when it is possible to employ strategies that do not rely on knowledge about numbers.

Density

Vamvakoussi and Vosniadou (2004, 2007, 2010) investigated secondary students' (grades 7th and 9th, and 11th) understanding of the dense ordering of rational numbers.

Vamvakoussi and Vosniadou (2010), administered to Greek 7th graders, 9th graders, and 11th graders a multiple-choice questionnaire consisting of items asking "How many numbers are there between *x* and *y*?", varying systematically the type of the *x* and *y* end points over natural numbers ("How many numbers are there between 0 and 1?"), decimals ("How many numbers are there between 0.1 and 0.2?"), and fractions ("How many numbers are there between 1/3 and 2/3?"). As predicted by the framework theory approach, they found that the idea of discreteness, a fundamental background assumption of learners' initial framework theory of number, was very robust, leading to considerable percentages of responses of the type "there is a finite number of intermediates between two given numbers," even for the older participants. Furthermore, the type of interval endpoints had a large effect on learners' judgments regarding the number as well as the type of numbers in a given interval. Specifically, learners were more inclined to answer that there is an infinite number of intermediates between two natural numbers, but less so in the case of decimals and fractions. Moreover, learners were reluctant to accept that there can be decimals between two fractions or that there can be fractions between two decimals. In addition, some learners changed their answers depending on the intervals end points. For instance, infinitely many intermediates were accepted between decimals, but a finite number of intermediates between fractions. Vamvakoussi and Vosniadou

(2010) attributed these findings to the synthetic conception discussed previously of the rational numbers as a collection of distinct, unrelated sets of numbers.

Vamvakoussi et al. (2011) replicated the previous study with Flemish 9th graders, and compared the results with those of the 9th grade Greek students. Although the Flemish participants outperformed their Greek peers, their patterns of responses were very similar. This latter result provided further evidence for the framework theory approach to conceptual change. Indeed, this study indicated that learners coming from different educational systems face the same conceptual difficulties regarding the dense structure of rational numbers. Finally, the density items used in the study by Vamvakoussi et al. (2012) proved the most challenging for the college students who participated in the study, compared to the comparison and operation tasks. As predicted by the dual process perspective, congruency for density items had a significant effect on accuracy (incongruent items elicited more errors), and participants took longer to respond correctly to incongruent items.

How are the Three Aspects Related to Each Other?

Based on the existing literature and an analysis of the Flemish mathematics curriculum, Van Hoof et al. (2015c) constructed and validated a comprehensive test instrument that enabled them to compare the strength (i.e., the ratio of correctly solved congruent to correctly solved incongruent items) of the natural number bias for the three different aspects of this bias (size, operations, and density). By administering this test to a large group of Flemish 4th–12th graders, they found a clear manifestation of the natural number bias in terms of higher accuracy rates for congruent than for incongruent items. Second, this manifestation of the bias was equally strong in tasks with decimal numbers as with fractions. This is an interesting finding, because the available theoretical and empirical literature contains evidence that different kinds of natural number-based errors may occur in items involving these two representations (for more information, see Van Hoof et al., 2015c). Third, learners' natural number-based errors decreased across grades and the natural number bias declined over a period of at least four years (from 6th to 10th grade), however, without completely disappearing at the end of secondary education (see Fig. 5.1). Fourth, the strength of the natural number bias was weakest in size tasks, somewhat stronger in operations tasks, and by far the strongest in density tasks (Fig. 5.1). Fifth, while there was an improvement in learners' understanding of all three aspects, they continued to struggle particularly with the operations and density aspects

Moreover, in a study including a large group of Flemish 4th graders Van Hoof et al. (2015a) conducted item response theory (IRT) analyzes to determine if 4th graders' ability to inhibit the inappropriate use of natural number knowledge in rational number tasks can be considered a single construct. Learners were given the Rational Number Sense Test (Van Hoof et al., 2015c) that includes both congruent and incongruent items within the three aspects of the natural number

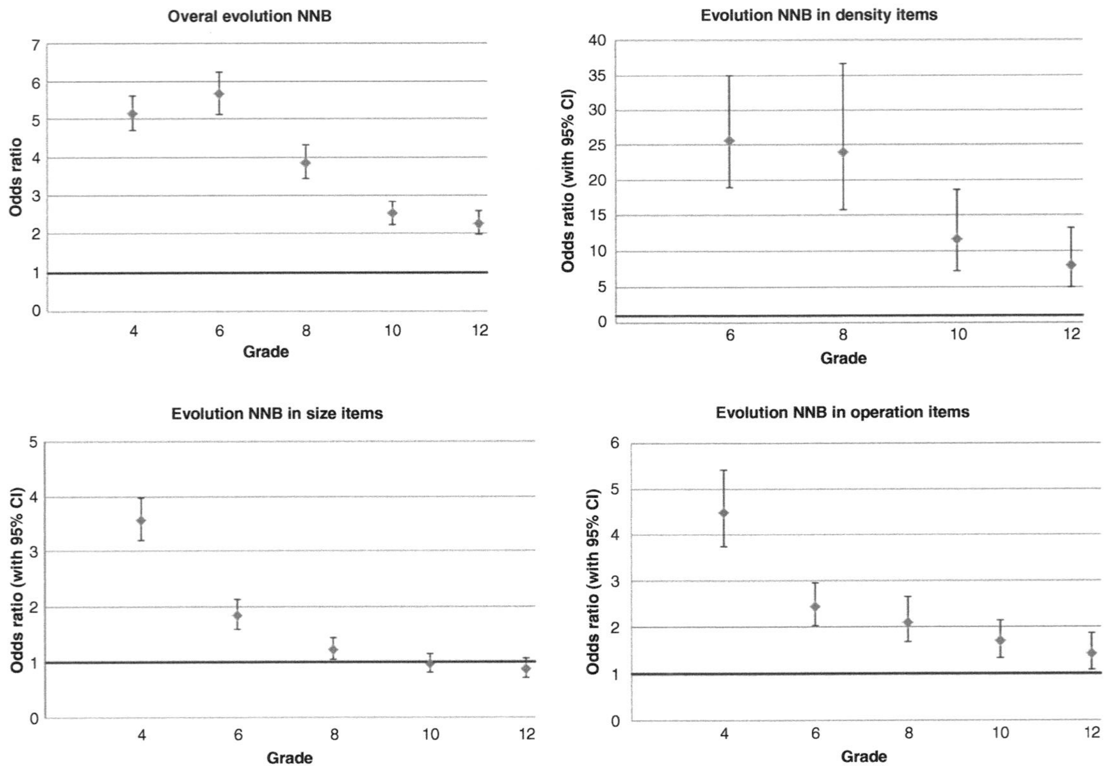

FIGURE 5.1 Overall change and change per aspect of the strength of the natural number bias as represented by the odds ratio (and 95% confidence interval) of accuracy for congruent and incongruent items (an odds ratio of 1 indicates that the observed accuracies for both types of items yield no evidence for the natural number bias).

bias. The finding that a Rasch scale fitted the data well is consistent with the ability to inhibit the natural number bias defining a single construct. If, on the contrary, results would have shown that 4th graders' ability was a multidimensional construct, this would have meant that learners develop an understanding of one aspect relatively independently of developing understanding of another aspect. However, this was not the case. A closer look at the distribution of the items on the IRT-scale revealed that most density items were located at the top of the scale, meaning that they had the highest difficulty level, whereas most size items were located at the bottom of the scale, implying that they had the lowest difficulty levels, and the operations items were distributed all over the scale. These findings indicated that a good understanding of the size of rational numbers forms a prerequisite for gaining understanding in the aspect of operations and that understanding of the aspects of size and operations is a prerequisite for gaining understanding in the dense structure of rational numbers. A similar finding was obtained by Vamvakoussi et al. (2012) with college students.

CONCLUSIONS AND FUTURE DIRECTIONS

In this chapter we presented a set of studies investigating the phenomenon of natural number knowledge interference in rational number learning and reasoning, that is, the natural number bias. We used two theoretical perspectives: the framework theory approach to conceptual change (Vamvakoussi et al., 2011, 2012, 2013b; Vamvakoussi & Vosniadou, 2010; Van Hoof et al., 2015a,b,c), and the dual process perspective on reasoning (Obersteiner et al., 2013, 2016; Vamvakoussi et al., 2013a; Van Hoof et al., 2013). The former predicts that the transition from the initial framework theory of numbers as natural numbers to the scientific perspective of numbers as rational number is a slow process, bugged by systematic errors. Initially, learners make mistakes that reflect the direct application of natural number knowledge to rational number tasks (e.g., longer decimals are bigger; the bigger the terms, the bigger the fraction; there is no other number between 2/5 and 3/5; etc.). As instruction on rational numbers continues, inconsistencies and synthetic conceptions appear.

From the second perspective, the dual process model of reasoning, the natural number bias is also evident when learners make correct inferences, through an increase in the time needed to solve rational number problems that are inconsistent with natural number properties. It is also predicted that learners might be inconsistent in the sense that the same kind of task sometimes elicit a correct response and other times an incorrect one. The latter case indicates a failure to inhibit the natural number bias.

The study by Van Hoof et al. (2015c) indicated that learners' rational number understanding develops over a period of at least four years, but not continuously. The development is characterized by periods of rapid development (for example, between 6th and 8th grade) and periods where it tends to stagnate (for example, between 10th and 12th grade). This finding cannot be entirely explained by the relative focus on rational numbers in the curriculum. If this were the case, then one would expect a clear evolution between 4th and 6th grade, which was not found. Mistakes that can be directly attributed to the natural number bias tended to disappear for size (for example, thinking that 0.12 is larger than 0.5, because 12 is larger than 5) at the beginning of secondary education. This result does not necessarily mean that learners are no longer affected by the bias. In fraction comparison tasks, secondary school students take more time to respond correctly to incongruent (10/27 vs. 10/21) than congruent (6/27 vs. 10/27) tasks (Van Hoof et al., 2013), and the same holds for college students (Vamvakoussi et al., 2012) and even experts (Obersteiner et al., 2013).

Furthermore, the fact that natural number based errors practically disappear overall does not mean that learners do not make mistakes. New kinds of mistakes become prevalent. For example, Vamvakoussi et al. (2012) reported that the college students participating in their study chose the "shorter" of two decimals as bigger (e.g., 2.6 vs. 2.899). It appears that the participants retained the idea that the magnitude of a decimal depends on the number of the digits (as with natural numbers), but also used information coming from instruction [possibly from instruction on fractions, as argued also by Peled and Awawdy-Shahbari (2009) and Resnick et al. (1989)]. In this sense, this is a synthetic conception from the perspective of the framework theory approach to conceptual change. Another example can be found in the study of Gómez and Dartnell (2015). They report a metaanalysis of five datasets of fraction comparison studies, and found, especially in the older age groups, a reverse bias in fraction comparison tasks for fractions with different numerators and denominators. Congruent items (e.g., 5/7vs. 1/3) elicited more errors than incongruent items (e.g., 11/25 vs. 8/13). Moreover, the participants needed more time to correctly solve the congruent items compared to the incongruent items. Gómez and Dartnell propose that the participants retained the idea that a smaller denominator means a larger fraction.

Van Hoof et al. (2015c) also showed that natural number bias based mistakes decrease throughout primary and secondary education for the aspect of arithmetical operations (e.g., 72 * 0.99 is larger than 72). However, such mistakes do not decrease in the last years of secondary school and remain present in college students, who also take longer to respond to incongruent tasks than to congruent tasks (Vamvakoussi et al., 2012, 2013a). It should be also taken into consideration that the overall increase in accuracy does not necessarily mean that individual learners respond correctly in a systematic way (e.g., Vamvakoussi et al., 2012). Interestingly, the college students in the latter study appeared biased also with respect to addition and subtraction, whereas the secondary school students in the Van Hoof et al. (2015b) study appeared biased only with respect to multiplication and division. Professional mathematicians on the other hand did not appear to be biased either in terms of accuracy, or in terms of reaction times (Obersteiner et al., 2016) for any of the four arithmetic operations, presumably because they do not draw on natural number-related knowledge but rather rely on knowledge of the solvability of inequalities.

By far the most challenging aspect is the density property of rational numbers (Vamvakoussi et al., 2012; Van Hoof et al., 2015c). A detailed account of the intermediate states of understanding this property was initially provided by Vamvakoussi and Vosniadou (2010), and further confirmed by the cross-cultural comparative study of Vamvakoussi et al. (2011). It appears that learners coming from different educational systems, regardless of an overall difference in performance, still face the same conceptual difficulties regarding the dense ordering of rational numbers.

The study of Van Hoof et al. (2015a) indicated that, at least in 4th graders, understanding the size of rational numbers forms a prerequisite for gaining understanding in the aspect of operations, for example, how arithmetic operations can differentially affect the outcomes for whole numbers (e.g., multiplication produces an answer that is greater than the multiplier and multiplicand) and rational numbers (e.g., multiplication can result in a smaller answer) and that both understanding of the aspect of size and operations is a prerequisite for gaining understanding in the dense structure of rational numbers. Vamvakoussi et al. (2012) found the same pattern with college students. This finding further is consistent with and expands the integrated theory of numerical development (Siegler et al., 2011), which states that understanding magnitudes forms a crucial step in the understanding of fractions. Taken together, these studies show that rational number learning is a slow and difficult process in which the natural number bias interferes. This bias shows up in terms of errors, of synthetic conceptions, but also in terms of reaction times, even when learners respond correctly, which is an indication of the intuitive character of the natural number bias. It is also persistent, even for experts, for certain tasks.

These findings confirmed the claim made by researchers nowadays that when conceptual change has occurred and learners have acquired a scientifically correct idea, this does not mean that the initial naïve idea no longer exists. On the contrary, the scientifically correct idea may coexist with the initial concept (DeWolf & Vosniadou, 2015). Therefore, when solving incongruent rational number tasks, learners need to first inhibit their natural number based idea of a rational number to come to the right answer, which lead to longer reaction times to give a correct answer to incongruent rational number tasks.

Future Directions

Several of the studies reported in this chapter attempted to capture the development of the natural number bias. However, these were mainly cross-sectional studies. The instrument developed by Van Hoof et al. (2015c) could be used in longitudinal studies to capture learners' rational number development over time and to investigate whether general developmental patterns can be found. As in the study of McMullen et al. (2015), data can be analyzed using latent profile analysis and latent transition analysis, because these analysis techniques allow us to map both gradual as well as abrupt changes in learners' conceptual structures. While these statistical techniques are novel in the research field of conceptual change theory, they have already proven useful in a few studies (see, for example, McMullen et al., 2015; Schneider & Hardy, 2013). Moreover, a more detailed analysis of learners' mistakes at the intermediate states of understanding size, operations, and density is necessary to provide an account of synthetic misconceptions predicted by the framework theory approach to conceptual change.

In the aforementioned studies, the intuitive character of learners' reasoning processes when they deal with rational number tasks was investigated. We note that the dual process perspective offers more methodological tools than have taken advantage of thus far (Gillard et al., 2009). Experimental manipulations that could be used to reveal the natural number bias and dual process mechanisms include restricting solution time and increasing working memory load, which are assumed to hinder analytic reasoning.

Another question that cannot be answered by reaction time studies is whether the learners experience an explicit conflict between their intuitions and the valid rational number knowledge when they make errors. This is an interesting question because—depending on the answer—errors due to the natural number bias can be attributed to a failure to detect the conflict between the intuitive, erroneous response and the mathematically correct answer, or to a failure to discard the initial, tempting intuition (De Neys, Moyens, & Vansteenwegen, 2010). Data about people's self-assessment of certainty, think-aloud protocols, but also methodologies coming from neuroscientific research, such as monitoring participants' skin conductance responses while they solve congruent and incongruent tasks could be valuable for establishing conflict detection.

In addition, more cross-cultural comparative studies are necessary in order to investigate the extent to which the development of learners' concept of rational number has certain universal features, or whether the development is qualitatively different when another kind of rational number instruction is given.

Moreover, as stated by Gómez et al. (2014), we still do not know enough about the extent to which instructional interventions can moderate the effect of the natural number bias, or even help learners overcome it (at least in terms of errors). There are, of course, studies aiming at remedying learners' natural number misconceptions with respect to one of the aspects discussed previously (see Vamvakoussi et al., 2013a); and there are also intervention studies that take into consideration the effects of prior natural number knowledge (e.g., Cramer et al., 2002; Moss, 2005). However, systematic, longitudinal interventions aiming at moderating the effects of the natural number bias are lacking.

There are, however, several principles for instruction stemming from the conceptual change perspective on learning that remain to be validated by further research (see, for example, Greer, 2004; Vosniadou, Ioannides,

Dimitrakopoulou, & Papademetriou, 2001; Vosniadou & Vamvakoussi, 2006). One of these principles is particularly relevant to this chapter, because it pertains to the distinction between intuitive and analytic reasoning: learners should be encouraged to become aware of their background assumptions regarding numbers; and the differences between natural number and rational numbers should be made explicit to learners, for example, through refutational texts. As stated by Tippett (2010), a refutational text consists of two parts: the first part is a statement of a commonly held misconception and the second part consists of an explicit refutation of that misconception while emphasizing the correct scientific explanation. There is a substantial body of literature indicating the effectiveness of refutational text for addressing learners' misconceptions (for an overview, see Tippett, 2010). These steps are necessary for learners to be able to "stop and think" when dealing with rational number tasks, that is, to critically evaluate their first, intuitive response, and engage in analytic reasoning.

However, current instructional practices, at least in Flanders and in Greece, do not include these techniques. For example, Debou and Verschetze (2012) found that the most commonly used textbooks in the Flemish classroom devote very little time to the (conceptual) differences between natural and rational numbers. On the contrary, they tend to capitalize on the continuity with natural numbers in their instruction about rational numbers by emphasizing the similarities between both types of numbers. For example, a common way to teach multiplication is through the model of repeated addition, and the common way to teach division is through the model of equal sharing, because these two models of operations cannot always be applied in tasks with rational numbers, a need for a stronger awareness of the possible negative consequences of introducing multiplication and division only using these two models is needed (Greer, 1994). Trying to build rational number knowledge on natural number ideas without emphasizing the differences between both types of numbers too may also have adverse results, because it validates and thus reinforces learners' reliance on natural number reasoning (Moss, 2005; Vamvakoussi, 2015). Careful consideration is required in order to decide when to emphasize the dissimilarities between the natural number system and the rational number system and when to point out the appropriate, deep similarities between both number systems to help learners to gain a stronger understanding of rational numbers.

One such similarity is that natural and rational numbers alike measure quantities, they have magnitudes, and they can be placed on number lines (Schneider & Siegler, 2010; Siegler et al., 2011; Vamvakoussi & Vosniadou, 2010). We found that a good understanding of size is a prerequisite for gaining understanding of operations and possibly of density. It appears that understanding of rational number magnitude is an essential step in the understanding of rational numbers in general. This claim is further corroborated by recent studies (see, for example, Gabriel et al., 2012, 2013).

REFERENCES

Asquith, P., Stephens, A. C., Knuth, E. J., & Alibali, M. W. (2007). Middle school mathematics teachers' knowledge of students' understanding of core algebraic concepts: Equal sign and variable. *Mathematical Thinking and Learning*, *9*, 249–272.

Babai, R., Levyadun, T., Stavy, R., & Tirosh, D. (2006). Intuitive rules in science and mathematics: A reaction time study. *International Journal of Mathematical Education in Science and Technology*, *37*, 913–924.

Bailey, D. H., Hoard, M. K., Nugent, L., & Geary, D. C. (2012). Competence with fractions predicts gains in mathematics achievement. *Journal of Experimental Child Psychology*, *113*, 447–455.

Booth, J. L., & Newton, K. J. (2012). Fractions: Could they really be the gatekeeper's doorman? *Contemporary Educational Psychology*, *37*, 247–253.

Booth, L. R. (1984). *Algebra: Children's strategies and errors*. Windsor, Berkshire: NFER-Nelson.

Christou, K. P., & Vosniadou, S. (2005). How students interpret literal symbols in algebra: A conceptual change approach. In B.G., Bara, L., Barsalou, M., Bucciarelli, (Eds.), *Proceedings of the XXVII Annual Conference of the Cognitive Science Society* (pp. 453–458). Stresa, Italy: Erlbaum.

Christou, K. P., & Vosniadou, S. (2012). What kinds of numbers do students assign to literal symbols? Aspects of the transition from arithmetic to algebra. *Mathematical Thinking and Learning*, *14*, 1–27.

Clarke, D. M., & Roche, A. (2009). Students' fraction comparison strategies as a window into robust understanding and possible pointers for instruction. *Educational Studies in Mathematics*, *72*, 127–138.

Cramer, K. A., Post, T. R., & delMas, R. C. (2002). Initial fraction learning by fourth- and fifth-grade students: A comparison of the effects of using commercial curricula with the effects of using the rational number project curriculum. *Journal for Research in Mathematics Education*, *33*, 111–144.

Dantzig, T. (1954). *Number: the language of science*. New York: Free Press.

De Corte, E., Verschaffel, L., & Van Coillie, V. (1988). The effect of type of number, problem structure and mode of response on children's solutions of multiplication word problems. *Journal for Mathematical Behavior*, *7*, 197–216.

De Neys, W. (2006). Automatic-heuristic and executive-analytic processing in reasoning: Chronometric and dual task considerations. *Quarterly Journal of Experimental Psychology*, *59*, 1070–1100.

De Neys, W., Moyens, E., & Vansteenwegen, D. (2010). Feeling we're biased: Autonomic arousal and reasoning conflict. *Cognitive, Affective, and Behavioral Neuroscience*, *10*, 208–216.

Debou, E., & Verschetze, L. (2012). *De overgang van natuurlijke naar rationale getallen. Een curriculumanalyse van drie Vlaamse wiskundemethoden* [The transition from natural to rational numbers. An analysis of three Flemish methods of mathematics] (Unpublished master's thesis). Katholieke Universiteit Leuven, Faculteit Psychologie en Pedagogische Wetenschappen, Belgium.

Depaepe, F., Torbeyns, J., Vermeersch, N., Janssens, D., Kelchtermans, G., Verschaffel, L., & Van Dooren, W. (2015). Prospective elementary and lower secondary teachers' content and pedagogical content knowledge in the rational number domain. *Teaching and Teacher Education*, *47*, 82–92.

DeWolf, M., & Vosniadou, S. (2015). The representation of the fraction magnitudes and the whole number bias reconsidered. *Learning and Instruction*, *37*, 39–49.

Evans, J. S. B. T., & Over, D. E. (1996). *Rationality and reasoning*. Hove, England: Psychology Press.

Fischbein, E. (1987). *Intuition in science and mathematics: An educational approach*. Dordrecht, The Netherlands: D. Reidel Publishing Company.

Fischbein, E., Deri, M., Nello, M., & Marino, M. (1985). The role of implicit models in solving problems in multiplication and division. *Journal for Research in Mathematics Education, 16*, 3–17.

Gabriel, F., Coché, F., Szucs, D., Carette, V., Rey, B., & Content, A. (2012). Developing children's understanding of fractions: An intervention study. *Mind, Brain, and Education, 6*, 137–146.

Gabriel, F., Coché, F., Szucs, D., Carette, V., Rey, B., & Content, A. (2013). A componential view of children's difficulties in learning fractions. *Frontiers in Psychology, 4*, 1–12.

Geary, D. (2008). An evolutionarily informed education science. *Educational Psychologist, 43*(4), 179–195.

Gillard, E., Van Dooren, W., Schaeken, W., & Verschaffel, L. (2009). Dual processes in psychology of mathematics education and cognitive psychology. *Human Development, 52*, 95–108.

Gómez, D., & Dartnell, P. (2015). Is there a natural number bias when comparing fractions without common components? A meta-analysis. *Proceedings of the 39th Conference of the International Group for the Psychology of Mathematics Education* (Vol. 3, pp. 1–8). Hobart, Australia: PME.

Gómez, D., Jiménez, A., Bobadilla, R., Reyes, C., & Dartnell, P. (2014). Exploring fraction comparison in school children. *Enero, 10*, 1–10.

Greer, B. (1994). Extending the meaning of multiplication and division. In G. Harel, & J. Confrey (Eds.), *The development of multiplicative reasoning in the learning of mathematics* (pp. 61–85). New York, NY: State University Press.

Greer, B. (2004). The growth of mathematics through conceptual restructuring. *Learning and Instruction, 14*, 541–548.

Hannula, M. S., Pehkonen, E., Maijala, H., & Soro, R. (2006). Levels of students' understanding on infinity. *Teaching Mathematics and Computer Science, 4*, 317–337.

Hasemann, C. (1981). On difficulties with fractions. *Educational Studies in Mathematics*, 171–187.

Janssen, R., Verschaffel, L., Tuerlinckx, F., Van den Noortgate, W., & De Fraine, B. (2010). *Tweede peiling wiskunde in het basisonderwijs*. Brussel, Belgium: Vlaamse Overheid [Second survey mathematics in primary education].

Kieran, C. (2006). Research on the learning and teaching of algebra: A broadening of sources of meaning. In A. Gutiérrez, & P. Boero (Eds.), *Handbook of research on the psychology of mathematics education: Past, present and future* (pp. 11–49). Rotterdam, The Netherlands: Sense Publishers.

Kilpatrick, J., Swafford, J., & Findell, B. (2001). *Adding it up. Helping children learn mathematics*. Washington, DC: National Academy Press.

Lortie-Forgues, H., Tian, J., & Siegler, R. S. (2015). Why is learning fraction and decimal arithmetic so difficult? *Developmental Review, 38*, 201–221.

Mazzocco, M. M. M., & Devlin, K. T. (2008). Parts and "holes": Gaps in rational number sense among children with vs. without mathematical learning disabilities. *Developmental Science, 11*, 681–691.

McMullen, J., Laakkonen, E., Hannula-Sormunen, M., & Lehtinen, E. (2015). Modeling the developmental trajectories of rational number concept(s). *Learning and Instruction, 37*, 14–20.

Meert, G., Grégoire, J., & Noël, M. -P. (2010). Comparing the magnitude of two fractions with common components: Which representations are used by 10- and 12-year-olds? *Journal of Experimental Child Psychology, 107*, 244–259.

Merenluoto, K., & Lehtinen, E. (2002). Certainty bias as an indicator of problems in conceptual change: the case of number line. In A.C., Cockburn, E., Nardi, (Eds.) PME, *Proceedings of the 26th Annual Conference* (Vol 3, pp. 337–344). University of East Anglia, Norwich, UK.

Merenluoto, K., & Lehtinen, E. (2004). Number concept and conceptual change: Towards a systemic model of the processes of change. *Learning and Instruction, 14*, 519–534.

Moss, J. (2005). Pipes, tubes, and beakers: New approaches to teaching the rational-number system. In M. S. Donovan, & J. D. Bransford (Eds.), *How students learn: Mathematics in the classroom* (pp. 121–162). Washington, DC: National Academic Press.

Ni, Y., & Zhou, Y. -D. (2005). Teaching and learning fraction and rational numbers: The origins and implications of whole number bias. *Educational Psychologist*, *40*, 27–52.

Nunes, T., & Bryant, P. (2008). Rational numbers and intensive quantities: Challenges and insights to pupils' implicit knowledge. *Anales de Psicologia*, *24*(2), 262–270.

Obersteiner, A., Van Dooren, W., Van Hoof, J., & Verschaffel, L. (2013). The natural number bias and magnitude representation in fraction comparison by expert mathematicians. *Learning and Instruction*, *28*, 64–72.

Obersteiner, A., Van Hoof, J., Verschaffel, L., & Van Dooren, W. (2016). Who can escape the natural number bias in rational number tasks? A study involving students and experts. *British Journal of Psychology*, *107*, 537–555.

Peled, I., & Awawdy-Shahbari, J. (2009). Journey to the past: Verifying and modifying the conceptual sources of decimal fraction knowledge. *Canadian Journal of Science, Mathematics and Technology Education*, *9*(2), 73–85.

Resnick, L. B., Nesher, P., Leonard, F., Magone, M., Omanson, S., & Peled, I. (1989). Conceptual bases of arithmetic errors: The case of decimal fractions. *Journal for Research in Mathematics Education*, *20*, 8–27.

Schneider, M., & Hardy, I. (2013). Profiles of inconsistent knowledge in children's pathways of conceptual change. *Developmental Psychology*, *49*, 1639–1649.

Schneider, M., & Siegler, R. S. (2010). Representations of the magnitudes of fractions. *Journal of Experimental Psychology: Human Perception and Performance*, *5*, 1227–1238.

Siegler, R. S., Thompson, C., & Schneider, M. (2011). An integrated theory of whole number and fractions development. *Cognitive Psychology*, *62*, 273–296.

Siegler, R. S., Duncan, G. J., Davis-Kean, P. E., Duckworth, K., Claessens, A., Engel, M., ... Chen, M. (2012). Early predictors of high school mathematics achievement. *Psychological Science*, *23*, 691-697.

Smith, C. L., Solomon, G. E. A., & Carey, S. (2005). Never getting to zero: elementary school students' understanding of the infinite divisibility of number and matter. *Cognitive Psychology*, *51*, 101–140.

Stafylidou, S., & Vosniadou, S. (2004). The development of students' understanding of the numerical value of fractions. *Learning and Instruction*, *14*, 503–518.

Stanovich, K. E., & West, R. F. (2000). Advancing the rationality debate. *Behavioral and Brain Sciences*, *23*, 701–726.

Tippett, C. D. (2010). Refutation text in science education: A review of two decades of research. *International Journal of Science and Mathematics Education*, *8*, 951–970.

Vamvakoussi, X. (2015). The development of rational number knowledge: Old topic, new insights. *Learning and Instruction*, *37*, 50–55.

Vamvakoussi, X., & Vosniadou, S. (2004). Understanding the structure of the set of rational numbers: A conceptual change approach. *Learning and Instruction*, *14*, 453–467.

Vamvakoussi, X., & Vosniadou, S. (2007). How many numbers are there in a rational numbers interval? Constrains, synthetic models and the effect of the number line. In S. Vosniadou, A. Baltas, & X. Vamvakoussi (Eds.), *Reframing the Conceptual Change Approach in Learning and Instruction* (pp. 265–282). Oxford: Elsevier.

Vamvakoussi, X., & Vosniadou, S. (2010). How many decimals are there between two fractions? Aspects of secondary school students ' understanding of rational numbers and their notation. *Cognition and Instruction*, *28*, 181–209.

Vamvakoussi, X., Christou, K. P., Mertens, L., & Van Dooren, W. (2011). What fills the gap between discrete and dense? Greek and Flemish students' understanding of density. *Learning and Instruction*, *21*, 676–685.

Vamvakoussi, X., Van Dooren, W., & Verschaffel, L. (2012). Naturally biased? In search for reaction time evidence for a natural number bias in adults. *The Journal of Mathematical Behavior, 31*, 344–355.

Vamvakoussi, X., Van Dooren, W., & Verschaffel, L. (2013a). Educated adults are still affected by intuitions about the effect of arithmetical operations: Evidence from a reaction-time study. *Educational Studies in Mathematics, 82*, 323–330.

Vamvakoussi, X., Vosniadou, S., & Van Dooren, W. (2013b). The framework theory approach applied to mathematics learning. In S. Vosniadou (Ed.), *International handbook of research on conceptual change* (2nd ed., pp. 305–321). New York, NY: Routledge.

Van Dooren, W., & Vamvakoussi, X. (2010). Greek and Flemish students' interpretation of the literal symbols as variables. In: Pinto, M., Kawasaki, T., (Eds.), *Proceedings of the 34th Conference of the International Group for the Psychology in Mathematics Education* (Vol. 4, pp. 257–264). Belo Horizonte, Brazil: PME.

Van Dooren, W., Lehtinen, E., & Verschaffel, L. (2015). Unraveling the gap between natural and rational numbers. *Learning and Instruction, 37*, 1–4.

Van Hoof, J., Lijnen, T., Verschaffel, L., & Van Dooren, W. (2013). Are secondary school students still hampered by the natural number bias? A reaction time study on fraction comparison tasks. *Research in Mathematics Education, 15*, 154–164.

Van Hoof, J., Janssen, R., Verschaffel, L., & Van Dooren, W. (2015a). Inhibiting natural knowledge in fourth graders: Towards a comprehensive test instrument. *ZDM Mathematics Education, 47*, 849–857.

Van Hoof, J., Vandewalle, J., Verschaffel, L., & Van Dooren, W. (2015b). In search for the natural number bias in secondary school students' interpretation of the effect of arithmetical operations. *Learning and Instruction, 37*, 30–38.

Van Hoof, J., Verschaffel, L., & Van Dooren, W. (2015c). Inappropriately applying natural number properties in rational number tasks: characterizing the development of the natural number bias through primary and secondary education. *Educational Studies in Mathematics, 90*, 39–56.

Vlaamse Overheid (2010). Peiling wiskunde in de eerste graad secundair onderwijs (A-stroom) [Survey mathematics in the first grade of secondary education]. Available from: http://www.ond.vlaanderen.be/curriculum/peilingen/secundair-onderwijs/brochures/brochure_peiling_wiskunde_1a.pdf

Vosniadou, S. (1994). Capturing and modelling the process of conceptual change. *Learning and Instruction, 4*, 45–69.

Vosniadou, S. (Ed.). (2013). *International handbook of research on conceptual change* (2nd ed.). New York, NY: Routledge.

Vosniadou, S., & Vamvakoussi, X. (2006). Examining mathematics learning from a conceptual change point of view: Implications for the design of learning environments. In L. Verschaffel, F. Dochy, M. Boekaerts, & S. Vosniadou (Eds.), *Instructional psychology: Past, present and future trends. Sixteen essays in honour of Erik De Corte* (pp. 55–70). Oxford, UK: Elsevier Science Ltd.

Vosniadou, S., & Verschaffel, L. (2004). Extending the conceptual change approach to mathematics learning and teaching. *Learning and Instruction, 14*, 445–451.

Vosniadou, S., Ioannides, C., Dimitrakopoulou, A., & Papademetriou, E. (2001). Designing learning environments to promote conceptual change in science. *Learning and Instruction, 11*, 381–419.

Vosniadou, S., Vamvakoussi, X., & Skopeliti, I. (2008). The framework theory approach to conceptual change. In S. Vosniadou (Ed.), *International handbook of research on conceptual change* (pp. 3–34). Mahwah, NJ: Erlbaum.

Chapter 6

Fraction Development in Children: Importance of Building Numerical Magnitude Understanding

Nancy C. Jordan*, Jessica Rodrigues*, Nicole Hansen**, Ilyse Resnick*
*University of Delaware, Newark, DE, United States; **Fairleigh Dickinson University, Teaneck, NJ, United States

Fractions are a crucial component of the US mathematics curriculum in elementary and middle school (Council of Chief State School Officers & National Governors Association Center for Best Practices, 2010). Yet fractions and their magnitudes are sources of great confusion for many students. For example, about half of eighth graders on the National Assessment of Education Progress (NAEP, 2004) experienced difficulties when asked to order three fractions with unlike denominators. Moreover, only a third of 11th graders can successfully translate decimals into fractions (Kloosterman, 2010) and many community college students still struggle when making fraction magnitude comparisons (Schneider & Siegler, 2010). Frequently students do not recognize that the numerator and denominator work together to form a fraction's magnitude. Persistent difficulties with fractions can have far-reaching consequences, both on professional and personal levels. Fraction knowledge supports learning of algebra (Booth & Newton, 2012), and algebra proficiency, in turn, prepares students for higher education and careers in science, technology, engineering, and mathematics (STEM) disciplines. Fraction knowledge also supports daily life functioning related to managing personal finances, understanding mortgage interest rates, doing home repairs, cooking, managing medical dosages, and so forth. Difficulties with fractions have even been suggested as a root cause of mathematics anxiety (Heitin, 2015).

This chapter has several aims. First, we situate fraction learning within the integrated theory of numerical development, proposed by Siegler, Thompson, and Schneider (2011). The integrated theory asserts that the unifying property of all real numbers is that they have magnitudes or numerical values that can

Acquisition of Complex Arithmetic Skills and Higher-Order Mathematics Concepts
http://dx.doi.org/10.1016/B978-0-12-805086-6.00006-0

be ordered on the number line (Siegler, Fuchs, Jordan, Gersten, & Ochsendorf, 2015). Next, we identify key areas of fraction knowledge and then chart fraction development from early childhood through middle school. In particular, we discuss research findings from our large longitudinal study of fraction learning from third through sixth grade, supported by the US Department of Education Institute of Education Sciences. The four-year study identified domain-general and domain-specific predictors and concomitants of fraction learning. Overall, we show that fraction development during this formal instructional period is fundamental to mathematics success more generally. Finally, we discuss implications for helping students who struggle with fractions, especially with respect to building numerical magnitude understandings.

INTEGRATED THEORY OF NUMERICAL DEVELOPMENT

Earlier theories of numerical development conceptualize whole number learning and fraction learning as two separate processes, with fractions secondary to whole numbers (Siegler et al., 2013; Siegler et al., 2011). Although whole numbers are characterized as being naturally acquired early in development, segmented theories assume fraction knowledge occurs much later and with great difficulty (e.g., Geary, 2004, 2006; Gelman & Williams, 1998). In fact, whole number understanding has been described as interfering with the learning of fractions (Geary, 2006; Vamvakoussi & Vosniadou, 2010). For example, children often overextend the principle of whole number ordinality (Ni & Zhou, 2005), identifying unit fractions with the larger denominator as having the larger magnitude (e.g., 1/6 > 1/5 because 6 > 5).

More recently, research has begun to emphasize how the numerical property of magnitude can unite number learning into a continuous process, one that includes fractions as well as whole numbers (Matthews, Lewis, & Hubbard, 2016; Siegler & Lortie-Forgues, 2014). Properties of whole numbers, however, are not invariant across all real numbers, and, in particular, differ from fractions in important ways. Whole numbers have unique successors, which are expressed as a single symbol and are finite within a given interval; they increase or stay the same with addition and multiplication and decrease or stay the same with subtraction and division. Properties of fractions, on the other hand, sometimes behave in the opposite manner. However, the one property that unites all real numbers is that they have magnitudes that can be located on a number line (Siegler, Fazio, Bailey, & Zhou, 2013; Siegler & Lortie-Forgues, 2014; Siegler et al., 2011).

According to the integrated theory, number learning involves continually expanding the size and type of number whose magnitudes can be accurately represented (Siegler & Lortie-Forgues, 2014). Children typically begin with nonsymbolic magnitude representations. The ability to discriminate between sets of one to four objects (the object tracking system) and more than four objects (the approximate number system) emerges in infancy (Feigenson, Dehaene, &

Spelke, 2004). For example, even infants can recognize the difference between 8 and 16 dots, a 2:1 ratio (Xu & Spelke, 2000). As infants prefer to look at novel objects, their looking time decreases, or habituates, to the repeated presentation of the same image (e.g., Groves & Thompson, 1970). In Xu and Spelke's (2000) study, after being habituated to an array of dots that either contained 8 or 16 dots, infants looked longer at an array of dots with the opposite amount (i.e., either 8 or 16 dots). Moreover, there is evidence that sensitivity to ratios is also present in infants (McCrink & Wynn, 2007). Matthews et al. (2016) argue there is a core cognitive system that processes nonsymbolic ratios, which is referred to as the "ratio processing system" (RPS). Analogous to whole number development, in which both formal and informal learning links nonsymbolic magnitudes (e.g., three dots) to symbolic representations (e.g., the numeral "3"), children's nonsymbolic understanding of ratios (e.g., the ability to recognize that the proportion of red to blue in a bar is the same or different as the ratio of red to blue in another bar) may underpin the learning of conventional fraction symbols (Boyer & Levine, 2012).

The ability to link nonsymbolic to symbolic magnitude representations develops gradually (Siegler & Lortie-Forgues, 2014). Comprehension of numerical magnitudes can be seen on a number line task, where children are presented with an image of a horizontal line flanked with two values (e.g., 0 on the left and 100 on the right) and asked to locate the position of a given number on the line (Siegler & Opfer, 2003). Accuracy on the number line task undergoes a developmental shift, with children becoming increasingly closer to the target number with age. On unfamiliar scales, children overestimate relatively smaller numbers and underestimate relatively larger magnitudes (Siegler & Booth, 2004; Siegler & Opfer, 2003). Magnitudes on a scale from 0 to 10 are typically understood linearly by first grade (Bertelletti, Lucangeli, Piazza, Dehaene, & Zorzi, 2010; Siegler & Booth, 2004), with 0–100 understood by second grade (Booth & Siegler, 2006), 0–1,000 by fourth grade (Siegler & Opfer, 2003), and 0–100,000 by sixth grade (Opfer & Siegler, 2007; Opfer & Thompson, 2008).

Eventually, children learn to represent fractions as well as whole numbers on the number line (Siegler & Lortie-Forgues, 2014). Children are first able to locate proper fractions (*a/b*, where *b* is larger than *a*) and then mixed numbers (a combination of a whole number plus a proper fraction) on a number line (e.g., 0 to 2), and later learn the magnitudes of improper fractions (*b/a*, where *b* is larger than *a*; Resnick et al., 2015). The integrated theory of numerical development emphasizes that fractions are a pivotal part of number learning more generally, with fractions representing the first opportunity for children to learn properties of whole numbers are not always true of all numbers (Siegler & Lortie-Forgues, 2014; Siegler et al., 2011).

There is growing evidence that magnitude understanding is foundational to mathematics learning. Accuracy in whole number line estimation (e.g., Booth & Siegler, 2006, 2008; Holloway & Ansari, 2008; Jordan et al., 2013; Sasanguie, Göbel, Moll, Smets, & Reynvoet, 2013) and fraction number line estimation

(Bailey et al., 2012; Resnick et al., 2015; Siegler et al., 2012; Siegler & Pyke, 2013; Siegler et al., 2011) both predict overall mathematics achievement.

UNDERSTANDING OF FRACTIONS INVOLVES BOTH CONCEPTUAL AND PROCEDURAL KNOWLEDGE

Any mathematical domain involves both concepts and procedures (Geary, 2004). Knowledge of mathematical concepts is often termed *conceptual knowledge*; knowing how to execute steps to reach a correct answer is termed *procedural knowledge*. In the domain of fractions, measures of conceptual understanding typically assess students' ability to estimate fraction magnitudes on a number line (as discussed previously), compare fraction magnitudes (e.g., Hecht, Close, & Santisi, 2003), find equivalent fractions, and recognize parts of a whole and parts of a set (e.g., Jordan et al., 2013). Measures of procedural knowledge require students to solve fraction arithmetic problems, with variations of addition, subtraction, multiplication, and division items (e.g., Siegler et al., 2011). Procedural knowledge may also be used to help students compare magnitudes and find equivalent fractions (e.g., cross multiplying the numerators with the denominators).

Facility with both fraction concepts and procedures is needed for students to develop fraction proficiency. Imagine, for example, a student who demonstrates near perfect accuracy when estimating fraction magnitudes on a number line. However, when given the addition problem 1/2 + 1/4, the student does not know the procedural steps for finding a common denominator (e.g., 1/2 = 2/4; 2/4 + 1/4 = 3/4). The student may nevertheless reach the correct answer by using his or her strong grasp of fraction magnitudes. That is, the student visualizes a mental number line and sees that adding 1/2 and 1/4 results in a location on the number line that is 3/4 of the whole number one. In this case, the student compensates for low procedural knowledge with high conceptual knowledge. However, the same student may struggle with the problem 2/7 + 4/21. Although he or she understands the conceptual underpinnings of the problem, the mental number line provides limited help for finding exact sums of two fractions with greater numerators and denominators. To solve this problem, the procedure of identifying a common denominator, which involves multiplicative skills, is more advantageous and accessible than a mental number line (2/7 = 6/21; 6/21 + 4/21 = 10/21), although magnitude understanding can help students evaluate the plausibility of their solutions (Bailey et al., 2015; Siegler & Lortie-Forgues, 2014). Conceptual knowledge of magnitudes is necessary but not sufficient for achieving with fractions. Likewise, students who execute procedures for solving fraction arithmetic problems without magnitude understanding will encounter limited success on problem solving tasks that require deeper knowledge.

Fraction concepts and procedures ideally develop hand over hand (Hansen et al., 2015b). Recent analyses exploring how each type of knowledge influences the other indicate that the relation between fraction concepts and procedures is bidirectional; knowledge of fraction concepts impacts the development of fraction procedures and in turn, knowledge of procedures affects concepts, although likely to a lesser extent (Bailey et al., 2015; Fuchs et al., 2013; Fuchs et al., 2014; Hecht & Vagi, 2010). Overall, examination of the development of both fraction concepts and fraction procedures provides a more accurate picture of students' fraction development.

FRACTION DEVELOPMENT IN EARLY CHILDHOOD

Before formal instruction, most preschoolers have a basic understanding of fractions. Young children successfully solve nonsymbolic calculations with fractions, make sense of fractions through equal sharing, and show an early sense of proportionality and partitioning. Despite these early foundational abilities, children also hold misconceptions about fractions that can persist throughout elementary school. Each of these areas is discussed in this section.

Early Fraction Calculation Ability

To assess fraction calculation in 3- to 7-year-old children, Mix, Levine, and Huttenlocher (1999) modified a previously used nonverbal calculation task where children are asked to view an array of black dots, which is then covered and transformed by adding dots to or subtracting them from the covered array; the child's task is point at a picture of an equivalent array of dots after the transformation (Jordan, Levine, & Huttenlocher, 1994). On Mix and coworkers' nonverbal fraction calculation task, an amount representing the first term of a fraction problem (represented as quarters of a circular sponge) was placed into a shallow hole (e.g., two quarters of the sponge). Then, the hole was hidden, and parts of the circular sponge (e.g., one quarter of the sponge) was either added to or subtracted from the amount that was hidden, but the child could not see the outcome (e.g., a quarter of the sponge; $1/2 - 1/4 = 1/4$). The experimenter then asked the child to choose the correct amount from an array of four pictures showing different fractions (pieces) of a circle. The researchers found that while nonverbal whole-number calculation skill emerges about a year earlier than nonverbal fraction calculation skill, children as young as four years of age perform significantly above chance on fraction calculations with totals less than or equal to one. Mix and coworkers suggest that this later development of nonverbal fraction calculation skills may be related to the complexity of the spatial skills required for fraction versus whole-number calculation. That is, whole number calculations simply require movements in and out of space whereas the fractions problems also required children to rotate, separate, and recombine amounts. Overall, the findings suggest that well before formal instruction in fractions begins, children can meaningfully calculate with fractions, most likely through their experience with parts and wholes in the environment as well as—to some extent—through their fundamental nonsymbolic understandings of number and proportions.

Equal Sharing

One way children make sense of fractions is through equal sharing experiences (Davis & Pepper, 1992; Empson, 1999). Children learn about fractions through real-world experiences with sharing snacks and toys, and perhaps unsurprisingly, many become concerned with the "fairness" or "evenness" of the sizes of the fractions when working on equal sharing problems (Brizuela, 2006; Empson, 1999). In a descriptive study of fraction learning in a first-grade classroom, Empson (1999) showed that children develop fraction concepts using their own representations in equal sharing situations. Before instruction, the majority of children successfully solved equal sharing problems that involved halving or repeated halving (e.g., when asked to share nine apples equally between two horses, children understood that both horses will receive four apples and the remaining apple must be split into two equal parts). By the end of the study period, nearly all of the children showed knowledge of at least one fraction other than one-half and could solve equal sharing problems that involved partitioning other than repeated halving after instruction based on children's informal understanding of fractions.

Early Knowledge of Proportionality

Proportional reasoning is foundational to understanding fractions (Boyer & Levine, 2012). Building on the idea that children have intuitive understanding of proportionality that is not dependent on formal instruction, Singer-Freeman and Goswami (2001) administered a proportional analogy task to preschool children. Two models of familiar foods were used in the task: one represented a continuous quantity (pizza) and the other a discontinuous quantity (box of chocolates). During each trial, the experimenter presented two models to the child (e.g., two pizzas) and then removed a proportion from one of the models (e.g., one-half of a pizza). Children were asked to take away the same amount from the other model. Proportional reasoning with continuous quantities (e.g., pizzas) was relatively easy for the young children. On the other hand, children encountered greater difficulty reasoning about proportions when the task involved discrete quantities (e.g., individual chocolates), most likely because the presence of discrete, countable quantities encourages counting (Boyer, Levine, & Huttenlocher, 2008; Singer-Freeman & Goswami, 2001). Rather than thinking of a fraction as one magnitude (e.g., "one-third of the candy" or "less than half of the candy"), children may focus only on the size of the numerator or denominator or the total amount (e.g., "one piece of candy" or "three pieces total").

Early Misconceptions

The early fraction knowledge that young children demonstrate is not without misunderstandings. Children even struggle with the concept of "half." Although children can solve concrete problems that involve halving (e.g., half of the pizza), they do not seem to have a complete understanding of the term.

Some kindergartners and first graders see "half" as any fraction, regardless of whether they were halves, thirds, or quarters (Brizuela, 2006; Empson, 1999). Still other children are uncomfortable naming fractions, instead preferring to refer to fractions of pizzas or cookies as "pieces" (Brizuela, 2006; Mack, 1990). Some children do not understand what a half is. For example, when asked to identify "half," they may point to a partitioning line that separates a figure and call that "the half," rather than one of the pieces themselves (Brizuela, 2006). Interestingly, young students sometimes assert that half means "a little more than," particularly in the context of age. For example, children might assert that on their next birthday, they will be seven, and on the day after the birthday, they will be seven and a half (Brizuela, 2006). These conflicting sets of beliefs indicate that young children are thinking about fractions; however, full understanding of seemingly simple concepts, such as half, develops later. As children live in a world that uses fractions in colloquial language, they pick up on aspects of fractions that are salient to them and, as a result, may hold several differing, seemingly conflicting understandings of fractions (Brizuela, 2006).

Although children show early knowledge of fraction calculation, equal sharing, and proportional reasoning, this knowledge does not translate easily to symbolic fractions and their magnitudes. When children use written arithmetic, they become confused because the same symbols that they have used for whole numbers are used in a new way to represent fractions (Sophian, Garyantes, & Chang, 1997). For example, children have seen the number "1" used to represent one object and "2" used to represent two objects, but the fraction "1/2" can have multiple meanings (half of one object, or three out of six objects).

FRACTION DEVELOPMENT BETWEEN THIRD AND SIXTH GRADE: FINDINGS FROM THE DELAWARE LONGITUDINAL STUDY

The goal of our longitudinal project was to identify predictors of fraction learning from third through sixth grade. We examined the development of fraction concepts and procedures over multiple time points: before, during, and right after formal fraction instruction in school. We uncovered empirically distinct growth trajectory classes for fraction learning and analyzed student performance to help explain why so many children struggle with fractions.

We started the project with about 500 children in third grade. Participants were drawn from schools that primarily followed a curriculum sequence based on the US Common Core State Standards-Mathematics (Council of Chief State School Officers & National Governors Association Center for Best Practices, 2010). Although students represented a diverse range of ethnicities, SES, and ability levels, we oversampled in higher risk schools serving low-income communities. Based on the aforementioned integrated theory of numerical development, we hypothesized that accurate representations of numerical magnitudes would be uniquely important for acquisition of fraction knowledge in particular and mathematics skills more generally. To test this premise, we examined the extent to which general and number specific cognitive processes predict fraction outcomes.

Domain-general predictors included measures of classroom attention, working memory, language, nonverbal reasoning, and reading fluency. Our mathematics specific predictors included whole number line estimation, nonsymbolic proportional reasoning, and whole number calculation skills (i.e., addition and multiplication fluency and long division). Nonsymbolic proportional reasoning supports thinking about multiplicative relations, a crucial component of fractions (Boyer & Levine, 2012). Most of the predictor measures were administered in third and fifth grades, with the exception of proportional reasoning, multiplication, and long division, which were administered only fifth grade. Fraction outcomes, which were assessed over multiple time points between fourth and sixth grades, included a fraction number line estimation task, a curriculum-based fraction concepts measure using released items from the US National Assessment of Educational Progress (NAEP, 2007, 2009), and a measure of fraction arithmetic procedures. In all of our main analyses we controlled for the influence of background factors (i.e., gender, income status, and age).

Predictors of Fraction Knowledge

Overall, multiple regression analyses showed that both domain-specific knowledge and domain-general knowledge shape change in fraction learning, with numerical magnitude understanding being especially important (Jordan et al., 2013; Hansen et al., 2015a). The standardized beta coefficients (measure of effect size or the expected increase or decrease of the dependent variable in standard deviation units) of predictors, by grade, are presented in Table 6.1.

1. *Third grade predictors of fourth grade fraction outcomes* (Jordan et al., 2013). For fourth-grade fraction concepts, number line estimation, attentive behavior, addition fluency, language, nonverbal ability, and reading fluency, all made significant and unique contributions. For fourth-grade fraction procedures, number line estimation, working memory, attentive behavior, and addition fluency contributed reliably and independently to the model. Overall, the complete set of number-related and general processes predicted fourth graders' fraction concepts better than fraction procedures (56% vs. 30% of explained variance).
2. *Fifth grade predictors of sixth grade fraction outcomes* (Hansen et al., 2015a). Whole number line estimation, nonsymbolic proportional reasoning, attentive behavior, long division, and working memory all contributed uniquely to fraction concepts in sixth grade. On a sixth-grade measure of fraction procedures, attentive behavior, whole number line estimation, multiplication fluency, and long division made unique and significant contributions. Overall, the entire set of fifth grade predictors accounted for 58% of the explained variance for fraction concepts and 40% for fraction procedures

TABLE 6.1 Multiple Regression Standardized Beta Coefficients of Predictors of Fraction Concepts and Procedures by Grade

	Fourth grade		Sixth grade	
Variable	**Concepts**	**Procedures**	**Concepts**	**Procedures**
Whole number line estimation[a]	−.271***	−.237***	−.361***	−.183***
Attention	.225***	.135*	.170***	.254***
Language	.196***	.081	—	—
Addition fluency	.169***	.126*	—	—
Nonverbal reasoning	.111**	.097	—	—
Reading fluency	.086*	−.037	.048	.011
Working memory	.044	.137**	.114**	.011
Nonsymbolic proportional reasoning	—	—	.198***	.080
Multiplication fluency	—	—	.037	.171**
Long division	—	—	.113*	.171**

Note: Predictors of fourth-grade outcomes were assessed in third grade; predictors of sixth-grade outcomes were assessed in fifth grade. Empty cells indicate variables that were not included in the multiple regression analysis.
*p < *.05.*
**p < *.01.*
***p < *.001.*
[a]*Percentage of absolute error with lower score indicating more accurate estimation.*

in sixth grade. The lower overall predictability of our measures for fraction procedures compared to fraction concepts could indicate that instruction in rote computation algorithms may be used more in current practice than instruction that emphasizes understanding of what fractions are.

In the aforementioned study of sixth grade predictors (Hansen et al., 2015a), number line estimation is more important to fraction concepts than to procedures, which differs from our earlier finding that third-grade number line estimation is equally predictive of fraction concepts and procedures at fourth grade (Jordan et al., 2013). A possible explanation for this developmental difference might be that fourth grade fraction computation involves relatively simple procedures involving like denominators for addition and subtractions. In later grades, students must be facile with multiple procedures and operations. Attention and whole number computational abilities (multiplication fluency and division) emerge as key predictors of sixth grade fraction procedures.

Interestingly, our nonsymbolic proportional reasoning measure contributed independently to students' proficiency with fraction concepts, over and above symbolic math-related and general predictors. The data are in keeping with

Matthews et al.'s (2016) conjecture, discussed earlier, that a core cognitive system that processes nonsymbolic ratios underpins fraction learning. Although proportional reasoning and whole number line estimation are moderately correlated, the independent contribution of proportional reasoning shows that fundamental understandings of scale relations and multiplicative reasoning (e.g., 1/3 is the same as 3/9; Boyer & Levine, 2012; Gunderson, Ramirez, Beilock, & Levine, 2012) are important for learning fraction concepts, particularly in the context of area models for representing fractions that are widely used in the United States.

We also examined students' performance on individual items on our fraction concepts and procedures measures in sixth grade (Hansen et al., 2015b). There were substantial gaps between lower and higher performing children on items requiring them to compare and order fractions. Lower performing students struggled relative to their higher performing peers on tasks that required flexible thinking, such as ones where the denominator of the fraction did not correspond directly to the pieces shown (e.g., "Shade 4/5 of 10 circles"). It should be noted that almost all students experienced difficulties when presented with fraction estimation problems (e.g., "Estimate the sum: 7/8 + 12/13"), suggesting that generally more attention should be devoted to this instructional topic. On procedures, we found that low performing sixth graders tend to misapply whole number strategies when adding and subtracting fractions with unlike denominators (e.g., 2/5 + 3/4 = 5/9). This approach reflects weak understanding or attention to the bipartite structure of fractions, that is, that the numerator and denominator of the fraction together represent one magnitude.

Growth in Fraction Magnitude Understanding

We assessed the development of children's fraction number line estimation over five time points between fourth and sixth grade (Resnick et al., 2015). No previous studies had examined the development of fraction magnitude estimation during this time, which is when the majority of fraction instruction takes place in American schools. At each time point, students were asked to estimate where fractions should be placed on number lines ranging from 0 to 1 and 0 to 2. See Fig. 6.1, for examples, of task items.

Over the course of the study, most students increased their estimation accuracy. However, latent class growth analyses revealed three empirically distinct growth trajectory classes: (1) students who were highly accurate from the start and became even more accurate; (2) students who were inaccurate at first but showed steep growth; and (3) students who started inaccurate and showed minimal growth. Disturbingly, about 42% of students left sixth grade with only a minimal grasp of fraction magnitudes. Another 26% of students seemed to benefit a great deal from their fractions instruction in later elementary and early middle school but still were not as accurate as roughly one third of the students who started the study with considerable strength in estimating fraction magnitudes. Growth trajectory class membership was highly predictive of performance on a high-stakes, statewide mathematics test at the end of sixth grade. That is, 67% of

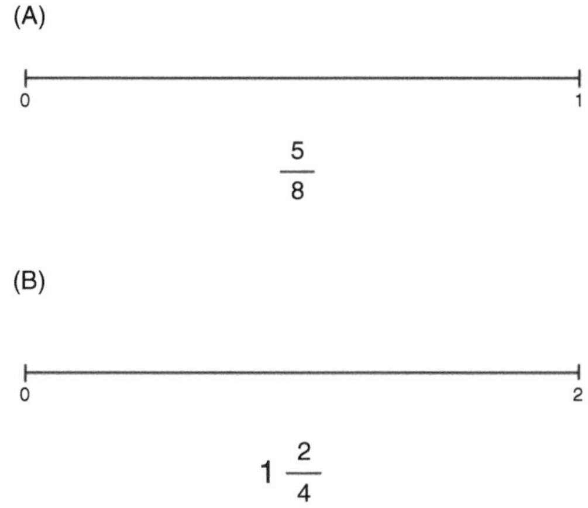

FIGURE 6.1 Example items for (A) 0–1 and (B) 0–2 fraction number line estimation tasks. Students are asked to mark where the fraction should be placed on the line.

the inaccurate, minimal growth group did not meet state mathematics standards versus only 5% of the highly accurate group and 17% of the steep growth group.

Students with poor multiplication fluency, weak classroom attention, and inaccurate whole number line estimation skill at the start of the study in fourth grade were much more likely to fall into the low-accuracy, minimal growth group. Analyses of performance showed that younger grades and low-accuracy/minimal-growth students typically estimated both proper and improper fractions as being less than one, indicating they were not attending to the relation between numerator and denominator. These students seemed to define fractions as "really small" or "less than one." Emphasis on proper fractions in early fraction instruction may lead students to view all fractions (*a*/*b*) as numbers between zero and one (Vosniadou, Vamvakoussi, & Skopeliti, 2008).

Overall analysis by fraction type revealed that students' estimation skills with improper fractions were less accurate and developed later than their skill with proper fractions and mixed numbers. This finding supports the integrated theory of numerical development, described earlier, which asserts that numerical development involves increasingly widening the range and type of number understood as magnitudes that can be accurately located on a number line (Siegler & Lortie-Forgues, 2014).

In sum, findings to date from the Delaware Longitudinal Study show that although a range of domain-general and number-specific competencies predict fraction outcomes, the ability to estimate numerical magnitudes on a number line is a key marker of fraction and mathematics success. Children with mathematics difficulties may have fundamental problems related to whole number magnitude representations that are further complicated by the introduction of fractions into the curriculum. In addition, attentive behavior, nonsymbolic proportional reasoning, and calculation fluency all were consistently important predictors of fraction learning and growth.

HELPING STUDENTS WHO STRUGGLE WITH FRACTIONS

According to a recent report by the US Department of Education National Center for Educational Statistics, only 32% of eighth graders nationwide scored at or above proficiency on the 2015 nationwide NAEP assessments (NAEP, 2015). Moreover, the situation is most urgent for minority and low-income students, who generally perform more poorly than their nonminority and nonlow income peers. As such, current research findings on the importance of numerical magnitude understanding for developing fraction knowledge in particular, and to mathematics achievement more generally, should have particular resonance for educators and policymakers.

Research findings strongly support an approach to teaching fractions as well as whole numbers that centers on a number line. Students who develop an understanding that all real numbers, including fractions, are assigned to their own location on a number line have an advantage not only in learning fractions but also in learning algebraic concepts (Booth, Newton, & Twiss-Garrity, 2014). Until recently, a part-whole interpretation of fractions has been dominant in the US mathematics curriculum (Siegler et al., 2015). Yet instructional activities using number lines successfully support student learning of whole numbers (Ramani & Siegler, 2008) and fractions (Fuchs et al., 2013; Fuchs et al., 2014).

Part-whole instruction defines *a/b* as *a* parts out of *b* parts. Even low-performing students are successful on part-whole problems where the denominator matches the number of partitions in the whole (Hansen et al., 2015b); while low-performing students can shade 3/4 of a rectangle when the rectangle is separated into four equal parts, they are unable to when it is partitioned into 8 equal parts. These same students, however, may be able to show that 3/4 is equivalent to 6/8 by performing the operation: $3/4 \times 2/2 = 6/8$. Such disconnects between conceptual and procedural knowledge makes it difficult for students to use fraction procedures in every day situations because they do not understand the concepts that support all procedures. A number line approach has the potential to remedy some of these difficulties by emphasizing that fractions are numbers that can be ordered and placed on the number line (Fuchs et al., 2013; Fuchs et al., 2014). For example, a number line could be used to illustrate fraction equivalence (3/4 is the same as 6/8 because they are actually the same location on a number line), the inversion property of fractions (fractions with the same numerator become smaller as the denominators increase), and the density of fractions (between any two fractions, there are an infinite number of fractions).

To promote learning of fraction magnitudes, interventions should also address students' attention. Attentive behavior allows students to stay on task and to acquire relevant knowledge and skills in their mathematics classrooms. Attentive behavior also facilitates effective strategy application on number line tasks; in order to understand the magnitude of a given fraction, students need to attend to the numerator and denominator simultaneously and to inhibit ineffective whole number strategies (Bonato, Fabrri, Umilta, & Zorzi, 2007; Meert, Gregoire, & Noel, 2010). Further, students must attend to the varying end points of the number line (e.g., 0 to 1, 0 to 2, 0 to 5) when placing fractions;

low-performing students tend to always place 1/2 in the midpoint of the line rather than the absolute value of 1/2 (Dyson, Rodrigues, & Jordan, 2016).

The importance of multiplication fluency should not be underestimated for fraction learning. Fast and accurate calculation skills facilitate reasoning about fractions (Seethaler, Fuchs, Star, & Bryant, 2011; Hecht et al., 2003). For example, it helps students see, with minimal cognitive effort, multiplicative relations between equivalent fractions (1/4 is the same as 2/8) as well as between improper fractions and mixed numbers (6/4 is the same as $6 \times 1/4$, or 1 1/2). Unfortunately, many US curricula do not support multiplication fluency practice after fifth grade.

Additionally, as noted earlier in this chapter, typical instruction focuses on proper fractions (i.e., fractions less than 1; Vosniadou et al., 2008), even if a number line approach is used. Focusing on proper fractions facilitates to the common misunderstanding that all fractions are small numbers less than one. This type of reasoning makes sense given the general use of the word "fraction" to mean a tiny part, such as a fraction of a second or a fraction of an inch. Children should work with number lines showing more than one from the start, although this assertion requires further investigation. Placing proper and improper fractions on number lines from 0 to 2 encourages reasoning about the relationship between the numerator and the denominator as students move from 0 to 1 and beyond 1. Number line approaches help children see when $a < b$, the fraction is less than 1; when $a = b$, the fraction is equal to 1; and when $a > b$, the fraction is greater than one.

According to the US Department What Works Clearinghouse Practice Guide on assisting students who are struggling with mathematics, in depth treatment of rational number should be the focus of middle school math interventions (Gersten et al., 2009). Research supported instruction emphasizing fraction magnitudes and centered on a number line must be implemented to stem cascading difficulties when algebra becomes a primary focus (Siegler et al., 2012; Siegler & Pyke, 2013).

ACKNOWLEDGMENT

This work was supported by Award R324C100004 from the US Institute of Education Sciences.

REFERENCES

Bailey, D. H., Hoard, M. K., Nugent, L., & Geary, D. C. (2012). Competence with fractions predicts gains in mathematics achievement. *Journal of Experimental Child Psychology*, *113*(3), 447–455.

Bailey, D. H., Zhou, X., Zhang, Y., Cui, J., Fuchs, L. S., Jordan, N. C., Gersten, R., & Siegler, R. S. (2015). Development of fraction concepts and procedures in U.S. and Chinese children. *Journal of Experimental Child Psychology*, *129*, 68–83.

Bertelletti, I., Lucangeli, D., Piazza, M., Dehaene, S., & Zorzi, M. (2010). Numerical estimation in preschoolers. *Developmental Psychology*, *46*, 545–551.

Bonato, M., Fabbri, S., Umiltà, C., & Zorzi, M. (2007). The mental representation of numerical fractions: Real or integer? *Journal of Experimental Psychology: Human Perception and Performance*, *33*(6), 1410–1419.

Booth, J. L., & Newton, K. J. (2012). Fractions: could they really be the gatekeeper's doorman? *Contemporary Educational Psychology*, *37*(4), 247–253.

Booth, J. L., & Siegler, R. S. (2006). Developmental and individual differences in pure numerical estimation. *Developmental Psychology*, *42*(1), 189.

Booth, J. L., & Siegler, R. S. (2008). Numerical magnitude representations influence arithmetic learning. *Child Development*, *79*(4), 1016–1031.

Booth, J. L., Newton, K. J., & Twiss-Garrity, L. K. (2014). The impact of fraction magnitude knowledge on algebra performance and learning. *Journal of Experimental Child Psychology*, *118*, 110–118.

Boyer, T. W., & Levine, S. C. (2012). Child proportional scaling: Is 1/3 = 2/6 = 3/9 = 4/12? *Journal of Experimental Child Psychology*, *111*(3), 516–533.

Boyer, T. W., Levine, S. C., & Huttenlocher, J. (2008). Development of proportional reasoning: Where young children go wrong. *Developmental Psychology*, *44*(5), 1478–1490.

Brizuela, B. M. (2006). Young children's notations for fractions. *Educational Studies in Mathematics*, *62*(3), 281–305.

Council of Chief State School Officers & National Governors Association Center for Best Practices. (2010). Common core state standards for mathematics. Common Core State Standards Initiative. Available from: http://www.corestandards.org/assets/CCSSI_Mathematics%20Standards.pdf

Davis, G. E., & Pepper, K. L. (1992). Mathematical problem solving by pre-school children. *Educational Studies in Mathematics*, *23*(4), 397–415.

Dyson, N., Rodrigues, J., & Jordan, N. C. (2016). Research-based interventionfor middle school students who struggle with fraction understanding. *Paper presented at the meeting of Council for Exceptional Children (CEC)*, St. Louis, MO.

Empson, S. B. (1999). Equal sharing and shared meaning: The development of fraction concepts in a first-grade classroom. *Cognition and Instruction*, *17*(3), 283–342.

Feigenson, L., Dehaene, S., & Spelke, E. (2004). Core systems of number. *Trends in cognitive sciences*, *8*(7), 307–314.

Fuchs, L. S., Schumacher, R. F., Long, J., Namkung, J., Hamlett, C. L., Cirino, P. T., Jordan, N. C., Siegler, R. S., Gersten, R., & Changas, P. (2013). Improving at-risk learners' understanding of fractions. *Journal of Educational Psychology*, *105*(3), 683–700.

Fuchs, L. S., Schumacher, R. F., Sterba, S. K., Long, J., Namkung, J., Malone, A., Hamlett, C. L., Jordan, N. C., Gersten, R., Siegler, R. S., & Changas, P. (2014). Does working memory moderate the effects of fraction intervention? An aptitude-treatment interaction. *Journal of Educational Psychology*, *106*(2), 499–514.

Geary, D. C. (2004). Mathematics and learning disabilities. *Journal of Learning Disabilities*, *37*(1), 4–15.

Geary, D. C. (2006). Development of mathematical understanding. In W. Damon, R.M. Lerner (Series, Eds.), D. Kuhn & R.S. Siegler (Vol. Eds.), *Handbook of child psychology: Volume 2 Cognition, perception, language* (6th ed., pp. 777–810). Hoboken, NJ: Wiley.

Gelman, R., & Williams, E. (1998). Enabling constraints for cognitive development and learning: Domain specificity and epigenesis. In W. Damon, D. Kuhn, & R. S. Siegler (Eds.), *Handbook of child psychology*. New York: Wiley.

Gersten, R., Beckmann, S., Clarke, B., Foegen, A., Marsh, L., Star, J. R., & Witzel, B. (2009). Assisting students struggling with mathematics: Response to Intervention (RtI) for elementary and middle schools (NCEE 2009-4060). Washington, DC: National Center for Education Evaluation and Regional Assistance, Institute of Education Sciences, US Department of Education. Available from: http://ies.ed.gov/ncee/wwc/publications/practiceguides/

Groves, P. M., & Thompson, R. F. (1970). Habituation: a dual-process theory. *Psychological Review*, *77*(5), 419.

Gunderson, E. A., Ramirez, G., Beilock, S. L., & Levine, S. C. (2012). The relation between spatial skill and early number knowledge: the role of the linear number line. *Developmental Psychology*, *48*(5), 1229–1241.

Hansen, N., Jordan, N. C., Siegler, R. S., Fernandez, E., Gersten, R., Fuchs, L., & Micklos, D. (2015a). General and math-specific predictors of sixth-graders' knowledge of fractions. *Cognitive Development, 35*, 34–49.

Hansen, N., Jordan, N. C., & Carrique, J. (2015b). Identifying learning difficulties with fractions: a longitudinal study of student growth from third through sixth grade. *Contemporary Educational Psychology*.

Hecht, S. A., & Vagi, K. J. (2010). Sources of group and individual differences in emerging fraction skills. *Journal of Educational Psychology, 102*(4), 843–859.

Hecht, S., Close, L., & Santisi, M. (2003). Sources of individual differences in fraction skills. *Journal of Experimental Child Psychology, 86*(4), 277–302.

Heitin, L. (2015). 'Fraction phobia': the root of math anxiety? [Web blog post]. Available from: http://blogs.edweek.org/edweek/curriculum/2015/02/fractionphobia_the_root_of_mat.html

Holloway, I. D., & Ansari, D. (2008). Domain-specific and domain-general changes in children's development of number comparison. *Developmental Science, 11*(5), 644–649.

Jordan, N. C., Levine, S. C., & Huttenlocher, J. (1994). Development of calculation abilities in middle- and low-income children after formal instruction in school. *Journal of Applied Developmental Psychology, 15*(2), 223–240.

Jordan, N. C., Hansen, N., Fuchs, L. S., Siegler, R. S., Gersten, R., & Micklos, D. (2013). Developmental predictors of fraction concepts and procedures. *Journal of Experimental Child Psychology, 116*(1), 45–58.

Kloosterman, P. (2010). Mathematics skills of 17-Year-Olds in the United States: 1978 to 2004. *Journal for Research in Mathematics Education, 41*(1), 20–51.

Mack, N. K. (1990). Learning fractions with understanding: building on informal knowledge. *Journal for Research in Mathematics Education, 21*(1), 16–32.

Matthews, P. G., Lewis, M. R., & Hubbard, E. M. (2016). Individual differences in nonsymbolic ratio processing predict symbolic math performance. *Psychological Science, 27*(2), 191–202.

McCrink, K., & Wynn, K. (2007). Ratio abstraction by 6-month-old infants. *Psychological Science, 18*(8), 740–745.

Meert, G., Grégoire, J., & Noël, M. P. (2010). Rational numbers: componential versus holistic representation of fractions in a magnitude comparison task. *The Quarterly Journal of Experimental Psychology, 62*(8), 1598–1616.

Mix, K., Levine, S. C., & Huttenlocher, J. (1999). Early fraction calculation ability. *Developmental Psychology, 35*(1), 164–174.

Ni, Y., & Zhou, Y. (2005). Teaching and learning fraction and rational numbers: the origins and implications of whole number bias. *Educational Psychologist, 40*(1), 27–52.

Opfer, J. E., & Siegler, R. S. (2007). Representational change and children's numerical estimation. *Cognitive Psychology, 55*(3), 169–195.

Opfer, J. E., & Thompson, C. A. (2008). The trouble with transfer: insights from microgenetic changes in the representation of numerical magnitude. *Child Development, 79*(3), 788–804.

Ramani, G. B., & Siegler, R. S. (2008). Promoting broad and stable improvements in low-income children's numerical knowledge through playing number board games. *Child Development, 79*(2), 375–394.

Resnick, I., Hansen, N., Carrique, J. Rajan, V., Jordan, N. C. (2015). Developmental growth trajectories in fraction magnitude from fourth through sixth grade. Poster presented at the Cognitive Developmental Society Biennial Meeting, Columbus, OH.

Sasanguie, D., Göbel, S. M., Moll, K., Smets, K., & Reynvoet, B. (2013). Approximate number sense, symbolic number processing, or number–space mappings: what underlies mathematics achievement? *Journal of Experimental Child Psychology, 114*(3), 418–431.

Schneider, M., & Siegler, R. S. (2010). Representations of the magnitudes of fractions. *Journal of Experimental Psychology: Human Perception and Performance, 36*(5), 1227–1238.

Seethaler, P. M., Fuchs, L. S., Star, J. R., & Bryant, J. R. (2011). The cognitive predictors of computational skill with whole versus rational numbers: an exploratory study. *Learning and Individual Differences*, *21*(5), 536–542.

Siegler, R. S., & Booth, J. L. (2004). Development of numerical estimation in young children. *Child development*, *75*(2), 428–444.

Siegler, R. S., & Lortie-Forgues, H. (2014). An integrative theory of numerical development. *Child Development Perspectives*, *8*(3), 144–150.

Siegler, R. S., & Opfer, J. E. (2003). The development of numerical estimation: evidence for multiple representations of numerical quantity. *Psychological Science*, *14*(3), 237–250.

Siegler, R. S., & Pyke, A. A. (2013). Developmental and individual differences in understanding of fractions. *Developmental Psychology*, *49*(10), 1994–2004.

Siegler, R. S., Thompson, C. A., & Schneider, M. (2011). An integrated theory of whole number and fractions development. *Cognitive Psychology*, *62*(4), 273–296.

Siegler, R. S., Fazio, L. K., Bailey, D. H., & Zhou, X. (2013). Fractions: The new frontier for theories of numerical development. *Trends in Cognitive Science*, *17*(1), 13–19.

Siegler, R. S., Fuchs, L., Jordan, N. C., Gersten, R., & Ochsendorf, R. (2015). The Center for Improving Learning of Fractions: A progress report. In S. Chinn (Ed.), *The Routledge international handbook ofdyscalculia and mathematical learning difficulties,* (pp. 292–303). New York: Routledge.

Siegler, R. S., Duncan, G. J., Davis-Kean, P. E., Duckworth, K., Claessens, A., Engel, M., Susperreguy, M. I., & Chen, M. (2012). Early predictors of high school mathematics achievement. *Psychological Science*, *23*, 691–697.

Siegler, R. S., Fazio, L. K., Bailey, D. H., & Zhou, X. (2013). Fractions: the new frontier for theories of numerical development. *Trends in Cognitive Science*, *17*(1), 13–19.

Siegler, R., Fuchs, L., Jordan, N., Gersten, R., & Ochsendorf, R. (2015). The center for improving learning of fractions: A progress report. In S. Chinn (Ed.), *The Routledge international handbook of dyscalculia and mathematical learning difficulties* (pp. 292–303). New York: Routledge.

Singer-Freeman, K. E., & Goswami, U. (2001). Does half a pizza equal half a box of chocolates? Proportional matching in an analogical task. *Cognitive Development*, *16*(3), 811–829.

Sophian, C., Garyantes, D., & Chang, C. (1997). When three is less than two: early developments in children's understanding of fractional quantities. *Developmental Psychology*, *33*(5), 731–744.

US Department of Education, Institute of Education Sciences, National Center for Education Statistics, National Assessment of Educational Progress (NAEP) (2007). Mathematics Assessment. Available from: http://nces.ed.gov/nationsreportcard/

US Department of Education, Institute of Education Sciences, National Center for Education Statistics, National Assessment of Educational Progress (NAEP), 2004 Mathematics Assessment. Available from: http://nces.ed.gov/nationsreportcard/

US Department of Education, Institute of Education Sciences, National Center for Education Statistics, National Assessment of Educational Progress (NAEP) (2009). Mathematics Assessment. Available from: http://nces.ed.gov/nationsreportcard/.

US Department of Education, Institute of Education Sciences, National Center for Education Statistics, National Assessment of Educational Progress (NAEP) (2015). Mathematics Assessment. Available from: http://nces.ed.gov/nationsreportcard/

Vamvakoussi, X., & Vosniadou, S. (2010). How many decimals are there between two fractions? Aspects of secondary school students' understanding of rational numbers and their notation. *Cognition and Instruction*, *28*(2), .

Vosniadou, S., Vamvakoussi, X., & Skopeliti, I. (2008). The framework theory approach to the problem of conceptual change. *International handbook of research on conceptual change*, 3–34.

Xu, F., & Spelke, E. S. (2000). Large number discrimination in 6-month-old infants. *Cognition*, *74*(1), B1–B11.

Chapter 7

Numbers as Mathematical Models: Modeling Relations and Magnitudes with Fractions and Decimals

Melissa DeWolf*, Miriam Bassok**, Keith J. Holyoak*
**Department of Psychology, University of California, Los Angeles, CA, United States;*
***Department of Psychology, University of Washington, Seattle, WA, United States*

UNDERSTANDING RATIONAL NUMBERS

Introduction

The introduction of rational numbers to elementary school students (typically, fractions, and then decimals) creates a situation in which students must make use of prior knowledge about natural numbers, but must construct coherent new models of concepts that diverge from their prior knowledge in important ways (see also Jordan, Chapter 6; Van Hoof et al., Chapter 5). For example, unlike whole numbers, fractions do not have unique successors, are not able to be ordered and counted, and have infinite ways to express the same magnitude. In addition, fractions differ from whole numbers perceptually in format due to their bipartite (*a*/*b*) structure. As we elaborate in the next section, various studies document misconceptions about rational numbers that reflect negative transfer from natural numbers. Such misconceptions may persist into adulthood, and are even prevalent among community colleges students (Givvin, Stigler, & Thompson, 2011; Stigler & Hiebert, 2009).

Most research that documents children's and adults' difficulties in understanding rational numbers, as well as many recent educational interventions that aim to alleviate these difficulties, focus on the fact that both natural and rational numbers represent magnitudes (e.g., Siegler, Fazio, & Bailey, 2013; Siegler, Thompson, & Schneider, 2011). This focus on the commonality between natural and rational numbers is obviously important. However, it neglects the fact that numbers, just as other mathematical expressions, serve

Acquisition of Complex Arithmetic Skills and Higher-Order Mathematics Concepts
http://dx.doi.org/10.1016/B978-0-12-805086-6.00007-2

as modeling tools. For example, both algebraic expressions and fractions represent a relation between entities (e.g., $4w$, 3/4) as well as the product of some operation ($4w$ = multiply 4 * w; 3/4 = 3 divided by 4; see Sfard & Linchevski, 1994). In particular, as we elaborate in this chapter, the bipartite structure of fractions (a/b) is well suited to model relations between sets [e.g., part (a)/ whole (b)], whereas the one-dimensional structure of decimals of the same magnitude is not.

After a brief review of research on misconceptions about fractions, and research on assessment of fraction magnitudes, this chapter describes a set of recent studies from our laboratories in which we examined how children and adults use fractions and decimals to model, respectively, relations between various quantities and magnitudes of various quantities. We first review research on modeling relations in mathematical word problems, and discuss how mathematical notations interact with the structures of the to-be-modeled world. We then review our work indicating that fractions are readily used to model relations between discrete, countable sets, whereas decimals are used more naturally for measurement of the magnitudes of continuous quantities. Finally, we discuss the implications of understanding the modeling role of rational numbers for acquisition of higher-level concepts in mathematics, particularly algebra (see also Booth, Chapter 13).

PRIOR RESEARCH ON MAGNITUDE ASSESSMENT AND MISCONCEPTIONS ABOUT RATIONAL NUMBERS

Student Misconceptions

As described, research on acquisition of rational numbers documents that knowledge of natural numbers leads to misconceptions about both fractions and decimals, involving basic understanding of arithmetic and relative magnitudes for rational numbers (e.g., that multiplication by a decimal results in a smaller value, not a larger one as with whole numbers). These and related conceptual differences between rational number types make it difficult for children to integrate fractions into their already well-established understanding of natural numbers (Ni & Zhou, 2005; Mack, 1995; Siegler, Fazio, Bailey, & Zhou, 2013). Stafylidou and Vosniadou (2004) found that many 5th–8th grade students used the whole number parts of fractions, or other derivative heuristics, to approximate the relative sizes of different fractions. For example, some students believed that the value of a fraction increases when either the value of the numerator or the denominator increase. Conversely, some students believed that the value of a fraction increases when both the value of numerator and denominator *decrease*. Yet others evaluated the size of the fraction based only on the size of the numerator. Vamvakoussi and Vosniadou (2004) found that some high-school students believe that there are no numbers between 5/7 and 6/7 (because 5 directly precedes 6).

Magnitude Representations for Rational and Natural Numbers

Research on student misconceptions of rational numbers has influenced work on adults' understanding of rational numbers. The central question has been: do well-educated adults represent the magnitudes of rational numbers much like whole numbers? Whole numbers are thought to be represented along an analog "mental number line" (Dehaene & Changeux, 1993; Gallistel, 1993). This hypothesis is supported by evidence for a *symbolic distance effect* in magnitude comparisons between whole numbers: speed and accuracy in comparing two numbers increases as the numerical distance between the two numbers increases (Moyer & Landauer, 1967).

There has been some controversy regarding whether adults with high education levels (e.g., college students) eventually acquire magnitude representations for rational numbers that are similar to their representations of natural numbers. Importantly, Siegler, Thompson and Schneider (2011) have argued that a crucial part of number development is being able to represent all types of numbers on a number line. This raises the question of whether fractions are eventually mentally represented the same way as other number types (i.e., on a mental number line), or if there are differences in how the magnitudes of fractions are accessed relative to other number types.

A number of studies have addressed the issue of whether the mental representation of fractions is the same as or different from representations of natural numbers (e.g., Bonato, Fabbri, Umiltà, & Zorzi, 2007; Meert, Gregoire, & Noel, 2010; Schneider & Siegler, 2010). Bonato et al. (2007) found no evidence for a distance effect related to the overall magnitudes of the fractions being compared. These investigators argued that adults process fractions in a piecemeal manner, focusing on the separable parts of the fractions (the numerator and denominator), and using heuristics based on these parts to determine the relative sizes of the fractions. However, Schneider and Siegler (2010) found that a distance effect can be obtained for fractions if the stimuli are designed so as to require processing of the holistic magnitude of the fraction, rather than just its parts. For example, Bonato et al. (2007) examined unit fractions in the form $1/n$, which do not require adults to compare the integrated magnitudes of the fractions. Instead, people could use a simple heuristic: fractions that have larger denominators are smaller overall. When Schneider and Siegler tested comparisons based on fractions for which this heuristic could not be used, they found a pattern consistent with the hypothesis that adults do in fact access the holistic magnitude of the fraction to perform comparisons.

Interestingly, the work on misconceptions and on children's and adults' understanding of fraction magnitudes is disconnected from an earlier line of research on rational numbers, which pointed out that fractions can be used in a variety of roles. Ohlsson (1988) summarized various definitions and uses of fractions as described by many researchers in math education and psychology (e.g., Kieren, 1975, 1980; Behr, Lesh, Post, & Silver, 1983; Vergnaud, 1983). The compiled list included representing part/whole relationships, ratio, or multiplicative comparison between two quantities, probability, operator, quotient, and points on a number line. Ohlsson argued that fractions can be understood by any of three basic interpretations: comparisons (e.g., ratio), partitioning (e.g., division), and composite operations (e.g., rates). He argued that these types of interpretations are applied when students use fractions to model physical quantities in the real world. Thus, not only is it difficult for students to interpret the symbolic notation of fractions due to their competing conceptions of other number types, but in addition fractions themselves may vary in meaning and purpose depending on the context. However, we argue and show that this variety of uses and contexts may actually be advantageous when rational numbers are considered as mathematical modeling tools.

Relational Affordances of the Fraction Notation

Whereas previous research seems to indicate that the bipartite format of fractions may impede access to magnitude representations, our own recent research has extended the work of Ohlsson (1988) and Kieren (1975), focusing on the relational importance of fractions. Our work has investigated the extent to which the bipartite format of fractions affords advantages in some tasks that require relational reasoning about quantities. When mathematics is viewed as a type of conceptual modeling, the notation serves a dual purpose: first as a bridge between quantitative concepts and the structure of the world being modeled, and second as a platform for performing mathematical procedures to solve problems defined in relation to the structure of the world. Procedures then derive meaning from their place in the conceptual structure, rather than standing in isolation as rote routines (e.g., English & Halford, 1995; Hinsley, Hayes, & Simon, 1977; Kintsch & Greeno, 1985; Mochon & Sloman, 2004). In the case of rational numbers, the bipartite structure of fractions (*a/b*) can be viewed as a type of *relational expression* that models several kinds of relations between two distinct quantities (e.g., quotient, ratio, multiplicative comparison, part-whole; Kieren, 1975). Decimals, in contrast, share the base-10 structure of natural numbers. Thus, they can be considered a more natural model of one-dimensional magnitude.

USING MATHEMATICS TO MODEL RELATIONS

Mathematical understanding fundamentally involves grasping that mathematics is a system of quantitative relations among concepts (e.g., the commutative relation between addends). Importantly, math educators aim to teach students that the formal quantitative relations of mathematics can be used to model real-life situations (see also Lee, Ng & Bull, Chapter 9; Thevenot, Chapter 3). Students are presented with mathematical word-problems, for which they are expected to construct a representation of the situation described in the problem and map it onto an appropriate mathematical representation (Kintsch & Greeno, 1985; Martin & Bassok, 2005). For example, solving a problem about the total of

Sam's stickers and Leo's stickers, students could add the two sets to find how many stickers the two boys have.

Various studies document that, instead of engaging in mathematical modeling, students often solve word problems using a variety of nonmodeling shortcut heuristics (e.g. Sherin, 2001; Silver, Shapiro, & Deutsch, 1993). For example, Silver et al. (1993) presented 13- and 14-year old students with the following division problem: "an army bus holds 36 soldiers. If 1128 soldiers are being bussed to their training site, how many buses are needed?" They found that only 63% of the students gave the sensible answer, 32 busses (31 plus an extra one to hold the remaining 12 people). The other students answered "31", or "31 remainder 12" (see also Thevenot, Chapter 3). These latter answers did not take into account the real-world constraints of the problem. Although students may avoid mathematical modeling in various situations, most students understand that mathematical representations (e.g., equations) should correspond to the relations between the entities described in the problem texts.

Semantic Alignment

Bassok and her coworker have shown that, in dealing with word problems, children and adults are guided by *semantic alignment*: an analogical mapping between the default semantic relation between the objects in a problem statement and the relation between the arguments of arithmetic operations (Bassok, 2001). For example, Bassok, Chase and Martin (1998) have shown that college students make systematic use of relational symmetry to align semantic and arithmetic relations (Fig. 7.1). When asked to construct addition or division word problems for various pairs of object sets, students preferred to construct addition problems for categorically-related (hence symmetrical) sets (e.g., "6 *tulips* plus 2 *roses* gives you how many flowers?"), but preferred to construct division problems for asymmetric functionally-related sets (e.g., "How many tulips per vase do you have with 21 *tulips* and 3 *vases*?).

Martin and Bassok (2005) found that middle school, high school, and college students are sensitive to semantic alignments when solving division word problems. Students committed fewer errors when the objects in the problems were aligned in terms of real-world relations with division (e.g., *cupcakes* and *trays*) than when the objects in the problems were misaligned with division (e.g., *cupcakes* and *muffins*). Fisher, Borchert and Bassok (2011) found similar results in an algebraic modeling task, using variants of the "students and professors" problem (Clement, Lochhead, & Soloway, 1979). In this task, participants are presented which a multiplicative-comparison statement (e.g., "There are six times as many students as professors") and are asked to construct an algebraic equation that represents the comparison relation in the statement. Many college students construct an incorrect (reversed) algebraic equation ($6S = P$, rather than as the correct $6P = S$), indicating reliance on the left-to-right direct-translation heuristic. Importantly, however, Fisher et al. found that students who constructed correct

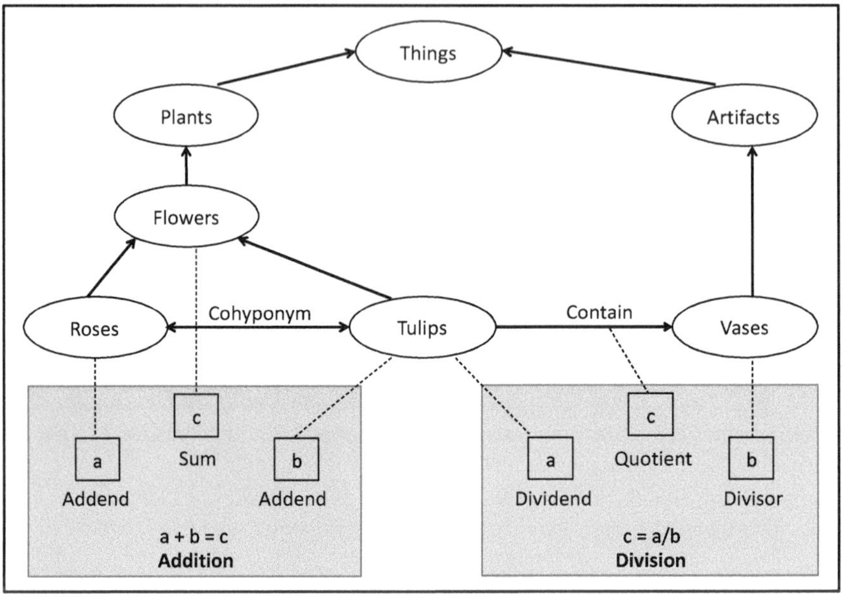

FIGURE 7.1 Semantic alignments between addition and division arithmetic relations and conceptual structures of items in the world as explained by Bassok et al. (1998). (Source: *Reprinted with permission from Guthormsen, A. M., Fisher, K. J., Bassok, M., Osterhout, L., DeWolf, M. & Holyoak, K. J. (2015). Conceptual integration of arithmetic operations with real-world knowledge: evidence from event-related potentials,* Cognitive Science, 40, *723–757).*

algebraic equations (i.e., engaged in modeling) were sensitive to semantic alignment: they took significantly longer to construct the correct equation when the statements related misaligned, semantically symmetric sets (e.g., "There are 6 times as many *cupcakes* as *muffins*"), than when the statements related aligned, semantically asymmetric sets (e.g., "There are 6 times as many *cupcakes* as *trays*").

Furthermore, there is evidence that alignment of semantic and mathematical relations is highly automatic for adults. Bassok, Pedigo, and Oskarsson (2008) found that adults retrieve addition facts (e.g., 6 + 2 = 8) when primed with categorically related objects (tulips and roses), but not when primed with functionally related objects (tulips and vases) or with unrelated objects (bears and thumbs). Guthormsen et al. (2016) also found evidence for the fairly automatic nature of alignments using event-related potentials (ERPs) measured using electroencephalography. In this set of studies, adults were asked to solve or affirm the validity of addition and division equations. Importantly, the equations were either aligned (3 tulips + 5 roses = ?; 12 tulips ÷ 4 vases = ?) or misaligned (3 tulips + 5 *vases* = ?; 12 tulips ÷ 4 *roses* = ?). Notice that the second object word was either aligned or misaligned. In the equation-solving task, participants showed an N400 effect for misaligned addition problems. Such an effect is typical when people encounter semantic violations. In an equation-verification task, participants showed a P600 effect (a signature of rule violation) when reading the second object label for misaligned problems. Together, these findings indicate that adults have strong expectations about systematic correspondences between real-world object relations and their mathematical representations. Any such correspondence can sometimes facilitate problem solving, but at other times may lead to mathematical errors (see Thevenot, Chapter 3).

Modeling with Rational Numbers

We recently proposed that viewing mathematical competence in terms of modeling and semantic alignment may help to understand the psychological representation of rational numbers, particularly fractions and decimals (DeWolf, Son, Bassok, & Holyoak, 2015a). As we have mentioned earlier, previous research on understanding of rational numbers had largely focused on their function as notations for expressing numerical magnitudes. Indeed, fractions and decimals are typically viewed as simply alternative notations, which (other than rounding error) are equivalent in magnitude (e.g., 3/8 vs. 0.375). Differences between notational variants have generally been treated as conceptually inconsequential (but see Cohen, Ferrell, & Johnson, 2002, and Goldstone, Landy, & Son, 2010, for important exceptions). What is often overlooked is the fact that, in addition to representing magnitudes, the core definition of a fraction (i.e., a rational number that can be expressed as the quotient of two natural numbers, a/b, where $b \neq 0$) is inherently *relational*. Specifically, the bipartite structure of the a/b notation expresses a relational model for the quantities corresponding to the numerator and denominator (Fig. 7.2). The form a/b expresses a division relation between natural numbers, which has the form of a fraction; whereas the

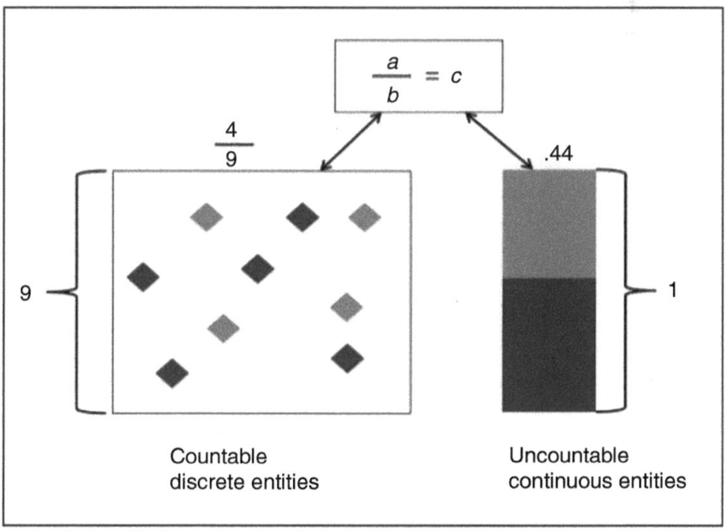

FIGURE 7.2 The bipartite structure of a fraction maps to countable subsets in a visual display (left), whereas the decimal equivalent represents a one-dimensional magnitude (right).

output of the division, *c*, expresses the magnitude of that relation and can be a decimal (with magnitude less than one when $a < b$). As illustrated in Fig. 7.2, the bipartite structure of the fraction provides a direct mapping to countable subsets in a visual display (e.g., a subset of 3 white objects within a set of 8). In contrast, the decimal equivalent loses the two-dimensional structure of the fraction. Instead, it provides a one-dimensional measure of a portion of a continuous unit quantity. (For discussions of the relationship between dimensionality of conceptual structures and relational complexity, see Halford, Wilson, & Phillips, 1998.)

ALIGNMENTS BETWEEN RATIONAL NUMBERS AND QUANTITY TYPES

Previous work on mathematics as a form of modeling has focused on numbers as defined by a unidimensional concept of magnitude (e.g., Dehaene & Changeux, 1993; Moyer & Landauer, 1969; Pinel, Piazza, Bihan, & Dehaene, 2004), or numbers as a system of relations involving mathematical expressions, such as equations (e.g., Bassok et al., 1998; Landy & Goldstone, 2007; Larkin, 1983; Kintsch & Greeno, 1985). However, the two formats of rational numbers—fractions and decimals—present a unique opportunity to examine how people understand symbolic notations that are equivalent in magnitude but represent different relationships (e.g., a two-part relationship versus a unidimensional magnitude). Importantly, rational numbers are used to convey information about parts of either discrete or continuous quantities, and the nature of these quantities may have different implications for which symbolic notation would be more appropriate to model them.

Discrete/Continuous Ontological Distinction

A particularly important ontological distinction that may be relevant to the use of rational numbers pertains to the *discrete* or *continuous* nature of quantities. Roughly, some entities are viewed as comprising a set of individual objects (e.g., a number of apples in a basket), whereas others are viewed as a continuous mass without individuation (e.g., a bucket of water). Continuous quantities can be subdivided into equal-sized units (i.e., discretized) for the purpose of measurement, but the divisions are arbitrary in the sense that they do not isolate conceptual parts (e.g., one can distinguish between specific subsets of red and green apples in a basket by noting that 2/3 of the apples are red and 1/3 are green, but there is no psychological difference between the water contained in 2/3 of a bucket and in the complementary 1/3 of the bucket).

The contrast between discreteness versus continuity (closely related to the count versus mass distinction in linguistics) (see Bloom, 1994; Bloom & Wynn, 1997) is based fundamentally not on physics, but on psychology. For example, a pile of sand is viewed as a continuous quantity even though it is composed of individual grains, because those units are too small and interchangeable to be typically viewed as "important". The impact of this basic ontological distinction has been documented both in young babies (e.g., Spelke, Breilinger, Macomber, & Jacobson, 1992) and in adults (Bassok & Olseth, 1995). For example, Bassok and Olseth found that college students viewed an increase in attendance at an annual conference as discrete (since it is based on a change between magnitudes associated with two discrete events well-separated in time), but viewed an annual increase in a city's population growth as continuous (since it is based on changes stemming from the psychologically-constant process of gaining and losing undifferentiated individual residents).

Infants and young children are able to distinguish the magnitudes of different continuous quantities (Clearfield & Mix, 2001; Fiegenson, Carey & Spelke, 2002). Dehaene and coworkers have argued that the ability to make these distinctions reflects an approximate number sense, which is evolutionarily more primitive than exact calculations (Dehaene, 1997; Feigenson, Dehaene, & Spelke, 2004; for a recent review see Geary, Berch, & Mann Koepke, 2015). However, school-aged children have an advantage when performing operations (e.g., counting) with discrete quantities (Gelman, 1993) over performing measurement operations using matched continuous quantities (Nunes, Light, & Mason, 1993). Measurement of continuous quantities depends on the introduction of arbitrary equal-sized units, which serve to parse a continuous whole into countable subparts (e.g., a continuous length can be broken down into some number of inches or centimeters).

Modeling Discrete and Continuous Quantities with Fractions and Decimals

Both fractions and decimals can represent unidimensional magnitudes of portions of quantities. However, as we have mentioned earlier, these two mathematical notations differ in format: fractions specify a two-part relation expressed in a bipartite (*a*/*b*) format, whereas decimals are one-dimensional. This difference has implications for what kinds of quantities each number type is best suited to represent. Fractions can model relations between the cardinality of two sets. Specifically, they can be used to model relations between discrete entities, such as a people or a balls, that can be counted but not partitioned. Therefore, the notation must adapt to fit the unit size of the entities being represented. For example, 3/7 might specify that 3 out of 7 children are girls. Using this notation, it is clear that the total number of children is 7, and that this total consists of 4 boys and 3 girls. An equivalent decimal (~0.429) does not provide such information—it merely conveys the fact that the proportion of girls is smaller than half of the total number of children. Therefore, fractions are more versatile because they can be used to represent discrete sets with arbitrary unit sizes (e.g., children), or describe continuous entities that can be partitioned (e.g., 3/5 of a pie).

Hiebert, Wearne and Taber (1991) argued that decimals have both discrete (base 10) and continuous features (e.g., can be represented on a number line), and are more specific than fractions because they have a specified unit size: base 10. Decimals are discrete in some sense because they can be used to meaningfully represent any discrete set of base 10. However, they are most useful for representing entities that can be partitioned into base-10 units, such as continuous entities. For example, a pie can be thought of as a continuous entity. If you think of .60 of a pie, this value does not specify the total number of slices of the pie you may need. However, you could easily think of the pie as portioned into 10 pieces, with 6 of them selected. Therefore, decimals are most useful for representing continuous entities because they make it possible to flexibly think of a proportion of the whole continuous entity (.60), or translate this to the number of partitioned pieces (6 out of 10 slices). Further, decimals are especially meaningful when standard metric units (e.g., centimeters, grams) are used for unidimensional measurement because of their mutual use of a base-10 measuring system.

There is evidence that educators and students honor an alignment between discrete (countable) entities with fractions, and continuous entities with decimals. Recently we completed a textbook analysis that showed textbook writers construct problems that associate fractions with discrete entities and decimals with continuous entities (Rapp, Bassok, DeWolf, & Holyoak, 2015). Fractions are used with both continuous and discrete entities, but are used far more often with discrete entities. When they are used with continuous entities, it is generally in cases where the entity is measured by imperial measures, such as feet or inches, which do not use a standardized base-10 system. Conversely, decimals are used overwhelmingly with continuous entities. When continuous entities are measured with metric base-10 units, they are paired almost exclusively with decimals. Further, research has shown that college students also show a similar alignment when using fractions and decimals to describe and create problems that use continuous and discrete entities (Rapp et al., 2015).

DeWolf et al. (2015b) gave college students pictures of various continuous, discretized, and discrete displays (Fig. 7.3). The displays contained red and green subsets. Importantly, both the discretized and discrete displays contained countable subsets while the continuous display had no individuating information between

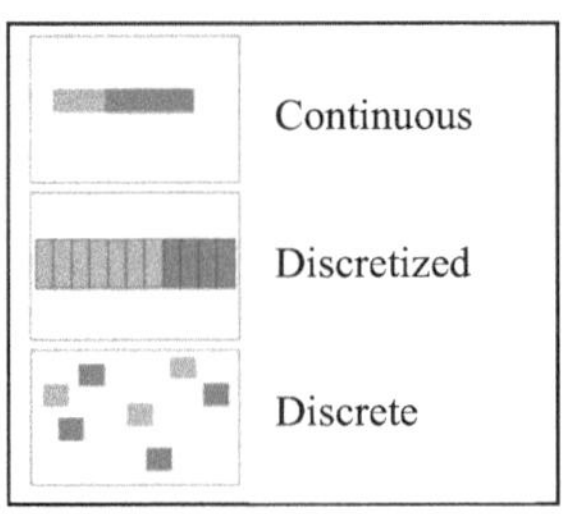

FIGURE 7.3 **Examples of continuous, discretized, and discrete displays used by DeWolf, Bassok and Holyoak (2015b).**

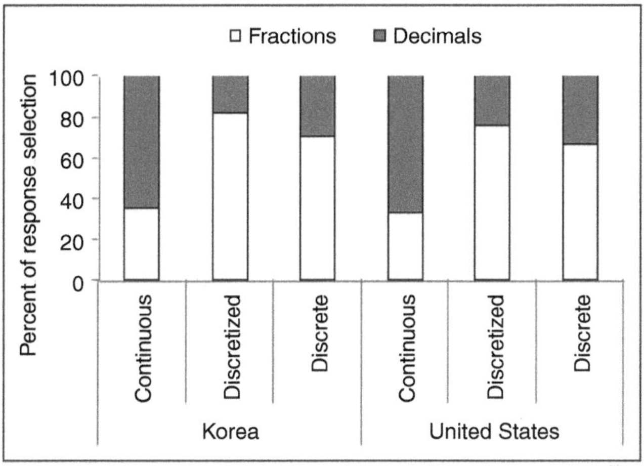

FIGURE 7.4 Preference for fractions vs. decimals to describe relations between continuous, discretized and discrete subsets in South Korea and the United States. The United States results are from DeWolf et al. (2015b, Experiment 1) and the South Korean results are from Lee et al. (2016).

the red and green portions. Participants were then asked whether they would prefer to use a fraction or a decimal or represent the relation between the red and green portions of the display. They were asked to make this decision "conceptually", and thus did not have to literally calculate what such a fraction or decimal would be. As hypothesized, students showed an overwhelming preference for fractions for both of the countable displays (discretized and discrete). Decimals were instead the preference for nonparsed continuous displays (Fig. 7.4, right panel).

Recently, we found the same pattern of alignment for Korean students, both in their textbooks and among college students (Lee, DeWolf, Bassok, & Holyoak, 2016). This similarity of alignment, despite important differences in language, culture, and educational practices, suggests that our findings capture a basic conceptual alignment between mathematic symbolic notations (fractions and decimals) and quantity type (discrete and continuous, respectively). Lee et al. (2016) replicated the findings of DeWolf, Bassok, & Holyoak (2015b), (Experiment 1), showing that Korean college students also prefer to use decimals to describe relations between continuous sets and fractions to describe relations between discrete sets (Fig. 7.4, left panel).

These findings may have important implications for learning about fractions and decimals. Hiebert et al. (1991) found that different types of physical representations can be used to highlight important features of different number types. For example, providing fourth-grade students with base-10 blocks helps them understand decimal notation, by highlighting the importance of the place-value system and the meaning of the decreasing sizes of decimals (e.g., .1 > .01 > .001). A similar intervention could be developed for fractions, by creating various unit sizes. The evidence reviewed previously suggests that there is an important mapping between fractions and decimals and the types of entities that they represent. Exploiting this mapping may help children attend to important differences between these symbolic notations, and the unique affordances of each (Dienes, 1964; Saxe, Taylor, McIntosh, Gearhart, & Saxe, 2012).

Modeling Magnitude with Decimals

As already described, decimals are preferred for modeling continuous quantities, because decimals are one dimensional, whereas fractions are two dimensional, the former number type should be more easily used as a measurement tool. Although a great deal of research has investigated whether fractions are well-suited for magnitude representation, considerably less research has addressed this question for decimals.

Findings from a few recent studies suggest the clear advantages that decimals have over fractions in magnitude representation. Iuculano and Butterworth (2011) assessed the accuracy of adults and children in locating natural numbers, decimals, and fractions on a number line. They found that performance was quite similar for decimals and natural numbers, both yielding good performance. In contrast, adults and children (aged 10 years) were much less accurate when evaluating the magnitudes of fractions by placing a mark on the number line at a certain fraction value, or generating a fraction value when shown a mark on a number line. Such findings suggest that, at least for well-educated adults, magnitudes are easier to grasp using decimals than fractions, and that processing decimals may be more similar to processing natural numbers.

DeWolf, Grounds, Bassok and Holyoak (2014) compared adults' ability to perform magnitude comparisons with matched fractions, decimals, and multidigit natural numbers (e.g., 3/5, .600, 600). The results (Fig. 7.5) showed that while the pattern of response times for comparing decimals was almost identical to that for comparing multidigit natural numbers, the pattern for fraction comparisons was dramatically slower, and showed a more exaggerated distance effect, indicative of less precise magnitude representations for fractions. This pattern of results held even when decimals were made more difficult by including a leading 0 (e.g., .050), varying the length of the decimal between 1 and 4 digits, and including a terminating 0 (e.g., .7140). Thus, decimals appear much more effective than fractions in conveying information about magnitudes.

More recently, we extended this line of work by using functional magnetic resonance imaging (fMRI) to compare how different types of numbers are processed and represented during a magnitude comparison task (DeWolf, Chiang, Bassok, Holyoak, & Monti, 2016). Previous research using neuroimaging and psychophysiological methods had provided insight into how numbers are processed and, more specifically, how numbers are processed in relation to the quantities that they are meant to represent. Magnitude processing appears to be dependent on the intraparietal sulcus (IPS), as the horizontal segment of the IPS has been implicated in the processing of quantities, and is associated with a "mental number line" (Dehaene, Piazza, Pinel, & Cohen, 2003). However, no previous studies had

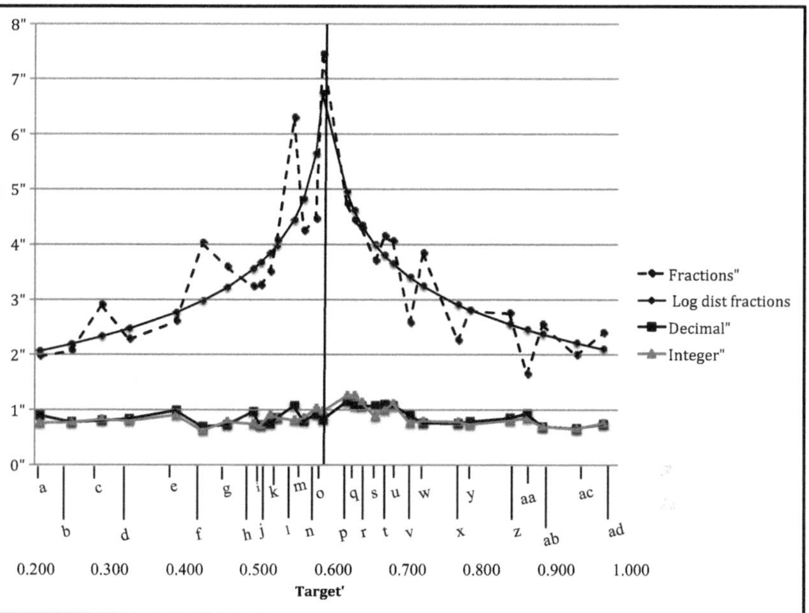

FIGURE 7.5 Mean response times across target magnitudes for fractions, decimals, and integers from DeWolf et al. (2014). The target magnitudes were each compared against a referent magnitude (3/5, .600, 600). There is a significant distance effect for all three number types but the fraction effect has a dramatically larger slope and is overall slower. Decimal and integer response times are nearly identical. *(Source: Reprinted with permission from DeWolf, M., Grounds, M. A., Bassok, M., & Holyoak, K. J. (2014). Magnitude comparison with different types of rational numbers.* Journal of Experimental Psychology: Human Perception and Performance, 40, *71–82.)*

examined processing of decimals in the brain, and no studies had compared activations for fractions, decimals, and integers within the same study.

The goal of our study was to investigate whether the processing for each of these number types was the same or different within well-educated individuals. Specifically, based on the results of DeWolf et al. (2014), we expected fractions to show a different pattern of results from both decimals and integers. However, whether decimals would behave the same as integers was less clear, as they are smaller in size (on an absolute scale) but conceptually distinct. The results of DeWolf et al. (2016) indicated that fractions recruit more activation in the IPS, and that this activation pattern differed from that of either decimals or integers. In contrast, decimals and integers showed no difference in level of activation, and produced nearly identical patterns of activation. These findings suggest that adults incorporate decimals into the neural system for representing magnitudes of whole numbers, but that fractions remain neurally distinct.

Modeling relations with fractions. One fundamental relationship that fractions represent is division. Formally, a fraction represents both a division relationship (numerator divided by denominator) and the outcome or quotient of that division. Understanding the magnitude of the fraction requires the latter interpretation. However, a true relational understanding of fractions should move flexibly between either of these two interpretations depending on task demands. In a magnitude task, such as magnitude comparison or number-line estimation, the magnitude interpretation (quotient) is required (e.g., DeWolf et al., 2014). But in a relational reasoning task, understanding the magnitude may be irrelevant.

For example, DeWolf et al. (2015b) administered an analogy task, in which participants had to identify one of two possible relations in pictures depicting two separate sets of items, and a decimal or fraction value corresponding to the target relation. There were two possible relations depicted: part-to-part ratio and part-to-whole ratio. To illustrate, a participant might see a picture with 2 red circles and 5 green squares paired with either the fraction 2/7 or the equivalent decimal .29. Participants needed to identify that this was a part-to-whole ratio (the part-to-part ratio would be 2/5 or .40). Then, they would be shown another picture with two different corresponding values, one that represented the part-to-part ratio and one that represented the part-to-whole ratio. Participants were instructed to choose which of the two values had the same relationship for the new picture as the one they had identified in the source picture. In this example, they would have to select the fraction or decimal that represented the part-to-whole ratio. Participants performed significantly better when the picture was paired with a fraction, which could be readily mapped onto the two sets of items, than when it was paired with a decimal expressing the magnitude of this relation. In addition to showing the advantage of the fraction format for relational reasoning, this finding suggests that adults can use fractions flexibly and do not automatically interpret the magnitudes of the fractions. Rather, they interpret the fraction in a way that is most advantageous given the task at hand.

CONNECTIONS BETWEEN RATIONAL NUMBERS AND OTHER MATH CONCEPTS

Multiplicative Reasoning and Fraction Understanding

Another important relational role of fractions is to highlight multiplicative and division relationships between whole numbers. Children learn multiplicative and division relations between whole numbers before they learn about fractions, when they are memorizing their multiplication and division tables. For example, children learn early on that any even number is divisible by 2; and conversely, that multiplying any number by 2 results in an even number. Fractions often utilize and highlight these relationships. For example, it can be determined that 6/8 is reducible because both the numerator and denominator are divisible by 2, and therefore the same relationship can be expressed by an equivalent fraction (3/4). Conversely, it can be determined that 6/7 is not reducible because the numerator and denominator do not share any common factors.

Because fractions inherently express a division relationship, using them when computing a multiplication or division problem involves understanding not only the relations within the multiplication and division problem itself, but also how these relate to the division relation within the fraction. Over the course of learning, students and teachers often seem to have trouble with multiplication and division with fractions. Students may get confused because of the differing procedures between fraction multiplication and fraction addition: in the latter case it is necessary to find a common denominator, but in the former case this is not necessary (Siegler et al., 2011, 2013). Even community-college students seem to have difficulty determining whether to obtain common denominators or multiply each number individually (Stigler, Givvin, & Thompson, 2010). Similarly, a well-known procedure for fraction division is the "invert and multiply" technique, in which one keeps the first factor, changes the division sign to a multiplication sign, and inverts the second factor. Prospective mathematics teachers show little understanding of this strategy and have difficulty explaining why this technique works (Tirosh, 2000).

At the essence of all of these difficulties is a lack of true conceptual understanding of multiplication and fractions. For example, applying the common adage "multiplication makes bigger" to the problem $31/56 \times 17/42$ might lead to the conclusion that the resulting product will be larger than either of the two multiplicands. And, in fact, middle-school students and preservice teachers reliably conclude that the product in such problems will be larger (Siegler & Lortie-Forgues, 2015). Correct understanding goes beyond simply conceptualizing multiplication as repeated addition (which does not fit well with the differences in procedures across arithmetic and multiplication with fractions). Rather, the reciprocal relation is key. Thompson and Saldanha (2003) proposed that understanding multiplication conceptually requires grasping multiple reciprocal relationships between the two factors, n and m, and their product nm (e.g., nm is n times as large as m; or conversely, n is $1/m$ times as large as nm). Thus, multiplication with fractions potentially can evoke a rich relational context. Further, Vergnaud (1988) argues that fractions have to be conceptualized as both operations (division) and as a resulting magnitude. However, during fraction multiplication and division, it can be fairly adaptive to think of fractions as an operation and not a single magnitude. Thinking of fractions as an instance of division allows the reasoner to take advantage of inverse relationships or simplifications that can be performed across an expression (e.g., $8 \times 1/4$ can be simplified to 2 by cross-canceling the 8 and 4—there is no need to multiply the 8 by 1).

DeWolf and Holyoak (2014) found that adults with a high level understanding of fractions and multiplication are able to exploit the reciprocal relation. Using a relational priming paradigm, we found that adults attended to reciprocal inverses in multiplication equations such as $8 \times 3/2 = 12$ and $12 \times 2/3 = 8$, where 3/2 and 2/3 are reciprocals. Time to solve the second problem was shorter when it included a reciprocal of the division relation (fraction) in the first problem, compared to when the second problem was preceded by one that was not reciprocally related. In contrast, when participants were presented with problems that used a decimal in place of the fraction multiplier ($8 \times 1.5 = 12$;

12 × .67 = 8), no priming effect was observed. Again, this finding suggests that adults use fractions flexibly—when processing multiplication problems, they manipulate the components of the fraction rather than estimate their magnitudes (as required when a decimal is used).

DeWolf et al. (2015a) used a similar priming paradigm to investigate whether this effect is consistent across levels of math achievement. Adult participants recruited from a wide range of mathematical backgrounds (noncompetitive and highly-competitive universities) were divided into low- and high-achieving math levels. Three groups of participants were given the multiplication priming task with either decimal problems (8 × 1.5 = 12; 12 × .67 = 8), high-similarity fraction problems (8 × 3/2 = 12; 12 × 2/3 = 8), or low-similarity fraction problems (8 × 6/4 = 12; 12 × 2/3 = 8) for which the successive fractions were constructed using different digits. Similar to the results of DeWolf and Holyoak (2014), neither low- nor high-achieving participants showed a priming effect for decimal problems, whereas both low- and high-achieving participants showed a priming effect for the high-similarity fraction problems. However, for the low-similarity fraction problems, low-performing participants were slower in general and did not show a priming effect, whereas high-performing participants continued to show priming. This finding suggests that high-performing participants may "see" fractions differently than do low-performing participants. The former may have an advantage in easily translating equivalent fractions based on ready access to common factors and reduced forms (Kellman et al., 2008). Low-performing participants, on the other hand, do not easily translate between alternative simplified forms of a fraction, and thus are not able to take advantage of the mapping between problems that are structurally but not visually similar.

Differential Contributions of Magnitude and Relational Knowledge to Learning Algebra

While it may seem as if fractions are difficult for students to master, attaining such mastery seems to be a necessary step to becoming a successful algebra student (National Mathematics Advisory Panel, 2008). Brown and Quinn (2007) found a link between fraction knowledge and knowledge of elementary and intermediate algebra. In a large sample of students from the United States and United Kingdom, Siegler et al. (2012) found that fraction knowledge in elementary school, measured by basic arithmetic and conceptual questions, uniquely predicted algebra knowledge and math achievement in general at age 16 (see also Torbeyns, Schneider, Xin, & Siegler, 2015). Further, Booth and coworkers found that 8th graders performance on a fraction number-line task (estimating the magnitudes of fractions on a 0–1 number line), along with questions measuring fraction declarative and procedural knowledge, was predictive of Algebra-1 performance (Booth & Newton, 2012; Booth, Newton, & Twiss-Garrity, 2014; Booth, Chapter 13). However, performance on a number-line estimation task with whole numbers was not predictive of Algebra-1 performance. This finding suggests that fraction magnitude estimation—rather than magnitude estimation in general—is linked to algebra understanding.

Although there seems to be an important link between fraction understanding and algebra performance, the nature of this link is not yet clear. Wu (2001) posited that the linkage between fractions and algebra may be due to the fact that both fractions and algebra are forms of generalized arithmetic. That is, algebra applies properties that are true of all numbers and thus involves abstraction of arithmetic properties. When students learn the rules and procedures that govern fractions, such as fraction addition, they must learn abstract procedures that apply for all types of fraction addition. Therefore, in 5th and 6th grade, when fraction addition is typically introduced, students must learn to abstract these general rules (even though they are usually taught with specific examples). Unlike learning whole-number arithmetic, for which students generally memorize the simple addition facts, fraction arithmetic "facts" often can't be memorized—rather, they require integration of specific procedures that must be understood on an abstract level.

Similar to this proposal, Empson and coworkers suggested that students' basic intuitions about arithmetic functions (across both whole numbers and fractions) can be exploited to build basic relational concepts that are common to arithmetic and algebra (Empson & Levi, 2011, Empson, Levi, & Carpenter, 2011). Specifically, students may have some implicit understanding of certain relations, such as the associative property. For example, to add 3/4 + 1/2 a student might reason that 3/4 is equal to 1/2 + 1/4. Then, one could add 1/2 + 1/2 to get 1, and have 1/4 left over (Empson, 1999). Empson argues that fraction learning is possible if the intuitive relational-reasoning skills of the students are fostered and can be connected to basic understanding of properties of algebra. For example, use of the distributive property of multiplication over addition to add $7a + 4a$ to get $11a$ is similar to the reasoning that one needs to understand how to add 70 + 40, or 7/5 + 4/5.

Wu (2009) and Siegler et al. (2011) have argued that an essential aspect of understanding fractions is grasping what unites fractions and other number types, as opposed to what makes each type unique. The main commonality is the fact that a fraction, like any other type of number, can be placed on a number line. This understanding requires an integration of procedural and conceptual knowledge about fractions and magnitudes. Together with the findings of Booth et al. (2014) indicating that fraction number-line estimation is predictive of algebra performance, this analysis suggests that magnitude understanding of fractions, and of how they relate to other number systems, may constitute important links between fractions and algebra.

Another possibility is that, as discussed earlier, the format of the fraction itself is unique in an important way. A fraction is at once a relationship between two quantities, expressed as a/b, and also the magnitude corresponding to the division of a by b. That is, fractions can either be interpreted as the result of some division operation (yielding a magnitude understanding) or as some specific comparison between two sets. Similar dualities arise in algebra (Sfard & Linchevski, 1994).

Fractions, then, are the first example of a *relational expression* to which students are exposed. This format has implications for the types of procedures that are appropriate for students to perform. Taking the example from Empson used previously, one could consider $7a$ to be an expression of 7 units each of size a. Similarly, one might interpret 7/5 as an expression of 7 units of 1/5. Such an understanding of 7/5 has implications for the appropriate ways to perform operations, such as the fact that "unlike terms" cannot be combined in algebra (e.g., $7a + 8b$ does not equal $15ab$, and 7/5 and 8/9 cannot be combined to get 15/45 or 15/14). As students become fluent with the concept of a relational expression, this understanding may help to bootstrap their learning about similar types of expressions within algebra.

To investigate the links between fraction understanding and later achievement in algebra and math in general, we performed a study with children engaged in early algebra learning (DeWolf, Bassok, & Holyoak, 2015c). Our study investigated how different components of rational-number conceptual understanding may contribute to algebra understanding for Algebra-I students in 7th and 8th grade. Students were given fraction and decimal number-line estimation tasks, as well as tests of procedural understanding of fractions and also relational understanding of fractions. Finally, they were asked to complete a set of basic algebra questions designed to measure their algebra performance. These questions covered basic equation solving, word problems, and interpretation of algebraic expressions. Consistent with previous findings, we found that fraction number-line estimation was significantly correlated with algebra performance. In fact, fraction number-line estimation, decimal number-line estimation, fraction relations, and fraction procedures were each significantly correlated with algebra performance. A multiple-regression analysis was then used to determine which of these aspects accounted for a unique portion of variance in algebra performance. The results showed that only decimal number-line estimation and relational fraction understanding each accounted for a unique portion of the variance in algebra performance. These findings connect well with our hypothesized conceptual distinction between fractions and decimals: decimals serve as representations of magnitude (important for overall math understanding), whereas fractions serve to prepare students for understanding and using relational representations (centrally important to understanding algebra).

CONCLUSIONS AND FUTURE DIRECTIONS

Prominent current approaches to understanding rational numbers focus on understanding student misconceptions involving fractions and decimals. Researchers have also examined the extent to which many adults ultimately overcome these misconceptions and incorporate fractions and decimals into their whole-number representational system. In particular, research has focused on whether adults show the behavioral signatures of representing fraction magnitudes on an analog mental number line (e.g., Siegler et al., 2011).

However, we propose that much of this research has failed to explore the affordances provided by fraction notation. If the test of adult understanding of fractions is whether they represent magnitudes in the same way as whole numbers, there is a danger of losing sight of the broader conceptual purposes for which fractions can be used. Fractions have many purposes outside of representing a magnitude: they can represent ratios, multiplicative comparisons, probabilities, operators, quotients, part/whole relationships, and points on a number line (Kieren, 1975, 1980; Ohlsson, 1988).

Accordingly, we have turned to areas of mathematical problem solving to explore potential uses of fractions and decimals. From this perspective, mathematical notation can be viewed as a type of conceptual modeling. In the case of fractions, the bipartite notation can be used to represent a relation between two distinct, countable quantities. By contrast, decimals provide a one-dimensional notation and therefore are best used to represent magnitudes and used as a measurement system for continuous quantities. Indeed, empirical tests show that well-educated adults and educators prefer to represent discrete sets with fractions and continuous quantities with decimals. Furthermore, adults show advantages in using fractions in relational contexts over decimals. Conversely, decimals are advantageous in contexts that involve magnitude representation.

Finally, we explored how differences between fractions and decimals affect other areas that connect to rational numbers, such as arithmetic with rational numbers, and algebra. In the case of multiplication with fractions, students must make use of the fraction as a division relationship, flexibly using the component parts of the fraction to simplify other whole-number components. This kind of relational use of a fraction can be transferred across structurally similar problems. Decimals, which do not share the same kind of relations between their components and whole numbers, cannot be simplified in such a way. In addition, based on our hypothesized alignments of fractions with relations, and of decimals with magnitudes, we found that decimal magnitude estimation and fraction relational understanding are the only two unique predictors of prealgebra performance in middle-school students (DeWolf et al., 2015c).

Many questions remain concerning how teaching of fractions might be improved (Jordan, Chapter 6), and how to leverage relational understanding of fractions to improve subsequent learning of more advanced topics, such as algebra. Further studies should investigate how to make connections between a student's understanding of fractions and of algebra more explicit, in order to bootstrap acquisition of algebra (Booth, Chapter 13). In addition, future research should focus on ways to overcome student misconceptions of rational numbers. It may be useful to first introduce decimals to teach the idea that rational numbers have magnitudes. The more complex notation of fractions might be introduced later (Moss & Case, 1999). More generally, this chapter has tried to relate research on number representation to work on higher-level relational reasoning, aiming to increase appreciation of the differential affordances of alternative mathematical notations.

REFERENCES

Bassok, M. (2001). Semantic alignments in mathematical word problems. *The Analogical mind: Perspectives from cognitive science*, 401–433.

Bassok, M., & Olseth, K. L. (1995). Object-based representations: transfer between cases of continuous and discrete models of change. *Journal of Experimental Psychology: Learning, Memory, and Cognition, 21*, 1522–1538.

Bassok, M., Chase, V. M., & Martin, S. A. (1998). Adding apples and oranges: alignment of semantic and formal knowledge. *Cognitive Psychology, 35*(2), 99–134.

Bassok, M., Pedigo, S. F., & Oskarsson, A. T. (2008). Priming addition facts with semantic relations. *Journal of Experimental Psychology: Learning, Memory, and Cognition, 34*, 343–352.

Behr, M. J., Lesh, R., Post, T. R., & Silver, E. A. (1983). Rational-number concepts. In R. Lesh, & M. Landau (Eds.), *Acquisitions of mathematics concepts and processes* (pp. 91–126). New York: Academic Press.

Bloom, P. (1994). Possible names: the role of syntax-semantics mappings in the acquisition of nominal. *Lingua, 92*, 297–329.

Bloom, P., & Wynn, K. (1997). Linguistic cues in the acquisition of number words. *Journal of Child Language, 24*, 511–533.

Bonato, M., Fabbri, S., Umiltà, C., & Zorzi, M. (2007). The mental representation of numerical fractions: real or integer? *Journal of Experimental Psychology: Human Perception and Performance, 33*, 1410–1419.

Booth, J. L., & Newton, K. J. (2012). Fractions: could they really be the gatekeeper's doorman? *Contemporary Educational Psychology, 37*(4), 247–253.

Booth, J. L., Newton, K. J., & Twiss-Garrity, L. K. (2014). The impact of fraction magnitude knowledge on algebra performance and learning. *Journal of Experimental Child Psychology, 118*, 110–118.

Brown, G., & Quinn, R. (2007). Fraction proficiency and success in algebra. *Australian Mathematics Teacher, 63*, 23–30.

Clearfield, M. W., & Mix, K. S. (2001). Infants use continuous quantity—not number—to discriminate small visual sets. *Journal of Cognition and Development, 2*, 243–260.

Clement, J., Lochhead, J., & Soloway, E. (1979). *Translating between symbol systems: Isolating a common difficulty in solving algebra word problems. A cognitive development project, department of physics and astronomy*. Amherst, Massachusetts: University of Massachusetts.

Cohen, D. J., Ferrell, J. M., & Johnson, N. (2002). What very small numbers mean. *Journal of Experimental Psychology: General, 131*(3), 424.

Dehaene, S. (1997). *The number sense*. NewYork: Oxford University Press.

Dehaene, S., & Changeux, J. -P. (1993). Development of elementary numerical abilities: a neuronal model. *Journal of Cognitive Neuroscience, 5*, 390–407.

Dehaene, S., Piazza, M., Pinel, P., & Cohen, L. (2003). Three parietal circuits for number processing. *Cognitive Neuropsychology, 20*, 487–506.

DeWolf, M. & Holyoak, K.J. (2014). Reciprocal and multiplicative relational reasoning with rational numbers. In B., Bello, M., Guarini, M., McShane, & B., Scassellati, (Eds.), *Proceedings of the 36th Annual Conference of the Cognitive Science Society*. Austin, TX: Cognitive Science Society.

DeWolf, M., Grounds, M. A., Bassok, M., & Holyoak, K. J. (2014). Magnitude comparison with different types of rational numbers. *Journal of Experimental Psychology: Human Perception and Performance, 40*, 71–82.

DeWolf, M., Son, J.Y., Bassok, M., & Holyoak, K.J. (2015). Implicit understanding of arithmetic with rational numbers: The impact of expertise. In R., Dale, C., Jennings, P., Maglio, T.,

Matlock, D., Noelle, A., Warfaumont, & J., Yoshimi, (Eds.), *Proceedings of the 37th Annual Conference of the Cognitive Science Society*. Austin, TX: Cognitive Science Society.

DeWolf, M., Bassok, M., & Holyoak, K. J. (2015b). Conceptual structure and the procedural affordances of rational numbers: relational reasoning with fractions and decimals. *Journal of Experimental Psychology: General, 144*(1), 127–150.

DeWolf, M., Bassok, M., & Holyoak, K. J. (2015c). From rational numbers to algebra: separable contributions of decimal magnitude and relational understanding of fractions. *Journal of Experimental Child Psychology, 133*, 72–84.

DeWolf, M., Chiang, J. N., Bassok, M., Holyoak, K. J., & Monti, M. M. (2016). Neural representations of magnitude for natural and rational numbers. *NeuroImage*.

Dienes, Z. P. (1964). *Learning mathematics: The cognitive science approach to mathematics education*. Norwood, NJ: Ablex.

Empson, S. B. (1999). Equal sharing and shared meaning: the development of fraction concepts in a first grade classroom. *Cognition and Instruction, 17*, 283–342.

Empson, S. B., & Levi, L. (2011). *Extending children's mathematics: Fractions and decimals*. Portsmouth, NH: Heinemann.

Empson, S. B., Levi, L., & Carpenter, T. P. (2011). The algebraic nature of fractions: Developing relational thinking in elementary school. In J. Cai, & E. Knuth (Eds.), *Early algebraization: A global dialogue from multiple perspectives* (pp. 409–428). Berlin: Springer.

English, L. D., & Halford, G. S. (1995). *Mathematics education: Models and processes*. Mahwah, NJ: Erlbaum.

Feigenson, L., Dehaene, S., & Spelke, E. (2004). Core systems of number. *Trends in Cognitive Sciences, 8*, 307–314.

Fiegenson, L., Carey, S., & Spelke, E. (2002). Infants' discrimination of number vs. continuous extent. *Cognitive Psychology, 44*, 33–66.

Fisher, K. J., Borchert, K., & Bassok, M. (2011). Following the standard form: effects of equation format on algebraic modeling. *Memory & Cognition, 39*(3), 502–515.

Gallistel, C. R. (1993). A conceptual framework for the study of numerical estimation and arithmetic reasoning in animals. In S. T. Boysen, & E. J. Capaldi (Eds.), *Development of numerical competence: Animal and human models* (pp. 211–223). Hillsdale, NJ: Erlbaum.

D. C.GearyD. B.BerchK.Mann Koepke*Evolutionary origins and early development of number processing, Mathematical Cognition and Learning* (Vol. 1). San Diego, CA: Elsevier.

Gelman, R. (1993). A rational-constructivist account of early learning about numbers and objects. In D. L. Medin (Ed.), *Learning and motivation* (pp. 61–96). (Vol. 33). New York, NY: Academic Press.

Givvin, K. B., Stigler, J. W., & Thompson, B. J. (2011). What community college developmental mathematics students understand about mathematics, Part 2: The Interviews. *MathAMATYC Educator, 2*(3), 4–18.

Goldstone, R. L., Landy, D. H., & Son, J. Y. (2010). The education of perception. *Topics in Cognitive Science, 2*(2), 265–284.

Guthormsen, A. M., Fisher, K. J., Bassok, M., Osterhout, L., DeWolf, M., & Holyoak, K. J. (2016). Conceptual integration of arithmetic operations with real-world knowledge: evidence from event-related potentials. *Cognitive Science, 40*, 723–757.

Halford, G. S., Wilson, W. H., & Phillips, S. (1998). Processing capacity defined by relational complexity: implications for comparative, developmental, and cognitive psychology. *Behavioral and Brain Sciences, 21*(06), 803–831.

Hiebert, J., Wearne, W., & Taber, S. (1991). Fourth grader's gradual construction of decimal fractions during instruction using physical representations. *The Elementary School Journal, 91*(4), 321–341.

Hinsley, D. A., Hayes, J. R., & Simon, H. A. (1977). From words to equations: meaning and representation in algebra word problems. *Cognitive Processes in Comprehension, 329.*

Iuculano, T., & Butterworth, B. (2011). Understanding the real value of fractions and decimals. *Quarterly Journal of Experimental Psychology, 64*, 2088–2098.

Kellman, P.J., Massey, C., Roth, Z., Burke, T., Zucker, J., Saw, A., ... & Wise, J.A. (2008). Perceptual learning and the technology of expertise: studies in fraction learning and algebra. *Pragmatics & Cognition, 16*(2), 356–405.

Kieren, T. (1975). On the mathematical, cognitive, and instructional foundations of rational numbers. In R. Lesh (Ed.), *Number and measurement: Papers from a research workshop* (pp. 101–144). Columbus, OH: ERIC/SMEAC.

Kieren, T. (1980). The rational number construct—its elements and mechanisms. In T. E. Kieren (Ed.), *Recent research on number learning* (pp. 125–149). Columbus, OH: ERIC/AMEAC.

Kintsch, W., & Greeno, J. G. (1985). Understanding and solving word arithmetic problems. *Psychological Review, 92*(1), 109.

Landy, D., & Goldstone, R. L. (2007). How abstract is symbolic thought? *Journal of Experimental Psychology: Learning, Memory, and Cognition, 33*, 720–733.

Larkin, J. H. (1983). The role of problem representation in physics. In D. Gentner, & A. L. Stevens (Eds.), *Mental models* (pp. 75–98). Hillsdale, NJ: Lawrence Erlbaum.

Lee, H. S., DeWolf, M., Bassok, M., & Holyoak, K. J. (2016). Conceptual and procedural distinctions between fractions and decimals: a cross-national comparison. *Cognition, 147*, 57–69.

Mack, N. (1995). Confounding whole-number and fraction concepts when building on informal knowledge. *Journal for Research in Mathematics Education, 26*, 422–441.

Martin, S. A., & Bassok, M. (2005). Effects of semantic cues on mathematical modeling: evidence from word-problem solving and equation construction tasks. *Memory & Cognition, 33*(3), 471–478.

Meert, G., Gregoire, J., & Noel, M. (2010). Comparing 5/7 and 2/9: adults can do it by accessing the magnitude of whole fractions. *Acta Psychologica, 135*, 284–292.

Mochon, D., & Sloman, S. A. (2004). Causal models frame interpretation of mathematical equations. *Psychonomic Bulletin & Review, 11*(6), 1099–1104.

Moss, J., & Case, R. (1999). Developing children's understanding of the rational numbers: a new model and an experimental curriculum. *Journal for Research in Mathematics Education, 30*, 122–147.

Moyer, R. S., & Landauer, T. K. (1967). Time required for judgments of numerical inequality. *Nature, 215*, 1519–1520.

National Mathematics Advisory Panel. (2008). *Foundations for Success: Final Report of the National Mathematics Advisory Panel*. Washington, DC: United States Department of Education.

Ni, Y., & Zhou, Y. D. (2005). Teaching and learning fraction and rational numbers: the origins and implications of the whole number bias. *Educational Psychologist, 40*, 27–52.

Nunes, T., Light, P., & Mason, J. (1993). Tools for thought: the measurement of length and area. *Learning and Instruction, 3*, 39–54.

Ohlsson, S. (1988). Mathematical meaning and applicational meaning in the semantics of fractions and related concepts. In J. Hiebert, & M. Behr (Eds.), *Number concepts and operations in the middle grades* (pp. 53–92). (Vol. 2). Reston, VA: Lawrence Erlbaum & National Council of Teachers of Mathematics.

Pinel, P., Piazza, M., Le Bihan, D., & Dehaene, S. (2004). Distributed and overlapping cerebral representations of number, size, and luminance during comparative judgments. *Neuron, 41*(6), 983–993.

Rapp, M., Bassok, M., DeWolf, M., & Holyoak, K. J. (2015). Modeling discrete and continuous entities with fractions and decimals. *Journal of Experimental Psychology: Applied, 21*(1), 47–56.

Saxe, G. B., Taylor, E. V., Mcintosh, C., Gearhart, M., & Saxe, B. (2012). Representing Fractions with Standard Notation: A Developmental Analysis. *Journal for Research in Mathematics Education*, *36*(2), 137–157.

Schneider, M., & Siegler, R. S. (2010). Representations of the magnitudes of fractions. *Journal of Experimental Psychology: Human Perception and Performance*, *36*(5), 1227–1238.

Sfard, A., & Linchevski, L. (1994). The gains and the pitfalls of reification—the case of algebra. In: *Learning Mathematics* (pp. 87–124). Springer, Netherlands.

Sherin, B. L. (2001). How students understand physics equations. *Cognition and Instruction*, *19*(4), 479–541.

Siegler, R. S., & Lortie-Forgues, H. (2015). Conceptual knowledge of fraction arithmetic. *Journal of Educational Psychology*, *107*(3), 909–918.

Siegler, R. S., Thompson, C. A., & Schneider, M. (2011). An integrated theory of whole number and fractions development. *Cognitive Psychology*, *62*, 273–296.

Siegler, R. S., Duncan, G. J., Davis-Kean, P. E., Duckworth, K., Claessens, A., Engel, M., et al. (2012). Early predictors of high school mathematics achievement. *Psychological Science*, *23*(7), 691–697.

Siegler, R. S., Fazio, L. K., Bailey, D. H., & Zhou, X. (2013). Fractions: the new frontier for. theories of numerical development. *Trends in Cognitive Sciences*, *17*, 13–19.

Silver, E. A., Shapiro, L. J., & Deutsch, A. (1993). Sense making and the solution of division problems involving remainders: an examination of middle school students' solution processes and their interpretations of solutions. *Journal for Research in Mathematics Education*, *24*(2), 117–135.

Spelke, E. S., Breinlinger, K., Macomber, J., & Jacobson, K. (1992). Origins of knowledge. *Psychological Review*, *99*(4), 605–632.

Stafylidou, S., & Vosniadou, S. (2004). The development of student's understanding of the numerical value of fractions. *Learning and Instruction*, *14*, 508–518.

Stigler, J. W., & Hiebert, J. (2009). *The teaching gap: Best ideas from the world's teachers for improving education in the classroom*. New York: Free Press.

Stigler, J. W., Givvin, K. B., & Thompson, B. J. (2010). What community college developmental mathematics students understand about mathematics. *MathAMATYC Educator*, *1*(3), 4–16.

Thompson, P. W., & Saldanha, L. A. (2003). Fractions and multiplicative reasoning. *Research companion to the principles and standards for school mathematics*, 95–113.

Tirosh, D. (2000). Enhancing prospective teachers' knowledge of children's conceptions: the case of division of fractions. *Journal for Research in Mathematics Education*, *31*, 5–25.

Torbeyns, J., Schneider, M., Xin, Z., & Siegler, R. S. (2015). Bridging the gap: fraction understanding is central to mathematics achievement in students from three different continents. *Learning and Instruction*, *37*, 5–13.

Vamvakoussi, X., & Vosniadou, S. (2010). Understanding the structure of the set of rational numbers: a conceptual change approach. *Learning and Instruction*, *14*, 453–467.

Vergnaud, G. (1983). Multiplicative Structures. In R. Lesh, & M. Landau (Eds.), *Acquisition of mathematics: Concepts and process* (pp. 127–174). New York: Academic Press.

Vergnaud, G. (1988). Multiplicative Structures. In J. Hiebert, & M. Behr (Eds.), *Number concepts and operations in the middle grades* (pp. 141–161). (Vol. 2). Reston, VA: Lawrence Erlbaum & National Council of Teachers of Mathematics.

Wu, H. (2001). How to prepare students for algebra. *American Educator*, *25*(2), 10–17.

Wu, H. (2009). What's sophisticated about elementary mathematics? *American Educator*, *33*(3), 4–14.

Part III

Algebraic, Geometric, and Trigonometric Concepts

Chapter 8

Understanding Children's Difficulties with Mathematical Equivalence

Nicole M. McNeil*, Caroline Byrd Hornburg*, Mary Wagner Fuhs**, Connor D. O'Rear*

**University of Notre Dame, Notre Dame, IN, United States; **University of Dayton, Dayton, OH, United States*

INTRODUCTION

To be complete, a theory of mathematics learning must explain not only how children learn new mathematical concepts and procedures, but also why some mathematical information is particularly difficult for children (and adults) to learn. This chapter focuses on math problems that are particularly difficult for children in the elementary school years—mathematical equivalence problems. They are equations that have operations on both sides of the equal sign (e.g., $3 + 4 + 5 = 3 + \_\_$; $7 + 4 + 5 = \_\_ + 5$) (Perry, Church, & Goldin-Meadow, 1988). Most elementary school children (ages 7–11) in the United States and other Western cultures solve these problems incorrectly. About 10 years ago, McNeil and Alibali (2005b) hypothesized that children's difficulties with math equivalence are due, at least in part, to knowledge that children *have* about mathematics. More specifically, they argued that children's difficulties are caused by inappropriate generalization of knowledge constructed from children's narrow experience with arithmetic in school. McNeil and Alibali were not the first to make this claim; the mathematics education community had long been warning about ineffective, overly procedural early instruction (e.g., Baroody & Ginsburg, 1983, 1986; Brownell, 1956; Cobb, 1987; DeCorte & Verschaffel, 1981; National Council of Teachers of Mathematics, 1989; Sáenz-Ludlow & Walgamuth, 1998). However, McNeil and Alibali (2005b) were the first to provide a mechanistic account that made specific, testable predictions, and their position diverged from several prevailing psychological theories at the time that tended to attribute math learning difficulties to something that children *lack* (e.g., domain-specific knowledge, executive function skills, proficiency with arithmetic facts).

Acquisition of Complex Arithmetic Skills and Higher-Order Mathematics Concepts
http://dx.doi.org/10.1016/B978-0-12-805086-6.00008-4

In this chapter, we will first define math equivalence problems, provide evidence that elementary school children have difficulties understanding them, and explain why these difficulties deserve our attention. Second, we will consider how three major psychological theories of children's mathematics learning difficulties might account for children's difficulties with math equivalence problems. Finally, we will describe the change-resistance account of children's difficulties with math equivalence and summarize the evidence supporting it. Although we extol the virtues of the change-resistance account, we acknowledge that multiple mechanisms likely play a role in children's difficulties with mathematical equivalence.

CHILDREN'S DIFFICULTIES WITH MATHEMATICAL EQUIVALENCE PROBLEMS

An *equation* is a mathematical statement in which the equal sign is used to indicate that two expressions are equivalent (or are defined as such). Elementary school children (ages 7–11) have little trouble understanding and solving simple equations for which the goal is to determine the numerical value of a combination of numbers and operations presented on the left side of the equal sign (e.g., 7 + 4 + 5 + 7 = __) (McNeil & Alibali, 2004). Indeed, most children understand how to solve simple addition problems (e.g., 7 + 4 = __) using informal strategies around age 5 years (Geary & Burlingham-Dubree, 1989; Siegler & Shrager, 1984). However, children have great difficulties solving math equivalence problems, which are equations that have operations on both sides of the equal sign (e.g., 7 + 4 + 5 = 7 + __; 7 + 4 + 5 = __ + 5) (Perry et al., 1988).

Math equivalence problems, as the name suggests, provide a means by which individuals' understanding of math equivalence can be assessed. In order to solve these equations correctly, individuals must mentally represent the problem accurately and understand at least two concepts: (1) for two numbers to be equal, they must have the same value and (2) the equal sign is a symbol used to indicate that the expressions on either side of it are (or are set to be) equal. Individuals also must understand the rules for how numbers and operations can be manipulated in order to come up with a solution that makes both sides of the equation to have the same value. Children who are successful at these equations look at the entire problem and identify the relation being expressed before calculating a solution (Jacobs, Franke, Carpenter, Levi, & Battey, 2007).

Any theory concerning children's performance on math equivalence problems will need to account for at least three findings. First, most elementary school children (ages 7–11) in Western cultures solve the problems incorrectly, whereas educated adults typically do not. Historically, children's difficulties with these problems were attributed to age-related changes in cognitive development, and the age of acquisition was thought to be around age 11 or 12 (Perry et al., 1988), but newer evidence suggests that children are capable of understanding the problems as early as age 7, maybe earlier (Capraro et al.,

2011; Carpenter, Levi, & Farnsworth, 2000). Moreover, understanding seems to decline a bit from first to third grade between the ages of 7 and 9 before starting to improve thereafter (Baroody & Ginsburg, 1983; McNeil, 2007; Mickey & McClelland, 2014). Second, children typically use one of the ten strategies shown in Table 8.1 when solving these equations, with two incorrect strategies being vastly more popular than the others. Third, children have poor retention and show limited transfer of newly learned correct strategies. Each of these findings will be discussed in turn.

Math equivalence problems are not typically included in traditional K-8 mathematics curricula in the United States (McNeil et al., 2006; Powell, 2012; Seo & Ginsburg, 2003). When elementary school children are presented with math equivalence problems, they rarely come up with a correct strategy for solving the problems in the absence of instruction. Although the specific numbers differ depending on the given study, most studies of children in grades 2–4 report that between 70% and 90% of children fail to solve the problems correctly.

When children are asked to solve math equivalence problems, they typically use one of the 10 strategies presented in Table 8.1, as noted. Children's use of these strategies has remained fairly stable since the late 1980s. The two most commonly used strategies are both incorrect—the add-all strategy

TABLE 8.1 Ten Documented Strategies for Solving the Problem $3 + 4 + 5 = 3 + \_\_$

Strategy	Solution	Verbal explanation of how solution was reached
1. Add-all	15	"I added 3 and 4 and 5 and 3."
2. Add-to-equal-sign	12	"I added 3 plus 4 plus 5, and that equals 12."
3. Carry	4	"There was a 4 over here, so I put 4."
4. Add-to-equal-sign and add-all	12 and 15	"I added 3 and 4 and 5, and I put 12, and then I added on the 3 and put 15."
5. Add-two-numbers	7	"I added 3 and 4."
6. Backward-add-to-equal-sign	2	"I put 2 because 2 + 3 equals 5."
7. Idiosyncratic	1	"I just guessed."
8. Equalize	9	"3 plus 4 plus 5 equals 12, and 3 plus 9 equals 12."
9. Add-subtract	9	"I added 3 plus 4 plus 5, and then I took 3 away from that."
10. Grouping	9	"I saw the two 3s, so I just added 4 and 5, and that's 9."

and the add-to-equal-sign strategy (McNeil, Fyfe, & Dunwiddie, 2015; Perry et al., 1988; Rittle-Johnson, 2006). One new incorrect strategy that we have observed in recent years is the backward-add-to-equal-sign strategy. Here, children create an equivalence relation between the expression on the right side of the problem and the one number immediately in front the equal sign. This incorrect strategy has been observed primarily in children using the Everyday Mathematics curriculum developed by the University of Chicago School Mathematics Project and published by McGraw-Hill Education. The equalize strategy is the most commonly used correct strategy. Here, children first add the numbers on the left side of the problem to get a total and then try to figure out what number to add to the known value on the right side of the problem to get that same total (e.g., Alibali, Phillips, & Fischer, 2009; Perry et al., 1988).

In several previous studies, children have been instructed on a correct strategy for solving math equivalence problems. Many children readily adopted these correct strategies and were able to use them to solve math equivalence problems correctly on a posttest. However, children failed to transfer their newly acquired knowledge to math equivalence problems that differed in terms of surface features (Alibali, 1999; Perry, 1991; Rittle-Johnson & Alibali, 1999). Moreover, when presented with identical math equivalence problems just two or four weeks after "learning" a correct strategy, many children do not retain their newly learned strategies and revert back to their incorrect strategies (Cook, Mitchell, & Goldin-Meadow, 2008; McNeil & Alibali, 2000).

Overall, the evidence indicates that elementary school children have substantial difficulties with math equivalence problems, and the academic community has known about these difficulties for at least 80 years (Renwick, 1932). However, the mechanisms underlying these difficulties have been less clear. In the next sections, we will discuss some psychological theories of learning and development in the domain of mathematics and consider how well each theory accounts for children's difficulties with math equivalence problems. Specifically, we will examine three proposed mechanisms of mathematical development: (1) changes in the understanding of mathematical symbols (Booth, Barbieri, Eyer, & Paré-Blagoev, 2014; Kieran, 1981; Knuth, Alibali, McNeil, Weinberg, & Stephens, 2005), (2) maturational changes in some parameter of the working memory system, such as total capacity, efficiency, or processing speed (Case, Kurland, & Goldberg, 1982; Cowan, 2014; McLean & Hitch, 1999), or attentional refreshing or inhibition (Barrouillet, Bernardin, & Camos, 2004; Conway, Cowan, & Bunting, 2001), and (3) increased proficiency with basic numerical knowledge (Anderson, 2002; Carnine, 1980; Haverty, 1999; Kaye, 1986). Although these prevailing theories differ in a number of ways, they have one thing in common—they attribute children's learning difficulties to something that children lack. Note that children's difficulties with math equivalence have also been attributed to age-related deficits in general logical/conceptual structures (Piaget & Szeminska, 1995; Sfard & Linchevski, 1994), but as this account has fallen out of favor, we will not consider it here.

POPULAR ACCOUNTS OF CHILDREN'S MATHEMATICS LEARNING DIFFICULTIES

The Symbol Misunderstanding Account

Some theorists argue that children's difficulties with mathematics are due to their misunderstanding of the symbol system of mathematics. In support of this view, research has shown that children have a poor understanding of important math symbols, such as the equal sign (Baroody & Ginsburg, 1983; Behr, Erlwanger, & Nichols, 1980; Booth et al., 2014; Kieran, 1981; McNeil & Alibali, 2005a), the decimal point (Wearne & Hiebert, 1989), the fraction bar (Mix, Levine, & Huttenlocher, 1999; Sophian & Wood, 1997), the negative sign (Vlassis, 2004), and variables (Knuth et al., 2005; McNeil et al., 2010b). According to this view, remediation hinges on instruction about the meaning of these and other abstract math symbols.

Research shows that children exhibit less advanced understanding of math concepts when working with abstract math symbols than with concrete quantities. For example, although many 8-year-old children understand the arithmetic principle of inversion as it applies to concrete quantities in a conservation task (Piaget & Szeminska, 1995), most children do not demonstrate their knowledge of the inversion principle in the context of math symbols until several years later (Bisanz & LeFevre, 1990; Siegler & Stern, 1998). The principle of inversion refers to the inverse relation between two mathematical operations, such as addition and subtraction or multiplication and division (see also Robinson, Chapter 2). In inversion problems children are presented with an equation of the form $a + b - b = \_\_$. Children with an understanding of the inversion principle will realize they do not need to add $a + b$ because $b - b$ is equal to zero, making the initial $a + b$ operation unnecessary.

Consistent with the symbolic misunderstanding account, children display an understanding of inversion with concrete amounts years before they begin to discover the same principle applies to symbolic arithmetic. In one study, 3-year-old children saw a set of blocks and then watched as a certain number of blocks was added and then removed, before being asked how many blocks were left (Sherman & Bisanz, 2007). Children were significantly more accurate on problems where the same number of blocks that was first added to a set was later removed ($a + b - b = ?$) than when a different number was removed ($a + b - c = ?$). The authors argued that this concrete presentation made it easier for children to reason about the quantities and utilize an understanding of the inversion principle (Sherman & Bisanz, 2007). When problems are presented symbolically (e.g., $3 + 5 - 5 = ?$) children do not start to demonstrate understanding until late elementary school (Siegler & Araya, 2005).

This advantage for concrete presentation can also be seen for math equivalence problems. In Sherman and Bisanz's (2009) study, second graders solved math equivalence problems presented either concretely (e.g., with blocks inside bins on either side of a divider) or symbolically (e.g., $5 + 2 = 4 + \_\_$). Children in the concrete condition outperformed those in the symbolic condition. Interestingly, children in the concrete condition were also more likely to display an understanding of math equivalence in their explanations. This suggests that second graders may have an intuitive, informal understanding of math equivalence that they do not readily link to symbolic arithmetic.

The math symbol most relevant to performance on math equivalence problems is the equal sign (=). Specifically, to solve a math equivalence problem correctly, solvers must interpret the equal sign in one of two ways. First, solvers can have a relational interpretation of the equal sign (e.g., the equal sign indicates that the expressions on either side of it have the same value and are interchangeable). Second, solvers can view the equal sign as a symbol that tells them to make both sides of the problem equal. Either conception will point solvers toward the correct solution. Unfortunately, most children hold neither conception.

It is typical for children to define the equal sign as "the total" or "the answer" (Baroody & Ginsburg, 1983; Behr et al., 1980; Kieran, 1981), even when they are asked to define it in the context of a math equivalence problem (McNeil & Alibali, 2005a). Children also tend to rate definitions, such as "the total" and "the answer" as smarter than definitions, such as "equal to" or "the same as" (McNeil & Alibali, 2000, 2005a; Rittle-Johnson & Alibali, 1999). This suggests that they lack a sophisticated understanding of the equal sign. As a further testament to children's poor understanding, when presented with an equation, such as "$3 + 4 = 5 + 2$," most say that the equation does not make sense (Behr et al., 1980; see also Baroody & Ginsburg, 1983; Rittle-Johnson & Alibali, 1999). Some go so far as to request that the equation be corrected to $3 + 4 + 5 + 2 = 14$, so that it corresponds to their operational interpretation (Behr et al., 1980).

If children's failure to understand the equal sign is responsible for their difficulties with math equivalence problems, then children should start solving math equivalence problems correctly once they understand the equal sign. However, at least three sources of evidence suggest that this is not the case. First, Rittle-Johnson and Alibali (1999) showed that 19% of children who defined the equal sign relationally failed to solve math equivalence problems correctly, whereas 70% of children who correctly solved the problems defined the equal sign as "the total" or "the answer." Second, the correlation between accuracy solving math equivalence problems and defining the equal sign relationally is only moderate (e.g., $r = .35$) (McNeil, Fyfe, Petersen, Dunwiddie, & Brletic-Shipley, 2011). Third, although accuracy tends to be better in nonsymbolic contexts, some children still fail to solve math equivalence problems correctly even when the problems are presented in contexts without the equal sign.

For example, across two of our current studies, 211 second graders were presented with two of Sherman and Bisanz's (2009) math equivalence problems in concrete, block form. Forty-three children (20%) solved both problems incorrectly, and another 31 (15%) were only successful on one problem. Additionally, when these children were presented with a contextualized word problem [e.g.,

"Bob has 9 dogs. If he gets 2 more cats, he will have as many cats as dogs. How many cats does Bob have?" (adapted from Fuson, Carroll, & Landis, 1996)], 19% incorrectly used the add-all strategy even though no equal sign was present (see also the "Joe and Jada" example in McNeil et al., 2011).

One additional piece of evidence comes from McNeil and Alibali's (2002) intervention study. One intervention targeted children's understanding of the equal sign. Children were presented with a math equivalence problem that had the correct answer written in the blank (e.g., 3 + 4 + 5 = 3 + 9). They were told to "notice that the equal sign means that the things on one side of it have to be *the same amount as* the things on the other side of it" (emphasis used). Most of the children (72%) who received this intervention demonstrated improvements in their understanding of the equal sign on a posttest. However, only 39% improved their performance solving math equivalence problems. Other interventions (e.g., asking children to notice where the equal sign is in a correctly solved problem) did not explicitly target children's understanding of the equal sign. Few of the children (26%) who received these interventions demonstrated improvements in their understanding of the equal sign. Nonetheless, 35% improved their performance solving math equivalence problems (an improvement rate that corresponds to that of children in the intervention that targeted equal-sign understanding). This suggests that children's performance on measures of equal-sign understanding is associated with—but does not directly correspond to—their performance solving math equivalence problems.

The symbol misunderstanding account places a strong emphasis on children's interpretation of the equal sign, and as a result, it accurately predicts the strategies that children use most often to solve math equivalence problems. If children interpret the equal sign as "the total" or "the answer," then they should be expected to solve problems by adding up all the numbers or adding up all the numbers before the equal sign. This account also provides a reasonable explanation for why children have limited transfer and poor retention of correct strategies. If children do not understand the equal sign as a relational symbol, then they will not be able to understand the correct strategies for solving math equivalence problems. Without a framework for understanding the strategies being taught, children's learning will be shallow, and children will be unable to transfer the strategies to novel problems or retain them for future use.

Thus, the symbol misunderstanding account accurately predicts the most common incorrect strategies and provides an explanation for the limited transfer and poor retention of correct strategies. However, as discussed, equal-sign understanding and problem-solving ability are not as tightly coupled as predicted by this account. Some children who perform well on measures of equal-sign understanding do not solve math equivalence problems correctly, and some children who perform poorly on measures of equal-sign understanding do solve math equivalence problems correctly. This contradicts the notion that children's difficulties with math equivalence problems can be distilled down to how they interpret the equal sign.

Equally important, it is not clear whether or not the symbol misunderstanding account would predict declines in understanding of math equivalence between the ages of 7 and 9 or dramatic improvements in children's understanding around age 11 or 12. Children's understanding of the equal sign likely improves gradually as their experience with the equal sign broadens across the elementary and middle school years. If children's lack of understanding of the equal sign determines performance on math equivalence problems, then performance should improve steadily as children gain a broader experience with the equal sign as it is presented in a variety of contexts (e.g., magnitude comparison problems; fraction to decimal conversion problems; prealgebra equations). However, the available evidence suggests that children's performance on math equivalence problems may dip around age 9 before improving dramatically around age 11 or 12 (McNeil, 2007). Even though most 12-year-olds can solve math equivalence problems correctly, about half still define the equal sign operationally (McNeil et al., 2006). Taken together with the evidence above, this finding reveals that children's difficulties with math equivalence problems cannot be reduced to their understanding of the equal sign.

The Deficient Working Memory System Account

Many theories link developmental and individual differences to variations in working memory capacity (see Lee, Ng, and Bull, Chapter 9). Each of several different variations of this approach focuses on different features of the working memory system. Some researchers attribute developmental gains as due, in part, to increases in the number of knowledge elements that can be held in mind at one time (Cowan, 2014; Cowan, Nugent, Elliot, Ponomarev, & Saults, 1999; Halford, Cowan, & Andrews, 2007; Pascual-Leone, 1970). Others have suggested that total capacity is fixed, but functional capacity increases as individuals gain familiarity with new information (Case et al., 1982). Still others have emphasized the role of processing speed (Hulme, Thomson, Muir, & Lawrence, 1984; Kail & Park, 1994; Kail & Salthouse, 1994), attentional refreshing (e.g., Barrouillet et al., 2004; see also Lee, Ng, and Bull, Chapter 9), or inhibition (e.g., Conway et al., 2001). Regardless of how working memory capacity is conceptualized, it is commonly assumed to be a general attribute of the cognitive system (but see MacDonald & Christiansen, 2002 for an alternative view).

Hitch's (1978) classic study of multicolumn addition problems (e.g., 628 + 931 = __) was one of the first demonstrations of working memory's involvement in calculations. Results suggested that working memory is crucial for remembering initial operands, keeping track of intermediate results, and executing additional operations. Since Hitch's study, many follow-up studies have sought to specify in more detail the relation between working memory and mathematics performance (see Raghubar, Barnes, & Hecht, 2010 for a review). Researchers have found associations between working memory performance and mathematics skills among children, including many studies with children in the age range of those who struggle with math equivalence problems (e.g., Bull & Lee, 2014; Friso-van den Bos, van der Ven, Kroesbergen, & van Luit, 2013; Li & Geary, 2013; Meyer, Salimpoor, Wu, Geary, & Menon, 2010; Rassmusen & Bisanz, 2005).

Many studies have followed Hitch's (1978) precedent by examining the role of working memory in mathematics performance on multicolumn arithmetic problems (e.g., 296 + 21 = __, 628 + 931 = __). Importantly, findings related to performance on these complex arithmetic problems can be extrapolated to explain performance on math equivalence problems because the solving of math equivalence problems involves many of the same component processes as the solving of complex arithmetic problems. For example, similar to the complex arithmetic problem 214 + 87 = __, the math equivalence problem 6 + 7 + 8 = 6 + __ requires a solver to construct or select a strategy and retrieve simple math facts. Moreover, depending on the strategy a solver uses to solve the math equivalence problem, he or she may need to perform carry operations and keep track of intermediate results. Thus, for our purposes here, findings concerning the role of working memory in performance on complex arithmetic problems are taken as evidence for the role of working memory in performance on math equivalence problems.

Most studies of the associations between working memory and mathematics skills have relied on the working memory model proposed by Baddeley and Hitch (1974), even though this model is no longer universally accepted by cognitive neuroscientists (Lewis-Peacock & Postle, 2008). The core components of this model involve a central executive and two domain-specific representational systems: the phonological loop and the visuospatial sketchpad. The phonological loop maintains rapidly decaying verbal information, and the visuospatial sketchpad maintains visual and spatial information. The episodic buffer was added to the model later and was modeled as an interface between the two representational systems to integrate information across different modalities (Baddeley, 2000). The central executive is a limited capacity system that supports a number of control functions, such as coordinating and integrating information from the representational systems and interacting with long-term memory.

Experiments with adults have used dual-task conditions designed to interfere selectively with the central executive or one of the two representational systems. Tasks thought to engage the central executive have been shown to interfere generally with arithmetic calculations (DeStefano & LeFevre, 2004) and specifically with the estimation of results (Logie, Gilhooly, & Wynn, 1994), the retrieval of simple facts (Rammelaere, Stuyven, & Vandierendonck, 2001), and the procedure of carrying (Fürst & Hitch, 2000; Imbo, Vandierendonck, & De Rammelaere, 2007). Tasks that engage the phonological loop have been shown to interfere with the maintenance of initial problem information (Fürst & Hitch, 2000; Logie et al., 1994) and the tracking of intermediate results (Logie et al., 1994). Evidence for the visuospatial sketchpad's role in arithmetic performance has been less consistent (Logie et al., 1994, but see Gathercole & Pickering, 2000). Only a few studies have examined dual-task performance in children, and they used simple arithmetic calculations (Imbo & Vandierendonck, 2007a,b; McKenzie, Bull, & Gray, 2003). McKenzie and coworkers found that visuospatial interruption affected younger children's (6- and 7-year-olds) simple arithmetic calculation skills, whereas phonological interruptions did not. Older children (8- and 9-year-olds) were affected by both types of disruption but were affected less by visuospatial disruption compared to younger children.

Given the evidence of a link between working memory and arithmetic calculation, it is not surprising that theorists have suggested that the development of the working memory system constrains the development of mathematical skills (see Adams & Hitch, 1997; Gathercole & Pickering, 2000). There is evidence that working memory components may differentiate students with a mathematics learning disability (performance on standardized math achievement test at or below the 25th percentile and performance on standardized reading achievement test at or above the 30th percentile) from their typically-developing peers (e.g., Andersson & Lyxell, 2007; Geary, Hoard, Nugent, & Bailey, 2012a). Swanson and Jerman (2006) conducted a metaanalysis and found that verbal working memory differentiated children with mathematics learning disabilities from typically-achieving children even under conditions of strict controls. This suggests that a working memory deficit may be contributing to deficits exhibited in children with mathematics learning disabilities.

Overall, the evidence for making causal statements about the role of working memory performance in poor mathematics performance is currently limited. There is not strong evidence from randomized trials reporting direct effects of working memory training on complex arithmetic calculation skills in children (Redick, Shipstead, Wiemers, Melby-Lervåg, & Hulme, 2015). For example, in their metaanalysis, Melby-Lervåg and Hulme (2013) found that although there was some evidence for short-term effects of working memory training, these effects did not transfer to other domains, such as arithmetic. However, because there is consistent evidence of strong longitudinal relations between working memory, specifically the updating component of the central executive, and mathematics skills more generally (e.g., Bull & Lee, 2014), we cannot rule out the possibility that causal effects could be found with alternative training methods. In other words, it remains possible that the reason for the lack of causal effects in randomized trials is due to limited knowledge about how to effectively train children's working memory skills in a way that would allow them to transfer specifically to mathematics skills. It should be noted that these experimental studies have generally used global mathematics achievement tests as the outcome of interest, and thus, there is currently no empirical data testing if working memory training specifically affects children's understanding of mathematical equivalence.

As presented, three findings come together to support the deficient working memory system account of elementary school children's difficulties on math equivalence problems: (1) working memory capacity is associated with mathematics skills, (2) tasks that engage working memory capacity interfere with mathematical calculation, and (3) children who have difficulties with mathematics have impaired performance on measures of working memory capacity. One advantage of the deficient working memory system account is that it provides an explanation for why children have trouble solving math equivalence problems before the age of 11 or 12 (i.e., they lack a mature working memory system). Consistent with this evidence, neuroimaging research suggests that children rely more heavily on the prefrontal cortex, which is associated with working memory skills, in early mathematics calculation but that children's reliance generally shifts from the prefrontal to the parietal cortex with age (Houdé, Rossi, Lubin, & Joliot, 2010, see also Qin et al., 2014). However, the mature working memory account does not readily account for the dip in performance between ages 7 and 9. Moreover, given that the evidence for the role of working memory is correlational, the validity of this account has yet to be established.

If deficiencies in the working memory system contribute to children's difficulties with math equivalence problems, then their performance on math equivalence problems should be inversely related to the amount of working memory capacity required for solving a given math equivalence problem, controlling for their fluency in executing fact retrieval. For example, children should be more likely to use a correct strategy on problems that involve three numbers (e.g., $7 + 4 = 7 + \_\_$) than on problems that involve four numbers (e.g., $7 + 4 + 5 = 7 + \_\_$), but data from Carpenter et al. (2000) and Byrd et al. (2015) do not support this prediction. In fact, in an unpublished sample of 153 children, children were significantly *less* accurate (4% on average) on three-number problems at pretest than four-number problems (8% on average), $p = .02$. Similarly, the percentage of children who solved at least one three-number problem correctly (11%) was comparable to the percentage of children who solved at least one four-number problem correctly (18%).

One drawback to an account that focuses exclusively on deficits in the working memory system is that it does not readily predict the specific strategies used by children on math equivalence problems. Although there are no available data on the working memory demands of each strategy shown in Table 8.1, it is fair to say that the add-all strategy is more demanding than the add-to-equal-sign strategy. When compared to the add-to-equal-sign strategy, the add-all strategy requires an additional intermediate result and an additional retrieval (or calculation) of a math fact. It also may require an additional carry operation. If capacity were the main factor determining children's strategy choices, then the add-to-equal-sign strategy would be chosen more often than the add-all strategy. However, the add-all strategy is more popular. Similarly, the equalize strategy is used more often than the grouping strategy, even though the grouping strategy likely requires less capacity to execute than the equalize strategy [on a four-number problem, such as $a + b + c = a + \_\_$, equalize requires at least 2 intermediate results and 3 retrievals (or calculations) of math facts, whereas grouping requires 0 intermediate results and 1 retrieval].

Proponents of the deficient working memory system account might counter by saying that working memory capacity is crucial for the construction of various correct strategies, such as grouping, in the first place. When working memory capacity is underdeveloped, children may not have resources available to discover new strategies (cf. Shrager & Siegler, 1998). This might help to explain the relative popularity of known strategies, such as the add-all strategy. However, by itself it still fails to explain why children do not retain and transfer the correct strategies they are taught. The grouping strategy is not capacity intensive, yet children do not continue to use it after it is taught to them (e.g., McNeil & Alibali, 2000). Thus, although deficiencies in the working memory system likely contribute to children's difficulties with math equivalence problems, an underdeveloped working memory system alone is unlikely to be a proximal cause of children's use of incorrect strategies on math equivalence problems.

The Poor Number Knowledge Account

Some theories emphasize the interface between working-memory-related processes and long-term memory representations (e.g., Anderson, Reder, & Lebiere, 1996; Ericsson & Kintsch, 1995). These theories are based, in part, on the skill acquisition approach (Anderson, 1982). They are related to the deficient working memory system account above, but they emphasize the role of domain experience and practice in overcoming the limitations of a fixed working memory system rather than on developmental and individual differences in general working memory itself. When a skill is unpracticed, the memory representations of domain facts and procedures required for its execution are weak and noisy, so it monopolizes working memory. In contrast, when a skill is highly practiced, it uses up less working memory, so working memory is free to be allocated to other processes.

Arithmetic calculation is one cognitive skill that is relevant to performance on math equivalence problems. Depending on the strategy a problem solver uses to solve a math equivalence problem, such as $7 + 4 + 5 = 7 + \_\_$, he or she will need to perform anywhere from one to three arithmetic operations (e.g., $7 + 4 = \_\_$, $11 + 5 = \_\_$, $16 - 7 = \_\_$). When a solver does not have sufficient practice with basic arithmetic facts, his or her working memory will likely be committed to controlling the step-by-step calculation of the operations. As a result, resources will be largely unavailable for the processing of other subtasks, such as encoding the novel problem structure or discovering novel problem-solving strategies. In contrast, when a problem solver has sufficient practice with basic facts, fewer demands will be placed on his or her working memory system, and resources can be allocated away from calculation to other processes. This view that fluency with basic arithmetic facts "frees up" working memory resources for higher-level problem solving is based on the Decomposition Thesis (Anderson, 2002). The basic argument is that a complex skill (e.g., solving math equivalence problems) can be decomposed into component "subskills" (e.g., performing calculations, encoding numbers, and operators in the problem), and that practice on those subskills facilitates learning and execution of the complex skill (Frensch & Geary, 1993; Kotovsky, Hayes, & Simon, 1985). This view also is consistent with Qin et al.'s (2014) recent finding that the engagement of the prefrontal cortex is high during early math facts learning, but then declines as people master those facts. According to this view, elementary school children's difficulties with math equivalence problems can be attributed, at least in part, to children's lack of proficiency with computing or retrieving basic arithmetic facts (cf., Carnine, 1980; Kaye, 1986).

As noted earlier, despite the logical structure of arithmetic, it takes children years of schooling to learn the concepts and skills necessary for calculating the arithmetic facts. There are two basic strategies for solving simple addition problems: retrieval and backup computation (Campbell & Timm, 2000; Siegler, 1987; Siegler & Shrager, 1984). When a problem, such as 4 + 5 = __ is presented, a problem solver may activate the long-term memory representation "4 + 5 = 9" and retrieve the answer 9. Alternatively, a problem solver may compute the answer using one of several back up counting strategies (Siegler, 1987, 1999), such as starting from the larger number and counting up by the smaller number (e.g., 5…6, 7, 8, 9). When children are able to use this "min" strategy, it does not automatically replace their less efficient strategies (e.g., "counting-all" or "sum" strategy), and even when children use the min strategy, they do not necessarily understand the commutative property (i.e., $a + b = b + a$) (Baroody & Gannon, 1984). In fact, it is a long developmental process for children to discover the "shortcut" strategies and regularities of arithmetic—beginning with counting all, then counting on from the first number, and then the min strategy (Baroody & Ginsburg, 1986). Once children have a good conceptual understanding of arithmetic, they can discover these shortcut strategies more rapidly (Siegler & Jenkins, 2014). Moreover, once children have committed some facts to memory, they may use a more sophisticated backup strategy known as decomposition, which involves translating a given problem into another known form to facilitate computation (Geary et al., 1996). For example, when solving 7 + 6, children may use their knowledge of the "tie" 7 + 7 = 14 and then subtract one to get 13, or when solving 4 + 7, they may use their knowledge that 4 + 6 = 10 and then add one to get 11 (Baroody & Tiilikainen, 2003; Siegler, 1987).

Children younger than age 10 or 11 frequently rely on backup computation strategies (the min model, Ashcraft & Fierman, 1982; Groen & Parkman, 1972). Backup strategies are more laborious than retrieval, and thus, require more working memory to be executed successfully. Even when children begin to retrieve basic arithmetic facts, the engagement of working memory resources during the act of retrieval continues for several years, as facts are consolidated in long-term memory (e.g., Geary, Hoard, & Nugent, 2012b; Qin et al., 2014;

Siegler, 1989). On top of the long consolidation process, learning of multiplication facts (typically around third grade) results in interference during the retrieval of both addition and multiplication facts such that correct retrieval now involves inhibiting irrelevant facts from intruding into working memory during problem solving (De Visscher & Noël, 2014b; Miller & Paredes, 1990) (Note: this provides a potential explanation for the dip in performance on math equivalence problems around age 9). Thus, when presented with a math equivalence problem, children's functional working memory capacity is reduced by the calculation of facts, and children may not have resources to carry out more advanced problem-solving behaviors.

Haverty (1999) specifically tested the relation between basic number knowledge and performance on more advanced mathematics problems. Specifically, she studied seventh graders' performance on function-finding problems (e.g., find a function to represent the relationship between x and y given $x = \{1, 2, 3, 4\}$ and $y = \{1, 4, 9, 16\}$). In two experiments, she trained children on specific arithmetic facts (the 17- or 19-times table) and then tested their ability to solve function-finding problems involving those facts. The training did not influence children's performance on "easy" problems (e.g., $y = 17x + 1$; $y = 19x - 1$), but it did influence children's performance on more difficult problems (e.g., $y = 17x - 34$; $y = 19x + 19$). On difficult problems, children's performance was significantly better on problems involving the facts on which they were trained than on problems involving the facts on which they were not trained. This suggests that proficiency with basic mathematics facts has the potential to influence performance on some types of more advanced mathematics problems (Geary & Widaman, 1992).

Children's lack of proficiency with basic arithmetic facts may contribute to their difficulties with math equivalence problems (cf. Carnine, 1980; Kaye, 1986). This is because, as noted, the solving of these problems necessarily involves the calculation or retrieval of several arithmetic facts. When children are presented with a math equivalence problem (e.g., $7 + 4 + 5 = 7 + \_\_$), they must devise a strategy for solving the problem. However, the embedded calculations require children to either retrieve facts or use backup strategies to calculate those facts, either of which will consume working memory resources, so resources may be largely unavailable for strategy discovery [see Shrager & Siegler's (1998) model].

Chesney et al. (2014) examined arithmetic strategy use and understanding of math equivalence and, contrary to the poor number knowledge account, found no evidence that use of retrieval strategies predicts understanding of math equivalence. They did, however, find that children who had decomposition in their strategy repertoires when solving simple addition problems exhibited a better understanding of math equivalence than those without decomposition in their repertoires, even after controlling for accuracy and reaction time on the addition problems. McNeil et al. (2011) found no significant differences in understanding of math equivalence between children who received extra practice with traditional arithmetic facts and children who did not receive extra practice. Thus, proficiency with arithmetic facts is not as strongly tied to understanding of math equivalence as might be expected by this account.

The poor number knowledge account is intuitively appealing. Mathematics is thought to be a domain in which knowledge of the "basics" is crucial for higher-level problem solving, with each new skill building on previously learned skills. Within such a system, it is reasonable to assume that difficulties with math equivalence problems can be traced back to difficulties with the basics. Indeed, the poor number knowledge account provides a specific explanation for why children have difficulties discovering a new strategy for solving math equivalence problems—their working memory resources are being consumed by basic fact calculation (or retrieval) and are unavailable for strategy discovery. In this sense, the account suggests a straightforward means by which accurate performance on math equivalence problems can be achieved—through experience and practice with basic facts (cf. Carnine, 1980; Haverty, 1999; Kaye, 1986). However, some available evidence actually shows the opposite—undergraduates from the United States are less likely to solve math equivalence problems correctly under speeded conditions after practicing simple addition facts than after participating in a control condition (McNeil, Rittle-Johnson, Hattikudur, & Petersen, 2010). Moreover, this effect is not due to simple priming of left-to-right processing strategies on math equivalence problems, as it carries over to more complex algebra word problems that require relational thinking. Undergraduates educated in high-achieving East Asian countries are not susceptible to these negative effects of facts practice (McNeil et al., 2010), presumably because they had never developed an operational view of equations as children (Capraro et al., 2011; Capraro et al., 2010; Watchorn, Lai, & Bisanz, 2009).

Additionally, the poor number knowledge account does not easily account for children's limited transfer and poor retention of newly learned correct strategies (e.g., Alibali, 1999; Cook et al., 2008; McNeil & Alibali, 2000). Although the account argues that children do not have working memory resources available to discover or generate a new strategy, it does not necessarily suggest that a new strategy should be difficult to maintain once it enters the repertoire. Overall, the evidence for and against the poor number knowledge account is mixed. Thus, future studies will need to elucidate the role of basic number knowledge on math equivalence problem performance.

THE CHANGE-RESISTANCE ACCOUNT

There are significant differences among the three prevailing accounts discussed thus far, but they share at least one thing in common—they attribute children's math learning difficulties to something that children lack. In contrast to this focus on what children lack, a number of scientists (e.g., Kuhl, 2000; Zevin & Seidenberg, 2002) consider the role of existing knowledge in learning difficulties. The general claim is that early learning constrains later learning (cf.

Munakata, 1998; Thelen & Smith, 1994). This view is derived from classic approaches to cognition that focus on "top-down" processes (e.g., Bruner, 1957; Rumelhart, 1980; Tolman, 1948). Contemporary theorists (e.g., Kuhl, 2000; Li, Zhao, & MacWhinney, 2007; Zevin & Seidenberg, 2002; see also Van Hoof et al., Chapter 5) explain the effects in terms of "entrenchment" of early-learned patterns. According to this view, the patterns with which people initially gain experience become entrenched, and difficulties arise when novel information shares commonalities with, but does not map directly onto, existing entrenched patterns.

Within this framework, McNeil and Alibali (2005b) proposed the *change-resistance account* of children's difficulties with math equivalence (see also McNeil, 2014). This account posits that these difficulties are due, at least in part, to their narrow experience with arithmetic in elementary school. According to this view, overly narrow practice with arithmetic operations in the early years hinders subsequent learning of more advanced topics, such as math equivalence. Thus, in contrast to accounts that focus on what children *lack*, the change-resistance account focuses on what children *have*—existing knowledge of mathematics.

In the United States, children's early math experience is weighted heavily toward arithmetic operations, such as addition, subtraction, multiplication, and division. This is because mathematics is considered to be a domain in which early competence with basic skills is necessary for advanced thinking and problem solving. Students learn arithmetic operations for many years before they are introduced to the conceptual meaning of the equal sign and learn to think relationally about equations. Indeed, data from the Third International Mathematics and Science Study (Beaton et al., 1996; Hiebert et al., 2005) show that, unlike students from higher-achieving countries, children in the United States spend substantial amounts of class time practicing and reviewing basic arithmetic skills throughout the elementary and middle school years (National Science Board, 2002; see also Valverde & Schmidt, 1997). Moreover, arithmetic practice tends to be focused narrowly on procedures and calculations rather than on the underlying concepts (Porter, 1989), and the problems are almost always presented with operations to the left of the equal sign and the "answer" to the right (e.g., $5 + 2 = \_\_$) (McNeil et al., 2006; Powell, 2012; Seo & Ginsburg, 2003). Limiting math problems to this left-to-right format obscures the interchangeable nature of the two sides of an equation.

Given children's early, narrow, and protracted experience with traditional arithmetic, it is likely that children are well versed in the patterns routinely encountered in arithmetic problems. Research has provided compelling evidence that humans possess powerful learning mechanisms that extract the statistical structure of patterns in their environment (Fiser & Aslin, 2002; Gomez, 2002; Saffran, 2003). Such mechanisms operate incidentally in a wide variety of domains and appear to be functional very early in life (Bulf, Johnson, & Valenza, 2011). Thus, it is likely that young learners pick up on the regularities present in the early mathematics problems they solve. These regularities eventually become entrenched long-term memory representations that are quickly and easily (often automatically) retrieved in the presence of related problems in the environment and guide how the problem solver encodes, interprets, and solves the presented problem (cf. Anderson, 1991; Bruner, 1957; Gray & Fu, 2001; Salthouse, 1991; Tolman, 1948).

McNeil and Alibali (2005b) argued that children in the United States (and other Western countries that teach arithmetic narrowly) extract at least three patterns from their experience with arithmetic operations that ultimately hinder their learning of and performance on math equivalence problems. First, children learn a strategy for solving math problems, namely "perform all given operations on all given numbers." For example, when presented with a typical addition problem, such as "2 + 4 + 3 + 5 = __," children learn to add up all the numbers and put the total in the blank. Second, children learn a perceptual pattern related to the structure of math problems, namely the "operations = answer" structure (McNeil & Alibali, 2004; see also Alibali et al., 2009; Cobb, 1987; Seo & Ginsburg, 2003). In a typical addition problem, such as "2 + 4 + 3 + 5 = __", the numbers and operations are to the left of the equal sign and the answer blank is to the right. Third, children learn a particular concept of the equal sign. Specifically, children infer that the equal sign is an operator (like plus or minus) that tells them to calculate the total (McNeil & Alibali, 2005a; see also Baroody & Ginsburg, 1983; Behr et al., 1980; Kieran, 1981). These patterns have been referred to as *operational patterns* in prior work.

Research has shown that children often adhere to these operational patterns, even when they encounter problems for which the patterns do not apply. For example, consider the math equivalence problem "7 + 4 + 5 = 7 + __." When children are asked to solve this problem, they frequently adhere to the "perform all given operations on all given numbers" strategy and put 23 (instead of 9) in the blank (McNeil & Alibali, 2000, 2004; Rittle-Johnson, 2006). When children are asked to reconstruct the problem after viewing it for a brief period of time, they frequently adhere to the "operations = answer" problem structure and reconstruct the problem as "7 + 4 + 5 + 7 = __" (McNeil & Alibali, 2004). And finally, when children are presented with the problem and asked to define the equal sign, they frequently adhere to the operational concept of the equal sign and say that it means, "add up all the numbers" or "the total" (McNeil & Alibali, 2005a).

Children who adhere strongly to these operational patterns on a pretest benefit less from a lesson on math equivalence than do children who are incorrect in ways that do not adhere to the operational patterns (McNeil & Alibali, 2005b). Moreover, undergraduates who have their representations of the operational patterns experimentally activated are less likely than participants in control conditions to encode and solve math equivalence problems correctly (Crooks & Alibali, 2013; McNeil & Alibali, 2005b). Importantly, undergraduates who solve math equivalence problems incorrectly after having their representations of the operational patterns activated tend to make errors similar to children and add up all the numbers.

McNeil et al. (2010) extended these findings by showing that practicing basic addition facts (e.g., 3 + 4) by itself is enough to activate participants' representations of the operational patterns and, thus, interfere with understanding of math equivalence and associated algebraic concepts. Undergraduates in this series of experiments were randomly assigned to an addition practice condition in which they practiced simple addition facts (e.g., 8 + 4) or to one of several control conditions before solving either math equivalence problems under speeded conditions, or algebra problems under no time constraints. Participants were less likely to solve problems correctly—both math equivalence problems and algebra problems—after practicing addition facts than after participating in a control condition. Note that the addition practice hurt subsequent performance on the higher-level math problems even when the higher-level problems were designed to incorporate the practiced facts (e.g., 6 + 8 + 4 = 7 + __ includes the addition fact 8 + 4). This finding suggests that the poor number knowledge account (mentioned earlier) does not generalize to all problem types and does not provide the best explanation of children's difficulties with math equivalence.

McNeil (2008) used a similar "experimental activation" method with children. In this study, children's representations of the operational patterns were activated (or not) by having them look at typical arithmetic problems (or not) during regular classroom lessons on the meaning of the equal sign. Children's classroom teacher taught two brief lessons on the relational meaning of the equal sign in front of the class, and children were randomly assigned to follow along either with booklets that contained typical arithmetic problems (e.g., 15 + 13 = 28), or with booklets that did not contain typical arithmetic problems (e.g., 28 = 28). All children in the study heard the same spoken lessons on the meaning of the equal sign. The only thing that differed across children was what they were looking at during the lessons. If children's knowledge of the operational patterns interferes with their understanding of math equivalence (as predicted by the change-resistance account), then children who are looking at typical arithmetic problems should learn less from lessons on the equal sign than children who are not looking at typical arithmetic problems. Result supported this prediction, and a follow-up experiment replicated the detrimental effect of the typical arithmetic context when compared to a condition in which the equal sign was presented alone with no accompanying numerical context (McNeil, 2008). Taken together with results from the undergraduate activation experiments, these results support the change-resistance account and suggest that activation of participants' representations of the operational patterns hinders understanding and learning of math equivalence.

Several more recent studies with children have employed longer-term interventions (on the scale of weeks or months) designed to prevent children from extracting, representing, activating, and applying the overly narrow operational patterns routinely encountered in arithmetic (Byrd et al., 2015; McNeil et al., 2012; McNeil et al., 2015; McNeil et al., 2011). These studies provided further support for the change-resistance account by showing that children construct a better understanding of math equivalence after receiving arithmetic practice that is modified to be less narrow. For example, McNeil et al. (2011) randomly assigned children (ages 7–8) to practice addition facts in one of three conditions: (1) traditional format (e.g., 9 + 8 = __), (2) nontraditional format (e.g., __ = 9 + 8), or (3) no extra practice. After approximately 100 min of practice distributed over the course of 3 weekly sessions and two homework assignments, children completed measures designed to assess their understanding of math equivalence. As predicted, children in the nontraditional format condition constructed a better understanding of math equivalence than did children in the other two conditions, and these gains were not accompanied by any detectable decrements in computational fluency.

Two additional modifications to traditional arithmetic practice have since been shown to improve children's understanding of math equivalence: (1) organizing problems into practice sets based on equivalent values (e.g., 2 + 5 = __, 3 + 4 = __, 6 + 1 = __) (McNeil et al., 2012) and (2) using relational phrases, such as "is equal to" and "is the same amount as" in place of the equal sign in practice problems (Chesney, McNeil, Petersen, & Dunwiddie, 2012). McNeil et al. (2015) incorporated all three modifications into a modified arithmetic practice workbook and compared it to an analogous traditional arithmetic practice workbook in several diverse second-grade classrooms. Children within classrooms were randomly assigned to use one of the two workbooks for 15 min/day, 2 days a week, for 12 weeks. Children completed measures to assess understanding of math equivalence and computational fluency at posttest and at a 6-month follow-up test.

As predicted, children who used the modified workbooks constructed a better understanding of math equivalence than children who used the traditional workbooks. This advantage persisted for approximately 5–6 months, and there were no observed trade-offs with computational fluency. Consistent with the change-resistance account, even when children were incorrect, children in the modified condition were less likely than those in the traditional condition to rely on the operational patterns. That is, they were less likely to add up all the numbers when solving math equivalence problems (e.g., 3 + 4 + 5 = 3 + <u>15</u>) and less likely to encode the problems as if they were typical addition problems (e.g., reconstruct 3 + 4 + 5 = 3 + __ as 3 + 4 + 5 + 3 = __). Further support for the theory of change was provided in an analysis of the arithmetic practice itself. Both workbook conditions contained sections where children had to generate and write their own arithmetic problems. In these sections, children could have written something that conformed to the "operations = answer" problem format like "3 + 4 = 7" or they could have written something else (e.g., "3 + 4" by itself, "7 = 3 + 4," or something nonsensical like "3 + 4 7"). The percentage of problems that conformed to the "operations = answer" format was much higher in the traditional workbook than in the modified workbook, and a mediation analysis showed that this reliance on the "operations = answer" format mediated the effect of workbook condition on children's understanding of math equivalence.

Recently, our research team used the change-resistance account as a framework for designing a more comprehensive intervention (Byrd et al., 2015). In addition to the modified arithmetic practice (described earlier), the intervention incorporated three components that had previously been shown to reduce children's reliance on the operational patterns: (1) lessons on the equal sign in nonarithmetic contexts (e.g., 28 = 28, McNeil, 2008), (2) "concreteness fading" activities to strengthen mappings between concrete relational contexts (e.g., sharing stickers, balancing a scale) and abstract math symbols (Fyfe, McNeil, & Borjas, 2015), and (3) lessons to prompt children to compare and explain different problem formats and problem-solving strategies (Carpenter et al., 2000; Hattikudur & Alibali, 2010; Rittle-Johnson, 2006). Evidence from two design experiments and a pilot study suggests that the comprehensive intervention improves children's understanding of math equivalence. Children who received the comprehensive intervention in a one-on-one design experiment greatly reduced their reliance on the operational patterns and improved their understanding of math equivalence. In support of the change-resistance account's theory of change, children reduced their reliance on the operational patterns. Specifically, fewer children used the add-all strategy (35% vs. 58% at pretest), fewer children converted equations to the typical "operations = answer" format (13% vs. 42% at pretest), and fewer children defined the equal sign operationally (13% vs. 88% at pretest). Similar improvements were observed in a classroom-based design experiment and in a small pilot study in which classrooms were randomly assigned to the comprehensive intervention or to modified arithmetic practice alone. These results provide further support for the change-resistance account.

Despite strong support in the literature for the change-resistance account, many accounts of children's difficulties with mathematics (including those discussed earlier) tend to overlook the dominant role of children's existing knowledge in math learning difficulties (but see De Visscher & Noël, 2014a; Van Hoof et al., Chapter 5), and that makes it difficult for them to account for children's performance on math equivalence problems. As stated at the outset, there are at least three findings for which any theory needs to account: (1) performance is poor between the ages of 7 and 11, bottoms out around age 9, and improves substantially around ages 11 or 12, (2) the add-all strategy is one of the most popular strategies, and (3) newly-learned correct strategies are not widely transferred or well retained. Accounts of children's math difficulties that focus on what children lack relative to adults are only able to explain one or two of these findings. The change-resistance account can explain all three.

First, the change-resistance account predicts that children's performance on math equivalence problems should get worse over the course of the early school years before starting to improve and then improving dramatically in the later elementary and middle school years (McNeil, 2007). This is because as children progress in a traditional math curriculum, they are gaining more narrow practice with traditional arithmetic, so they are strengthening the representations of operational patterns, which are hypothesized to hinder understanding. As children move into upper elementary and middle school, they begin to see arithmetic versions of magnitude comparison problems, equivalent fractions, and then ultimately prealgebra and algebra problems (around ages 11 and 12). As a result, their representations of the operational patterns start to lose predictive power, and thus, decrease in strength, freeing children up to successfully encode and solve math equivalence problems. Second, the change-resistance account predicts that the add-all strategy will be the most popular strategy because it is the default strategy derived from children's existing knowledge of the operational patterns, such as the "operations = answer" problem format, performing all given operations on all given numbers, and interpreting the equal sign as "the total." Third, it predicts that children should hold on to their preexisting strategies and resist changing to a new strategy even after instruction because previously learned strategies are entrenched and constantly reinforced in the day-to-day instruction and practice they receive in their math classrooms, and thus, easily overpower newly learned strategies.

CONCLUSIONS AND FUTURE DIRECTIONS

Prevailing accounts of why some mathematical information is particularly difficult to learn suggest that difficulties lie in learners' deficiencies: inadequate understanding of symbols, insufficient working memory capacity, or poor fluency with basic arithmetic facts. In contrast, the change-resistance account focuses on what children have—experience and practice with the overly narrow patterns encountered in traditional arithmetic instruction. This account suggests that difficulties lie in how new information maps onto previously experienced patterns (cf. Best, McRoberts, & Goodell, 2001; Zevin & Seidenberg, 2002), and the evidence to date suggests that it provides the best explanation of children's difficulties understanding mathematical equivalence.

In terms of practical implications, the change-resistance account challenges popular traditional instructional practices in the United States. Most schools teach arithmetic for years before introducing algebraic thinking. This makes perfect sense within the context of theories suggesting that young children lack the necessary knowledge, processes, or skills for learning math equivalence. In contrast, the change-resistance account agrees with math educators who suggest that important generalizable prealgebraic concepts can and should be taught in the early grades (Blanton et al., 2015; Carraher, Schliemann, Brizuela, & Earnest, 2006). In support of this view, several studies have suggested that young children—some as early as first- and second-grade—can understand math equivalence problems when their learning environment is refocused away from traditional arithmetic practice (Carpenter et al., 2000; DeCorte & Verschaffel, 1981; Jacobs et al., 2007; Sáenz-Ludlow & Walgamuth, 1998). According to McNeil et al. (2011), children will be less likely to become entrenched in overly narrow operational patterns and more prepared for the transition to algebra if they are exposed to a variety of problem formats early in arithmetic instruction.

Not only do American schools teach arithmetic for many years, but they also tend to emphasize practice and proficiency with those arithmetic operations. Children are encouraged to memorize the basic facts and strongly discouraged from relying on finger counting. This practice makes sense according to the poor number knowledge account. If working memory resources are being consumed by the calculation (or slow retrieval) of basic facts, then there will be few resources left over for higher-level problem solving. Although this may be true for some types of advanced problems (Haverty, 1999), educators need to be cautious because many higher-level problems, such as math equivalence problems, do not map directly on to the narrow patterns of traditional arithmetic. If the curriculum consistently reinforces narrow knowledge that does not generalize broadly, then students may have difficulties learning advanced mathematics. Indeed, it has been argued that an excessive focus on traditional arithmetic contributes to American students' below international average performance in middle school mathematics (Valverde & Schmidt, 1997).

Prior experience and practice have the potential to facilitate learning and memory (e.g., Chase & Simon, 1973), but as the findings reviewed in this chapter demonstrate, they also have the potential to interfere with learning and memory (e.g., Tolman, 1948; Wiley, 1998). Specifying the conditions under which interference occurs in mathematics and other domains will inform our understanding of the process of learning and the nature of knowledge.

ACKNOWLEDGMENTS

This research was supported by Grant R305A110198 to McNeil from the US Department of Education, Institute of Education Sciences and Grant DRL-1054467 to McNeil from the National Science Foundation DGE-1313583. Hornburg was supported by a Graduate Research Fellowship from the National Science Foundation. The opinions expressed are those of the authors and do not represent views of the Institute, the US Department of Education, or NSF. We would like to thank Martha Alibali for helpful discussions of the ideas in this chapter.

REFERENCES

Adams, J. W., & Hitch, G. J. (1997). Working memory and children's mental addition. *Journal of Experimental Child Psychology*, *67*, 21–38.

Alibali, M. W. (1999). How children change their minds: strategy change can be gradual or abrupt. *Developmental Psychology*, *35*, 127–145.

Alibali, M. W., Phillips, K. M., & Fischer, A. D. (2009). Learning new problem-solving strategies leads to changes in problem representation. *Cognitive Development*, *24*, 89–101.

Anderson, J. R. (1982). Acquisition of cognitive skill. *Psychological Review*, *89*, 369–406.

Anderson, J. R. (1991). Is human cognition adaptive? *Behavioral and Brain Sciences*, *14*, 471–517.

Anderson, J. R. (2002). Spanning seven orders of magnitude: a challenge for cognitive modeling. *Cognitive Science*, *26*, 85–112.

Anderson, J. R., Reder, L. M., & Lebiere, C. (1996). Working memory: activation limitations on retrieval. *Cognitive Psychology*, *30*, 221–256.

Andersson, U., & Lyxell, B. (2007). Working memory deficit in children with mathematical difficulties: a general or specific deficit? *Journal of Experimental Child Psychology*, *96*, 197–228.

Ashcraft, M. H., & Fierman, B. A. (1982). Mental addition in third, fourth, and sixth graders. *Journal of Experimental Child Psychology*, *33*, 216–234.

Baddeley, A. (2000). The episodic buffer: a new component of working memory? *Trends in Cognitive Science*, *11*, 417–423.

Baddeley, A. D., & Hitch, G. J. (1974). Working memory. In G. Bower (Ed.), *The psychology of learning and motivation: advances in research and theory* (pp. 47–90). (Vol. 8). New York: Academic Press.

Baroody, A. J., & Gannon, K. E. (1984). The development of the commutativity principle and economical addition strategies. *Cognition and Instruction*, *1*, 321–329.

Baroody, A. J., & Ginsburg, H. P. (1983). The effects of instruction on children's understanding of the "equals" sign. *Elementary School Journal*, *84*, 199–212.

Baroody, A. J., & Ginsburg, H. P. (1986). The relationship between initial meaningful and mechanical knowledge of arithmetic. In J. Hiebert (Ed.), *Conceptual and procedural knowledge: the case of mathematics* (pp. 75–112). Hillsdale, NJ: Erlbaum.

Baroody, A. J., & Tiilikainen, S. H. (2003). Two perspectives on addition development. In A. J. Baroody, & A. Dowker (Eds.), *The development of arithmetic concepts and skills: constructive adaptive expertise* (pp. 75–125). Mahwah, NJ: Erlbaum.

Barrouillet, P., Bernardin, S., & Camos, V. (2004). Time constraints and resource sharing in adults' working memory spans. *Journal of Experimental Psychology: General*, *133*, 83–100.

Beaton, A. E., Mullis, I. V. S., Martin, M. O., Gonzalez, E. J., Kelly, D. L., & Smith, T. A. (1996). *Mathematics achievement in the middle school years: IEA's Third International Mathematics and Science Study (TIMSS)*. Chestnut Hill, MA: Boston College.

Behr, M., Erlwanger, S., & Nichols, E. (1980). How children view the equals sign. *Mathematics Teaching*, *92*, 13–15.

Best, C. T., McRoberts, G. W., & Goodell, E. (2001). Discrimination of non-native consonant contrasts varying in perceptual assimilation to the listener's native phonological system. *Journal of the Acoustical Society of America*, *109*, 775–794.

Bisanz, J., & LeFevre, J. -A. (1990). Strategic and nonstrategic processing in the development of mathematical cognition. In D. F. Bjorklund (Ed.), *Children's strategies: Contemporary views of cognitive development*. Hillsdale, NJ: Erlbaum.

Blanton, M., Stephens, A., Knuth, E., Gardiner, A. M., Isler, I., & Kim, J. -S. (2015). The development of children's algebraic thinking: the impact of a comprehensive early algebra intervention in third grade. *Journal for Research in Mathematics Education*, *46*, 39–87.

Booth, J. L., Barbieri, C., Eyer, F., & Paré-Blagoev, E. J. (2014). Persistent and pernicious errors in algebraic problem solving. *Journal of Problem Solving*, *7*, 10–23.

Brownell, W. A. (1956). Meaning and skill-maintaining the balance. *The Arithmetic Teacher*, *3*, 129–142.

Bruner, J. S. (1957). On perceptual readiness. *Psychological Review*, *2*, 123–152.

Bulf, H., Johnson, S. P., & Valenza, E. (2011). Visual statistical learning in the newborn infant. *Cognition*, *121*, 127–132.

Bull, E., & Lee, K. (2014). Executive functioning and mathematics achievement. *Child Development Perspectives*, *8*, 34–41.

Byrd, C. E., McNeil, N. M., Carrazza, C., Matthews, J. M., Brletic-Shipley, H., & Celeste, E. (2015). Pilot test of a comprehensive intervention to improve children's understanding of math equivalence. In D. Francis (Organizer) and E. Bullock (Chair), *Explorations in mathematics in the elementary grades*. Paper presented at the Annual Meeting of the American Educational Research Association (AERA), Chicago, IL.

Campbell, J. I., & Timm, J. C. (2000). Adults' strategy choices for simple addition: effects of retrieval interference. *Psychonomic Bulletin & Review*, *7*, 692–699.

Capraro, R. M., Capraro, M. M., Yetkiner, A. E., Ozel, S., Kim, H. G., & Corlu, M. S. (2010). An international comparison of Grade 6 students' understanding of the equal sign. *Psychological Reports*, *106*, 49–53.

Capraro, R. M., Capraro, M. M., Yetkiner, E. Z., Corlu, S. M., Ozel, S., Ye, S., & Kim, H. G. (2011). An international perspective between problem types in textbooks and students' understanding of relational equality. *Mediterranean Journal for Research in Mathematics Education*, *10*, 187–213.

Carnine, D. (1980). Three procedures for presenting minimally different positive and negative instances. *Journal of Educational Psychology*, *72*, 452–456.

Carpenter, T. P., Levi, L., & Farnsworth, V. (2000). *Building a foundation for learning algebra in the elementary grades*. Madison, WI: National Center for Improving Student Learning and Achievement in Mathematics Education.

Carraher, D. W., Schliemann, A. D., Brizuela, B. M., & Earnest, D. (2006). Arithmetic and algebra in early mathematics education. *Journal for Research in Mathematics Education*, *37*, 87–115.

Case, R., Kurland, M., & Goldberg, J. (1982). Operational efficiency and the growth of short-term memory span. *Journal of Experimental Child Psychology*, *33*, 386–404.

Chase, W. G., & Simon, H. A. (1973). Perception in chess. *Cognitive Psychology*, *4*, 55–81.

Chesney, D. L., McNeil, N. M., Petersen, L. A., & Dunwiddie, A. E. (2012). *Arithmetic practice that includes relational words promotes conceptual understanding and computational fluency*. Chicago, IL: Poster presented at the annual convention of the Association for Psychological Science.

Chesney, D. L., McNeil, N. M., Matthews, P. G., Byrd, C. E., Petersen, L. A., Wheeler, M. C., … Dunwiddie, A. E. (2014). Organization matters: mental organization of addition knowledge relates to understanding math equivalence in symbolic form. *Cognitive Development*, *30*, 30–46.

Cobb, P. (1987). An investigation of young children's academic arithmetic contexts. *Educational Studies in Mathematics*, *18*, 109–124.

Conway, A. R. A., Cowan, N., & Bunting, M. F. (2001). The cocktail party phenomenon revisited: the importance of working memory capacity. *Psychonomic Bulletin & Review*, *8*, 331–335.

Cook, S. W., Mitchell, Z., & Goldin-Meadow, S. (2008). Gesturing makes learning last. *Cognition*, *106*, 1047–1058.

Cowan, N. (2014). Working memory underpins cognitive development, learning, and education. *Educational Psychology Review*, *26*, 197–223.

Cowan, N., Nugent, L. D., Elliot, E. M., Ponomarev, I., & Saults, J. S. (1999). The role of attention in the development of short-term memory: age differences in the verbal span of apprehension. *Child Development*, *70*, 1082–1097.

Crooks, N. M., & Alibali, M. W. (2013). Noticing relevant problem features: activating prior knowledge affects problem solving by guiding encoding. *Frontiers in Educational Psychology*, *4*, 1–10.

De Rammelaere, S., Stuyven, E., & Vandierendonck, A. (2001). Verifying simple arithmetic sums and products: are the phonological loop and the central executive involved? *Memory and Cognition*, *29*, 267–273.

De Visscher, A., & Noël, M. -P (2014a). Arithmetic facts storage deficit: the hypersensitivity-to-interference in memory hypothesis. *Developmental Science*, *17*, 434–442.

De Visscher, A., & Noël, M. -P. (2014b). The detrimental effect of interference in multiplication facts storing: typical development and individual differences. *Journal of Experimental Psychology: General*, *143*, 2380–2400.

DeCorte, E., & Verschaffel, L. (1981). Children's solution processes in elementary arithmetic problems: analysis and improvement. *Journal of Educational Psychology*, *73*, 765–779.

DeStefano, D., & LeFevre, J. (2004). The role of working memory in mental arithmetic. *European Journal of Cognitive Psychology*, *16*, 353–386.

Ericsson, K. A., & Kintsch, W. (1995). Long-term working memory. *Psychological Review*, *102*, 211–245.

Fiser, J., & Aslin, R. N. (2002). Statistical learning of higher-order temporal structure from visual shape-sequences. *Journal of Experimental Psychology: Learning, Memory, and Cognition*, *28*, 458–467.

Frensch, P. A., & Geary, D. C. (1993). The effects of practice on component processes in complex mental addition. *Journal of Experimental Psychology: Learning, Memory, and Cognition*, *19*, 433–456.

Friso-van den Bos, I., van der Ven, S. H. G., Kroesbergen, E. H., & van Luit, J. E. H. (2013). Working memory and mathematics in primary school children: a meta-analysis. *Educational Research Review*, *10*, 29–44.

Fürst, A. J., & Hitch, G. J. (2000). Separate roles for executive and phonological components of working memory in mental arithmetic. *Memory and Cognition*, *28*, 774–782.

Fuson, K. C., Carroll, W. M., & Landis, J. (1996). Levels in conceptualizing and solving addition and subtraction compare word problems. *Cognition and Instruction*, *14*(3), 345–371.

Fyfe, E. R., McNeil, N. M., & Borjas, S. (2015). Benefits of "concreteness fading" for children's mathematics understanding. *Learning and Instruction*, *35*, 104–120.

Gathercole, S. E., & Pickering, S. J. (2000). Working memory deficits in children with low achievements in the national curriculum at 7 years of age. *British Journal of Educational Psychology*, *70*, 177–194.

Geary, D. C., & Burlingham-Dubree, M. (1989). External validation of the strategy choice model for addition. *Journal of Experimental Child Psychology*, *47*, 175–192.

Geary, D. C., & Widaman, K. F. (1992). Numerical cognition: on the convergence of componential and psychometric models. *Intelligence*, *16*, 47–80.

Geary, D. C., Bow-Thomas, C. C., Liu, F., & Siegler, R. S. (1996). Development of arithmetical competencies in Chinese and American children: Influence of age, language, and schooling. *Child Development*, *67*(5), 2022–2044.

Geary, D. C., Hoard, M. K., Nugent, L., & Bailey, D. H. (2012a). Mathematical cognition deficits in children with learning disabilities and persistent low achievement: a five-year prospective study. *Journal of Educational Psychology*, *104*, 206–223.

Geary, D. C., Hoard, M. K., & Nugent, L. (2012b). Independent contributions of the central executive, intelligence, and in-class attentive behavior to developmental change in the strategies used to solve addition problems. *Journal of Experimental Child Psychology*, *113*, 49–65.

Gomez, R. L. (2002). Variability and detection of invariant structure. *Psychological Science*, *13*, 431–436.

Gray, W. D., & Fu, W. -T. (2001). Ignoring perfect knowledge in-the-world for imperfect knowledge in-the-head: implications of rational analysis for interface design. *CHI Letters*, *3*, 112–119.

Groen, G. J., & Parkman, J. M. (1972). A chronometric analysis of simple addition. *Psychological Review*, *79*, 329–343.

Halford, G. S., Cowan, N., & Andrews, G. (2007). Separating cognitive capacity from knowledge: a new hypothesis. *Trends in Cognitive Sciences*, *11*, 236–242.

Hattikudur, S., & Alibali, M. W. (2010). Learning about the equal sign: does comparing with inequality symbols help? *Journal of Experimental Child Psychology*, *107*, 15–30.

Haverty, L. A. (1999). *The importance of basic number knowledge to advanced mathematical problem solving*. Pittsburgh, PA: Carnegie Mellon University (Unpublished Doctoral Dissertation).

Hiebert, J., Stigler, J. W., Jacobs, J. K., Givvin, K. B., Garnier, H., Smith, M., ... & Gallimore, R. (2005). Mathematics teaching in the United States today (and tomorrow): Results from the TIMSS 1999 video study. *Educational Evaluation and Policy Analysis, 27*, 111–132.

Hitch, G. J. (1978). The role of short-term working memory in mental arithmetic. *Cognitive Psychology, 10*, 302–323.

Houdé, O., Rossi, S., Lubin, A., & Joliot, M. (2010). Mapping numerical processing, reading, and executive functions in the developing brain: an fMRI meta-analysis of 52 studies including 842 children. *Developmental Science, 13*, 876–885.

Hulme, C., Thomson, N., Muir, C., & Lawrence, A. (1984). Speech rate and the development of short-term memory span. *Journal of Experimental Child Psychology, 38*, 241–253.

Imbo, I., & Vandierendonck, A. (2007a). The development of strategy use in elementary school children: working memory and individual differences. *Journal of Experimental Child Psychology, 96*, 284–309.

Imbo, I., & Vandierendonck, A. (2007b). Effects of problem size, operation, and working-memory span on simple-arithmetic strategies: differences between children and adults? *Psychological Research, 72*, 331–346.

Imbo, I., Vandierendonck, A., & De Rammelaere, S. (2007). The role of working memory in the carry operation of mental arithmetic: number and value of the carry. *The Quarterly Journal of Experimental Psychology, 60*, 708–731.

Jacobs, V. R., Franke, M. L., Carpenter, T. P., Levi, L., & Battey, D. (2007). Professional development focused on children's algebraic reasoning in elementary school. *Journal for Research in Mathematics Education, 38*, 258–288.

Kail, R., & Park, Y. -S. (1994). Processing time, articulation time, and memory span. *Journal of Experimental Child Psychology, 57*, 281–291.

Kail, R., & Salthouse, T. A. (1994). Processing speed as a mental capacity. *Acta Psychologica, 86*, 199–225.

Kaye, D. B. (1986). The development of mathematical cognition. *Cognitive Development, 1*, 157–170.

Kieran, C. (1981). Concepts associated with the equality symbol. *Educational Studies in Mathematics, 12*, 317–326.

Knuth, E. J., Alibali, M. W., McNeil, N. M., Weinberg, A., & Stephens, A. S. (2005). Middle school students' understanding of core algebraic concepts: equality & variable. *Zentralblatt für Didaktik der Mathematik [International Reviews on Mathematical Education], 37*, 68–76.

Kotovsky, K., Hayes, J. R., & Simon, H. A. (1985). Why are some problems hard? Evidence from Tower of Hanoi. *Cognitive Psychology, 17*, 248–294.

Kuhl, P. K. (2000). A new view of language acquisition. *Proceedings of the National Academy of Science, 97*, 11850–11857.

Lewis-Peacock, J. A., & Postle, B. R. (2008). Temporary activation of long-term memory supports working memory. *Journal of Neuroscience, 28*, 8765–8771.

Li, Y., & Geary, D. C. (2013). Developmental gains in visuospatial memory predict gains in mathematics achievement. *PLoS ONE, 8*, e70160.

Li, P., Zhao, X., & Mac Whinney, B. (2007). Dynamic self-organization and early lexical development in children. *Cognitive Science, 31*, 581–612.

Logie, R. H., Gilhooly, K. J., & Wynn, V. (1994). Counting on working memory in arithmetic problem solving. *Memory and Cognition, 22*, 395–410.

MacDonald, M. C., & Christiansen, M. H. (2002). Reassessing working memory: comment on Just and Carpenter (1992) and Waters and Caplan (1996). *Psychological Review, 109*, 35–54.

McKenzie, B., Bull, R., & Gray, C. (2003). The effects of phonological and visual-spatial interference on children's arithmetical performance. *Educational and Child Psychology, 20*, 93–108.

McLean, J. F., & Hitch, G. J. (1999). Working memory impairments in children with specific arithmetic learning difficulties. *Journal of Experimental Child Psychology*, *74*, 240–260.

McNeil, N. M. (2007). U-shaped development in math: 7-year-olds outperform 9-year-olds on equivalence problems. *Developmental Psychology*, *43*, 687–695.

McNeil, N. M. (2008). Limitations to teaching children 2 + 2 = 4: typical arithmetic problems can hinder learning of mathematical equivalence. *Child Development*, *79*, 1524–1537.

McNeil, N. M. (2014). A change–resistance account of children's difficulties understanding mathematical equivalence. *Child Development Perspectives*, *8*, 42–47.

McNeil, N. M., & Alibali, M. W. (2000). Learning mathematics from procedural instruction: externally imposed goals influence what is learned. *Journal of Educational Psychology*, *92*, 734–744.

McNeil, N. M., & Alibali, M. W. (2002). A strong schema can interfere with learning: The case of children's typical addition schema. In C.D. Schunn, W. Gray, (Eds.), *Proceedings of the 24th Annual Conference of the Cognitive Science Society*. Mahwah, NJ: Erlbaum.

McNeil, N. M., & Alibali, M. W. (2004). You'll see what you mean: students encode equations based on their knowledge of arithmetic. *Cognitive Science*, *28*, 451–466.

McNeil, N. M., & Alibali, M. W. (2005a). Knowledge change as a function of mathematics experience: all contexts are not created equal. *Journal of Cognition and Development*, *6*, 285–306.

McNeil, N. M., & Alibali, M. W. (2005b). Why won't you change your mind? Knowledge of operational patterns hinders learning and performance on equations. *Child Development*, *76*, 883–899.

McNeil, N. M., Grandau, L., Knuth, E. J., Alibali, M. W., Stephens, A. C., Hattikudur, S., & Krill, D. E. (2006). Middle-school students' understanding of the equal sign: the books they read can't help. *Cognition and Instruction*, *24*, 367–385.

McNeil, N. M., Rittle-Johnson, B., Hattikudur, S., & Petersen, L. A. (2010a). Continuity in representation between children and adults: arithmetic knowledge hinders undergraduates' algebraic problem solving. *Journal of Cognition and Development*, *11*, 437–457.

McNeil, N. M., Weinberg, A., Stephens, A. C., Hattikudur, S., Asquith, P., Knuth, E. J., & Alibali, M. W. (2010b). A is for apple: mnemonic symbols hinder students' interpretation of algebraic expressions. *Journal of Educational Psychology*, *102*, 625–634.

McNeil, N. M., Fyfe, E. R., Petersen, L. A., Dunwiddie, A. E., & Brletic-Shipley, H. (2011). Benefits of practicing 4 = 2 + 2: nontraditional problem formats facilitate children's understanding of mathematical equivalence. *Child Development*, *82*, 1620–1633.

McNeil, N. M., Chesney, D. L., Matthews, P. G., Fyfe, E. R., Petersen, L. A., Dunwiddie, A. E., & Wheeler, M. C. (2012). It pays to be organized: organizing arithmetic practice around equivalent values facilitates understanding of math equivalence. *Journal of Educational Psychology*, *104*, 1109–1121.

McNeil, N. M., Fyfe, E. R., & Dunwiddie, A. E. (2015). Arithmetic practice can be modified to promote understanding of mathematical equivalence. *Journal of Educational Psychology*, *107*, 423–436.

Melby-Lervåg, M., & Hulme, C. (2013). Is working memory training effective? A meta-analytic review. *Developmental Psychology*, *49*, 270–291.

Meyer, M. L., Salinpoor, V. N., Wu, S. S., Geary, D. C., & Menon, V. (2010). Differential contribution of specific working memory components to mathematics achievement in 2nd and 3rd graders. *Learning and Individual Differences*, *20*, 101–109.

Mickey, K. W. & McClelland, J. L. (2014). A neural network model of learning mathematical equivalence. In P. Bello, M. Guarini, M. McShane, B. Scassellati, (Eds.), *Proceedings of the 36th Annual Conference of the Cognitive Science Society* (pp. 1012–1017). Austin, TX: Cognitive Science Society.

Miller, K. F., & Paredes, D. R. (1990). Starting to add worse: effects of learning to multiply on children's addition. *Cognition*, *37*, 213–242.

Mix, K. S., Levine, S. C., & Huttenlocher, J. (1999). Early fraction calculation ability. *Developmental Psychology*, *35*, 164–174.

Munakata, Y. (1998). Infant perseveration and implications for object permanence theories: a PDP model of the A-not-B task. *Developmental Science*, *1*, 161–184.

National Council of Teachers of Mathematics. (1989). *Curriculum and evaluation standards for school mathematics*. Reston, VA: NCTM.

National Science Board (2002). *Science and engineering indicators*. Arlington, VA: National Science Foundation.

Pascual-Leone, J. A. (1970). A mathematical model for the transition rule in Piaget's developmental stages. *Acta Psychologica*, *32*, 301–345.

Perry, M. (1991). Learning and transfer: instructional conditions and conceptual change. *Cognitive Development*, *6*, 449–468.

Perry, M., Church, R. B., & Goldin-Meadow, S. (1988). Transitional knowledge in the acquisition of concepts. *Cognitive Development*, *3*, 359–400.

Piaget, J., & Szeminska, A. (1995). The child's conception of number (C. Gattegno & F. M. Hadgson, Trans.). In H. E. Gruber, J. J. Voneche, (Eds.), *The essential Piaget* (pp. 298–341). Northvale, NJ: Jason Aronson (Original work published 1941).

Porter, A. (1989). A curriculum out of balance: the case of elementary school mathematics. *Educational Researcher*, *18*, 9–15.

Powell, S. R. (2012). Equations and the equal sign in elementary mathematics textbooks. *The Elementary School Journal*, *112*, 627–648.

Qin, S., Cho, S., Chen, T., Rosenberg-Lee, M., Geary, D. C., & Menon, V. (2014). Hippocampal-neocortical functional reorganization underlies children's cognitive development. *Nature Neuroscience*, *17*, 1263–1269.

Raghubar, K. P., Barnes, M. A., & Hecht, S. A. (2010). Working memory and mathematics: a review of developmental, individual difference, and cognitive approaches. *Learning and Individual Differences*, *20*, 110–122.

Rassmusen, C., & Bisanz, J. (2005). Representation and working memory in early arithmetic. *Journal of Experimental Child Psychology*, *91*, 137–157.

Redick, T. S., Shipstead, Z., Wiemers, E. A., Melby-Lervåg, M., & Hulme, C. (2015). What's working in working memory training? An educational perspective. *Educational Psychology Review*, *27*, 617–633.

Renwick, E. (1932). Children's misconceptions concerning the symbols for mathematical equality. *British Journal of Educational Psychology*, *2*, 173–183.

Rittle-Johnson, B. (2006). Promoting transfer: effects of self-explanation and direct instruction. *Child Development*, *77*, 1–15.

Rittle-Johnson, B., & Alibali, M. W. (1999). Conceptual and procedural knowledge of mathematics: does one lead to the other? *Journal of Educational Psychology*, *91*, 175–189.

Rumelhart, D. E. (1980). Schemata: The building blocks of cognition. In R. J. Spiro, B. C. Bruce, & W. F. Brewer (Eds.), *Theoretical issues in reading comprehension* (pp. 33–58). Hillsdale, NJ: Erlbaum.

Sáenz-Ludlow, A., & Walgamuth, C. (1998). Third graders' interpretations of equality and the equal symbol. *Educational Studies in Mathematics*, *35*, 153–187.

Saffran, J. R. (2003). Statistical language learning: mechanisms and constraints. *Current Directions in Psychological Science*, *12*, 110–114.

Salthouse, T. A. (1991). Expertise as the circumvention of human processing limitations. In K. A. Ericsson, & J. Smith (Eds.), *Toward a general theory of expertise*. Cambridge, MA: Cambridge University Press.

Seo, K. -H., & Ginsburg, H. P. (2003). You've got to carefully read the math sentence…": Classroom context and children's interpretations of the equals sign. In A. J. Baroody, & A. Dowker (Eds.), *The development of arithmetic concepts and skills: Constructing adaptive expertise* (pp. 161–187). Mahwah, NJ: Erlbaum.

Sfard, A., & Linchevski, L. (1994). The gains and pitfalls of reification—the case of algebra. *Educational Studies in Mathematics*, *26*, 191–228.

Sherman, J., & Bisanz, J. (2007). Evidence for use of mathematical inversion by three-year-old children. *Journal of Cognition and Development*, *8*, 333–344.

Sherman, J., & Bisanz, J. (2009). Equivalence in symbolic and non-symbolic contexts: benefits of solving problems with manipulatives. *Journal of Educational Psychology*, *101*, 88–100.

Shrager, J., & Siegler, R. S. (1998). SCADS: a model of children's strategy choices and strategy discoveries. *Psychological Science*, *9*, 405–410.

Siegler, R. S. (1987). The perils of averaging data over strategies: an example from children's addition. *Journal of Experimental Psychology: General*, *116*, 250–264.

Siegler, R. S. (1989). Mechanisms of cognitive development. *Annual Review of Psychology*, *40*, 353–379.

Siegler, R. S. (1999). Strategic development. *Trends in Cognitive Science*, *3*, 430–435.

Siegler, R. S., & Araya, R. (2005). A computational model of conscious and unconscious strategy discovery. In R. V. Kail (Ed.), *Advances in child development and behavior* (pp. 1–42). (Vol. 33). Oxford, England: Elsevier.

Siegler, R. S., & Jenkins, E. (2014). *How children discover new strategies*. New York, NY: Psychology Press.

Siegler, R. S., & Shrager, J. (1984). Strategy choices in addition: how do children know what to do? In C. Sophian (Ed.), *Origins of cognitive skills* (pp. 229–293). Hillsdale, NJ: Erlbaum.

Siegler, R. S., & Stern, E. (1998). Conscious and unconscious strategy discoveries: a microgenetic analysis. *Journal of Experimental Psychology: General*, *127*, 377–397.

Sophian, C., & Wood, A. (1997). Proportional reasoning in young children: the parts and the whole of it. *Journal of Educational Psychology*, *89*, 309–317.

Swanson, H. L., & Jerman, O. (2006). Math disabilities: a selective meta-analysis of the literature. *Review of Educational Research*, *76*, 249–274.

Thelen, E., & Smith, L. B. (1994). *A dynamic systems approach to the development of cognition and action*. Cambridge, MA: MIT Press.

Tolman, E. C. (1948). Cognitive maps in rats and men. *Psychological Review*, *55*, 189–208.

Valverde, G. A., & Schmidt, W. H. (1997). Refocusing US math and science education. *Issues in Science and Technology*, *14*, 60–67.

Vlassis, J. (2004). Making sense of the minus sign or becoming flexible in 'negativity'. *Learning and Instruction*, *14*, 469–484.

Watchorn, R., Lai, M., & Bisanz, J. (2009). *Failure on equivalence problems is not universal.* Paper presented at the meeting of the Cognitive Development Society, San Antonio, TX.

Wearne, D., & Hiebert, J. (1989). Cognitive changes during conceptually based instruction on decimal fractions. *Journal of Educational Psychology*, *81*, 507–513.

Wiley, J. (1998). Expertise as mental set: the effects of domain knowledge in creative problem solving. *Memory and Cognition*, *26*, 716–730.

Zevin, J. D., & Seidenberg, M. S. (2002). Age of acquisition effects in reading and other tasks. *Journal of Memory and Language*, *47*, 1–29.

Chapter 9

Learning and Solving More Complex Problems: The Roles of Working Memory, Updating, and Prior Skills for General Mathematical Achievement and Algebra

Kerry Lee, Swee Fong Ng, Rebecca Bull
National Institute of Education, Singapore, Singapore

INTRODUCTION

As children advance through the school years, they encounter mathematical problems that become increasingly complex. In the early primary years, arithmetic begins with addition, the teaching of which is often aided by the use of concrete exemplars. This is soon joined by subtraction, followed by multiplication and division. Difficulties and complexity are further increased when operands increase in size and different number systems (e.g., fractions) are introduced. Given the hierarchical nature of many mathematics curricula, by the end of high school, children would have experienced many such increases in conceptual or procedural complexity. In this chapter, we examine some of the factors that contribute to individual differences in mathematical performance as children advance through the school years and are asked to learn and solve more complex mathematical problems.

Prior achievement is generally one of the better predictors of later achievement. This is the case for mathematics in particular (Duncan et al., 2007), and great efforts are made to ensure that children have the basic numeracy and arithmetic skills needed for learning more complex mathematics. Some mathematical curricula go further and take steps to make explicit the relations between different domains of mathematics. In Singapore, for example, the curriculum is often

Acquisition of Complex Arithmetic Skills and Higher-Order Mathematics Concepts
http://dx.doi.org/10.1016/B978-0-12-805086-6.00009-6

described as having a spiral structure in which material taught earlier are revised and used again in the introduction of interconnected, but more complex material.

Although the importance of basic numeracy is not in dispute, questions have been raised concerning the extent to which some domains of mathematics are dependent on skills that are purportedly foundational (Bailey, Watts, Littlefield, & Geary, 2014). Algebra, for example, has long been assumed to be a more complex domain of mathematics than arithmetic and is generally not introduced to students until the secondary or high school years. However, recent works suggests that prealgebra or algebraic reasoning (e.g., in the form of mathematical pattern tasks) should be introduced much earlier to help ameliorate difficulties encountered by students, who are introduced to algebra only in secondary school (Kaput, Carraher, & Blanton, 2008). Indeed, mathematical pattern tasks have long been part of the Algebra Standards (NCTM, 2000). In contrast, the National Mathematics Advisory Panel (2008) argued that most countries that performed well in international comparisons of mathematics achievement do not devote a large amount of curriculum time to such tasks, and called for a relative reduction in attention to them.

In this chapter, we examine the extent to which successes with learning and solving more complex mathematics problems are dependent on both earlier achievement in mathematics and domain-general cognitive capabilities. We will first describe the theoretical rationale for expecting linkages. Because patterns tasks have not been examined widely in the educational psychology literature, we describe them in some detail, from both the mathematics education and cognitive developmental perspectives.

We adopted both a wider lens that examined the development of mathematical skills in general and a narrower lens that focused on the development of algebraic skills in particular. For algebraic skills, we examined the extent to which performance on algebraic word problems was associated with performance on relational tasks and arithmetic word problems. Regarding the domain-general cognitive capabilities, we focused on working memory (WM) capacity and the capacity of updating, a closely related executive function.

Our observations are based on a four-year longitudinal study that examined the development of executive functioning (Lee, Bull, & Ho, 2013) and their associations with children's developing mathematical skills. The study employed a cross-sequential design and involved children from four overlapping cohorts with starting grades that ranged from kindergarten to 6th grade. Here, we report findings on (1) the extent to which skill at solving more complex mathematical problems were predicted by skill at solving easier problems, both measured by a standardized achievement test, (2) the development of competencies in mathematical relational tasks and arithmetic word problems: domains that are believed to be closely related to and predictive of performance in algebra, and (3) the relations between children's developing domain-specific and domain-general capabilities and their performance on tests of general mathematics achievement and algebraic word problems.

ALGEBRA AND EARLIER MATHEMATICS SKILLS

Algebra is often regarded as a gateway to learning higher mathematics (National Mathematics Advisory Panel, 2008). The Algebra Standards specified by the NCTM (2000) identified four components of algebraic thinking. To think algebraically, students have to be able to understand and make use of mathematical relations, specifically (1) understand patterns, relations, and functions; (2) represent and analyze mathematical situations and structures using algebraic symbols; (3) use mathematical models to represent and understand quantitative relationships; and (4) analyze change in various contexts. Thus, algebra is more than just symbolic manipulation and solving for x and y; it is a way of thinking that forms the bedrock of mathematics. To design effective and well-founded learning approaches that cater to children with different abilities requires pedagogical content knowledge as well as a sound understanding of the cognitive demands of algebra.

A large number of studies in the mathematics literature have examined issues related to the teaching and learning of algebra (e.g., Bush & Karp, 2013; Cai et al., 2005; Kieran, 1997; Sfard, 1995; Stacey & MacGregor, 1999). In addition to components of algebraic thinking, learning algebra entails learning the meaning of letters and the conventions associated with the use of letters. Learning the symbolic conventions of algebra also involves learning to inhibit conventions that are acceptable in arithmetic (e.g., $2 + 3/4 = 2\ 3/4$), but not acceptable in algebra (e.g., $a + b$ cannot be concatenated) (see also Van Hoof et al., Chapter 5). It also requires the learning of new knowledge regarding the meanings of algebraic expressions and equations, including the structural meanings of the equal sign (i.e., supplementing the procedural meanings of the equal sign $3 + 4 = 7$ with the structural property of the equal sign $3 + 4 = 2 + 5$, which denotes equivalence) (see McNeil et al., Chapter 8).

In a study comparing the primary mathematics curriculum of five countries, Cai et al. (2005) showed that the Singapore primary mathematics curriculum, like those of China and South Korea, placed a heavy emphasis on the first three components specified in the Algebra Standard. The report of NCTM (2000) suggests such efforts should begin at the preschool level and continue into the secondary years. However, as Carraher, Schliemann, and Schwartz (2007) argued, an important point to note is that early algebra does not mean introducing algebra earlier. In Singapore, algebraic word problems identical in mathematical structure to those typically taught to secondary schools students are introduced early, in Grade 4. However, instead of formal letter-symbolic algebra, students are taught to use a schematic representation, known locally as the model method, to represent quantitative relationships depicted in the word problems (Ng & Lee, 2009). Solution is attained using heuristics that are arithmetical in nature. The purpose of introducing these problems early is to integrate algebraic thinking developmentally into the primary mathematics curriculum and to use it as a bridge between arithmetic thinking and formal letter-symbolic algebra.

Mathematical relational tasks, such as Number Sequences, Function Machine, and Function types tasks are used in this context to provide experiences of algebraic thinking (Kieran, 2004), which include analyzing relationships between quantities, noticing structure, studying change, and generalizing.

Relational Tasks

Pattern type tasks, geometric or number sequences, are a useful means to engage learners to look for regularities. Geometric core repeating patterns (e.g., □ ◊ ○ △ □ ◊ ○ △ □ ? ○ △) can be used to assess whether younger learners, who have yet to acquire number facts and means to operate on numbers, can identify structure inherent in a given sequence of shapes. Number sequence tasks are more demanding as they require knowledge of numbers, counting, and arithmetic operations. Given a sequence of even numbers 2, 4, 6, 8, ?, 12, 14, 16, for example, identification of the recursive rule—adding 2 to the preceding number—will provide the answer to the missing number. However, at a deeper level, if this sequence of even numbers is continued, there is another level of regularity. As the number sequence grows, the digit in the ones place (i.e., 0, 2, 4, 6, 8) returns with clockwork regularity. An awareness of this regularity will facilitate an understanding of the structure of not only the number sequence but the general base-10 structure of numbers: 0, 2, 4, 6, 8, 10, 12, 14, 16, 18 ... 30, 32, 34, 36, 38, 40, 42. At a more advanced level, with the introduction of letters at the secondary level, and the introduction of algebra as generalized arithmetic (Usiskin, 1988), the general formula, $T_n = T_{n-1} + d$, can be used to represent the *n*th term of the sequence where letters, *T*, are used as givens and as unknowns (Harper, 1987). Here, *d* refers to the absolute or functional difference between two consecutive terms.

Function tasks are a variation of Number Sequence tasks. Fig. 9.1 shows an example of how a core repeating geometric pattern varies with the growing number sequence. Guess and check strategy can be used to answer the questions. However, exploring the deeper structure underpinning such tasks brings new awareness of the relations between the numbers, shapes and the remainder when the numbers are divided by 3. This is demonstrated in Fig. 9.2. The remainders of the numbers return with clockwork regularity when the shapes

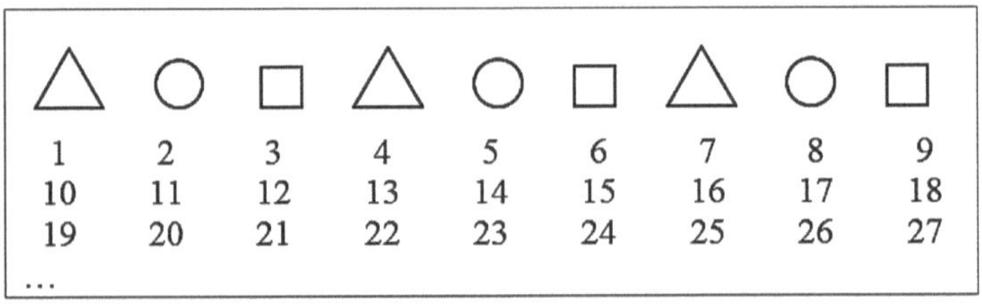

FIGURE 9.1 **Function task showing how numbers grow with a core repeating geometric pattern.**

Shapes	△	○	□	△	○	□	△	○	□	△	○	□	△	○	□
Number sequence	1	2	3	4	5	6	7	8	9	10	11	12	13	14	15
Remainder when divided by 3	1	2	0	1	2	0	1	2	0	1	2	0	1	2	0

FIGURE 9.2 The relations between shapes and numbers in the number sequence and their corresponding remainders when divided by 3. In reference to the question on which shape is above the number 96, 96 occurs at the 32nd repetition of the geometric sequence (96/3 = 32, remainder = 0 or third shape: square).

and the corresponding numbers are written in one row. Such function tasks are precursors to periodic functions expressed formally as $f(x) = f(x + a)$, $x \geq 0$, $a \in W$.

Function is the study of relationships. Authors of the Comprehensive School Mathematics Program (1975) argued that functions should be used as a vehicle through which letters as variables and algebra are introduced. When functions are first introduced at the lower grades the focus is on the covariation between the inputs and the outputs and the rule underpinning a particular set of inputs and outputs. The examples at the top of Fig. 9.3 show how a 1st grade child expressed the covariation between the inputs and the outputs. With the introduction of letters as variables at 6th grade, it is expected that students at this level

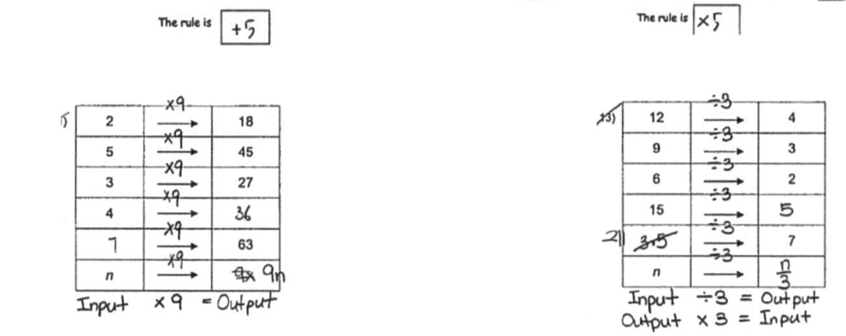

FIGURE 9.3 In the Function Machine task, children have to identify the numerical rules relating the input and output numbers. Children from higher grades were asked to use letters to state the rule relating the inputs with the outputs.

uses letters to state the rule underpinning the covariation between the inputs and the outputs and also to show how the expression, input ÷ 3 = output, could be further expressed as $n \div 3$. It could be said that such Function Machine tasks are instances of "seeing a generality through the particular and seeing the particular in the general" (Mason, 1996, p. 65). That is, (4, 9) is a specific example of the general rule $n + 5$.

Although there is a wealth of information on children's learning and use of mathematical patterns (e.g, Nathan & Kim, 2007; Orton & Orton, 1999; Rittle-Johnson, Fyfe, McLean, & McEldoon, 2013; Walkowiak, 2014), there is a dearth of empirical studies on the relation between individual differences in mathematical relational and algebraic skills. In a report of early findings from this study (Lee et al., 2012), we found performances on mathematics relational tasks (Function Machine and Number Sequence) at 5th grade predicted success with algebraic word problems at 6th grade. Individual differences on the relational tasks were associated with the children's WM and updating capacities. Of particular interest was that associations between WM, updating, and algebraic performances were fully mediated by performances on the relational tasks.

Algebra and Arithmetic

In the mathematics literature, algebra is sometimes regarded as generalized arithmetic. Nonetheless, the progression from arithmetic to algebra is surprisingly problematic and multifaceted (Stacey & Chick, 2004). Previous studies argued that success in algebra is dependent upon foundational knowledge that stems from early childhood and includes competency in arithmetic (English & Warren, 1998; Lannin, Barker, & Townsend, 2006; Orton & Orton, 1999). However, knowledge of how algebra problems can be solved using arithmetic methods can also interfere with the learning of algebraic methods (Khng & Lee, 2009).

Take, for example, two questions that are typical of standard arithmetic and algebraic word problems used in the 4th grade curriculum in Singapore.

Item 1: Men and women attended a parade. There are 500 men. There are 400 more women than men. How many people were there altogether?

Item 2: There were 1400 men and women at a parade. There were 400 more women than men. How many were women?

The first question requires the generation of an arithmetic equation $500 + (500 + 400)$. There is closure to the process as there is a unique answer of 1400. The question takes on a different complexion when it is posed with total number of attendants as a given. Solving this problem requires a shift from performing arithmetic operations to setting up an equation involving the unknown. In algebra, any unknown can be used to generate an appropriate algebraic expression. If x represents the number of men, then the algebraic expression $(x + 400)$

represents the number of women and the equation $x + (x + 400) = 1400$ specifies the total number of participants.

The solution to algebraic problems can be found by first generating appropriate algebraic expressions followed by algebraic equations. Generating such algebraic objects "is one of the key features that makes algebra different from arithmetic" (Lesh, Post, & Behr, 1987, p. 657). Solving the unknown, x, requires a transformational activity: rewriting the original equation in various equivalent forms: first as $x + (x + 400) - 400 = 1400 - 400$, and then in its equivalent form, $2x = 1000$. The arithmetic expression $500 + (500 + 400)$ differs significantly from the algebraic expression $x + (x + 400) = 1400$ in that there is closure in the former and a lack of it in the latter (Collis, 1975). Only when the letter x is treated as a specific unknown (Küchemann, 1981), can the expression $2x + 400 = 1400$ be solved. Persisting with an arithmetic approach in solving such problems will not result in any solution.

Thus, though arithmetic word problems may provide experiences with arithmetic computation and extracting quantitative information from a text based format, competencies in arithmetic alone are not likely to be sufficient for success in algebra. Kieran (1996) proposed that although much of secondary school algebra focused on the generational and transformational activities aforementioned, other mathematical activities are also necessary to support the learning of algebra. These activities include modeling, noticing structure, studying change, generalizing, and analyzing relationships, justifying, proving, and predicting, and the provision of supporting bridges between arithmetic and algebra.

Arithmetic and Algebraic Word Problems in the Singapore Curriculum

In addition to relational tasks, in Singapore, algebraic word problems are introduced in 4th grade, both to introduce children to algebraic thinking and to serve as a supporting bridge between arithmetic and algebra. The model method, described earlier as a heuristic to support primary students in solving algebraic word problem, is actually introduced even earlier, in 2nd grade. In the arithmetic context, rectangles represent specific numerical quantities and the part-whole nature of numbers, a concept that is fundamental to arithmetic and algebra (Davydov, 1962; Freudenthal, 1974). Fig. 9.4 depicts a model used to solve the questions titled Item 1 (previously).

In 4th grade, the model method is used to model information in arithmetic word problems and also simple algebraic word problems. The model method remains a heuristic of choice for algebraic word problems through the end of 6th grade. The affordance of the model method is such that it can be used to solve algebra word problems that normally require the construction of linear equations with up to two unknowns. When the model method is used to solve algebraic word problems, rectangles are used to represent numerical quantities

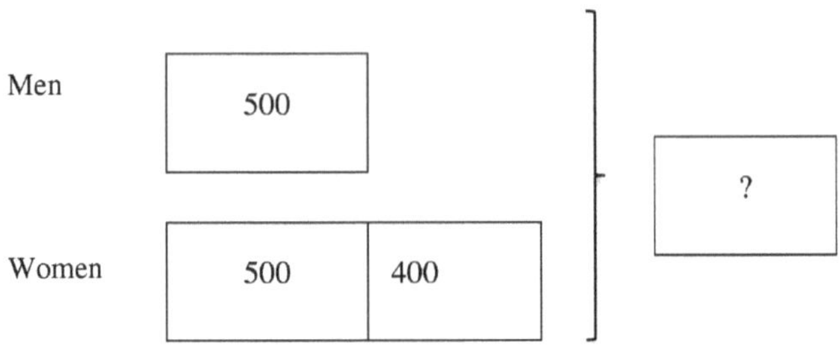

Total number of people = 500 + (500 + 400) = 1400

FIGURE 9.4 The model method used for solving Item 1, an arithmetic word problem.

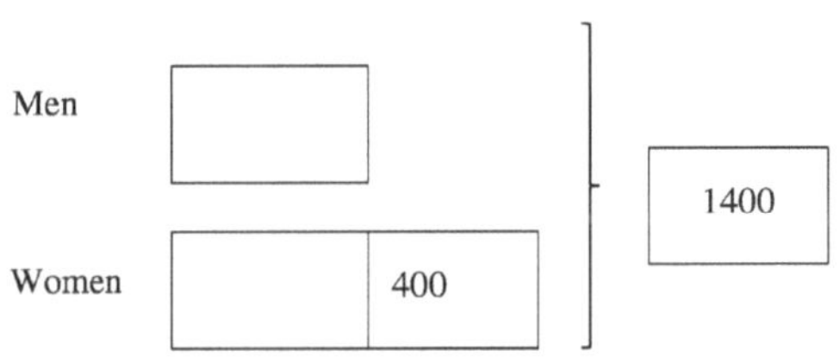

One rectangle is one unit.
2 units + 400 = 1400
2 units = 1400 – 400
2 units = 1000
1 unit = 500
There are 900 women.

Let m represent the number of men and w the number of women.

$w = m + 400$
$m + w = 1400$
$m + (m + 400) = 1400$
$2m + 400 = 1400$
$2m + 400 - 400 = 1400 - 400$
$2m = 1000$
$m = 500$
$w = 500 + 400 = 900$

There are 900 women.

FIGURE 9.5 The upper panel depicts the model method used for solving an algebraic word problem, Item 2. The lower panel shows the same question solved using letter-symbolic algebra.

as well as unknown values. Fig. 9.5 (see upper panel) depicts a model drawing used to solve Item 2 (previously). The solution in the lower panel of Fig. 9.5 shows how letter-symbolic algebra can be used to solve the same problem.

There is evidence to suggest that some 5th grade children already appreciate the dual nature of the rectangles (Ng & Lee, 2008). That is, that the rectangles can be used to depict specific and varying quantities. However, the way in which children in 5th grade drew the rectangles across questions with different operands suggested that they were still using the rectangles to depict specific numerical quantities, just as they did with arithmetic questions (Lee, Khng, Ng, & Ng Lan Kong, 2013).

In secondary schools, letters are used to replace the role played by the rectangles. Pedagogically, secondary school teachers are encouraged to demonstrate to students the dual role of the rectangles. The transition from the model method to letter-symbolic algebra is further supported by using algebraic problems that are similar in both structure and complexity. The same kinds of algebraic questions are used in primary and secondary schools, but in secondary schools, students are taught to use letter-symbolic algebra to solve such problems.

Of interest is the extent to which early experiences with algebraic word problems prepare students for letter-symbolic algebra. On the one hand, early experiences with these problems should give students opportunities to learn about the different types of quantitative relations depicted in these problems. On the other hand, learning to use algebra as a tool still requires learning the meaning of letters, the conventions associated with the use of letters and the structural relationships inherent in algebraic problems. Furthermore, having well-established routines may be detrimental in that learning to use letter-symbolic algebra requires updating information on conventions that are acceptable in arithmetic, but not acceptable in algebra (see also Van Hoof et al., Chapter 5). If this is the case, students' domain-general abilities may also play a role in their success in negotiating the transition from the model method to letter-symbolic algebra.

GENERAL MATHEMATICS ACHIEVEMENT, ALGEBRA, AND RELATIONS WITH DOMAIN-GENERAL AND DOMAIN-SPECIFIC INFLUENCES

In the last 20 years, many studies have examined the contributions of one particular aspect of cognitive ability to academic achievement: WM (for a review, see Bull & Lee, 2014; Raghubar, Barnes, & Hecht, 2010; see also Mammarella et al., Chapter 10). WM is involved in short-term memory storage, reasoning, problem solving, and other higher cognitive tasks that require simultaneous representation and manipulation of information. Here, we adopt the model and terminology first used by Baddeley and Hitch (1974). The latest version of Baddeley's WM model consists of four components: central executive, phonological loop, visuospatial sketchpad, and an episodic buffer (Baddeley, 2000). Both the phonological loop and the visual spatial sketchpad are short-term storage systems, responsible for storing and rehearsing auditory and visuospatial information respectively. The episodic buffer facilitates exchange of information between the central executive and long-term memory. The central executive was argued to be responsible for the simultaneous processing and storage of information.

More recent works suggest this ability to process and store information is just one aspect of executive functioning. Miyake et al. (2000) examined three aspects of executive functioning: inhibition (overriding of prepotent or dominant responses), shifting (shifting flexibly between tasks or mental sets), and updating (monitoring, and the addition or deletion of contents from WM). Amongst the three components, updating is most consistently related to mathematical achievement, with the relations between other executive functioning and mathematics often found to be nonsignificant when WM or updating is entered into explanatory models (Bull & Lee, 2014). For this reason, we give more attention throughout this chapter to the role of WM and updating in mathematical achievement.

Although the relation of WM and updating to mathematical achievement is relatively robust, few studies have considered how the nature of this relation varies from early schooling to the midadolescent years. Some researchers argue that while a certain level of WM is needed to support the performance of mathematical tasks, performance is largely predicted by domain-specific variables. Focusing on arithmetic fluency, Fuchs et al. (2010) found that only domain-specific skills significantly predicted simple arithmetic computation skills in 1st grade, and that domain-general abilities did not mediate this relationship. Based on this result, they argued that arithmetic fluency grows out of the development of meaningful and interconnected knowledge about numbers that may have their foundation in the early ability to subitize and to estimate magnitude (c.f., Fuchs, Geary, Fuchs, Compton, & Hamlett, 2014; Fuchs, Geary, Fuchs, Compton, & Hamlett, 2016).

For children in 2nd and 3rd grades, Cowan and Powell (2014) found that WM failed to explain variances in either calculation fluency or written arithmetic, but did explain variance in arithmetic word problems. For all three arithmetic skills, numeracy and magnitude knowledge regarding multidigit numbers accounted for the most variance. Zheng, Swanson, and Marcoulides (2011) similarly found that WM was a significant predictor of skill in arithmetic word problems in 7–10 year olds, but the direct relation of WM to word problem solving accuracy was fully mediated by reading and computational skills. Finally, Träff (2013) found that numeracy and magnitude comparison skills, but not WM, uniquely predicted performance in arithmetic fact retrieval in 10–13 year olds. These findings are consistent with those from the neuroimaging literature, which suggests that WM may be particularly important during the early stages of learning arithmetic. Its importance declines as procedures become automatized and solution becomes increasingly dependent on direct retrieval (Delazer, 2003; Qin et al., 2014; Rivera, Reiss, Eckert, & Menon, 2005).

Other researchers argued that WM and updating are needed at all stages of mathematical skills acquisition, from when children are first introduced to the new skills, through to understanding, interpreting, integrating new information with existing mathematical knowledge, and on to the generation of solutions by considering the problems in light of their existing mathematical knowledge. For example, even simple tasks like magnitude comparison and deciding whether a verbally presented word refers to a number, demand both simultaneous and sequential processes of perceiving, coding, interpreting, and comparing information in different modalities (Kolkman, Hoijtink, Kroesbergen, & Leseman, 2013; Noël, 2009). Even a simple arithmetic task will likely rely on counting and other effortful strategies that require children, who are inexperienced with such tasks, to keep track, update, and manipulate numbers at hand (Monette, Bigras, & Guay, 2011).

Many areas of mathematics have a cumulative or hierarchical nature (e.g., multiplication building on earlier competencies in addition), but in other areas, skills are relatively independent. This may be because of their foundational nature (e.g., magnitude comparison versus knowledge of number words) or because of the specialist knowledge required for solution (e.g., probability). By definition, for areas that are relatively independent, prior performance will not necessarily have a strong influence on subsequent performance. Even among areas of mathematics that are logically interconnected, though earlier material may become automatized, learning, and solving new problems should still require WM. Furthermore, learners may not perceive or understand interconnections between topics and treat them as discrete. If this is the case, WM and updating, being domain-general, will serve as a more consistent explanatory variable for mathematical performance. Indeed, Bailey et al. (2014) found that most of the long-term stability in children's performance on standardized achievement tests could be attributed to WM, reading achievement, SES and other trait-like factors.

Consistent with this line of reasoning, a number of cross-sectional studies, especially those conducted with younger children, suggest that WM explains mathematical performance more consistently than does related domain-specific variables. For example, in a report of data from earlier stages of this study (Lee et al., 2012), we found that while performance on numerical patterns was significantly related to arithmetical proficiency, this was rendered nonsignificant when WM and updating were included in the model. Toll, van der Ven, Kroesbergen, and Van Luit (2011) found that while WM, updating, and preparatory mathematical abilities provided equally good prediction of low mathematical achievement measured 1.5 years later, WM and updating predicted persistence of mathematical difficulties, over and above the predictive value of preparatory mathematical abilities. Monette et al. (2011) found that kindergarteners' WM predicted performance on mathematical reasoning one year later, even when prenumeracy abilities were included as a covariate. They argued that the lack of automaticity in simple arithmetic at this age might make performance on mathematical reasoning questions particularly demanding on WM. Geary (2004) argued that deficits in WM and other related executive functions may lead to more long-term difficulties in retrieving basic arithmetic facts.

In a longitudinal study, van der Ven, Boom, Kroesbergen, and Leseman (2012) found that updating explained mathematical achievement at each of four time-points across 1st and 2nd grades. Latent growth models showed significant individual differences in the rate at which children improved in mathematics, with minimal variation in growth rate for updating. They also found a significant correlation between attainment in mathematics and updating, and significant correlation in their rates of growth, but attainment was not related to rate of growth in either case. van der Ven et al. argued that updating plays an important role in the mathematical learning process, and that they influence each other's development. Working with a wider age range (1st–5th grade), Geary, Hoard, Nugent, and Bailey (2012) found WM measures mediated differences in the growth rates of mathematical achievement between children with versus without mathematical learning difficulties. Similarly, Li and Geary (2013) tracked children's performance on a standardized mathematical test and found WM measured at 1st grade predicted rate of growth in achievement between 1st and 5th grades. WM measured at 1st, but not that measured at 5th grade explained variance in mathematical achievement at 5th grade.

Other longitudinal studies indicate that the role of WM varies both with children's experience and the type of mathematical problems that they were asked to solve. Focusing on the use of addition strategies, Geary, Hoard, and Nugent (2012) concluded that individual differences in WM are particularly important for understanding variation in the rate of learning to use strategies. Once these strategies have been learned, WM largely affects rate of executing the strategies. In contrast, findings on children's performance on achievement tests showed that easier items did not require extensive WM engagement. WM became important with successive grades and more complex test items (Geary, 2011). Geary also found that mathematical cognition measured at 1st grade predicted mathematical achievement and growth beyond the contribution of domain-general abilities. Fuchs et al. (2014) found relations between WM assessed at 1st grade and mathematical performance at 3rd grade varied depending on the WM measure and also whether the criterion measure is numeration understanding or multidigit calculations. WM predicted understanding more strongly than it did calculation, but both understanding and calculation were predicted by more foundational numerical competencies. Similarly, Fuchs et al. (2015) found arithmetic calculation at 3rd grade was predicted by both WM and early mathematical competencies at 1st grade, but with WM explaining double the amount of variance than did early mathematical competencies.

In summary, previous findings point to three ways in which patterns of relations between WM, updating, and mathematical performance can be characterized. Studies conducted with younger children, in particular, suggest that WM predicts mathematical performance more consistently than does prior mathematical performance. Others suggest that the explanatory power of WM and updating is largely mediated by prior mathematical achievement. A third possibility is a dynamic model, which specifies that the relative importance of WM, updating, and prior mathematical performance vary both with age and the domain of mathematics under consideration.

Focusing more specifically on learning and solving algebra, few studies have examined the relations between WM and achievement in algebraic word problems. In a previous study, we examined whether WM predicted algebraic problem solving abilities in 5th grade students (Lee, Ng, & Ng, 2009). Even after controlling for differences in intelligence and English literacy, WM provided reliable and additional information on pupils' likelihood of success. The data showed that WM, together with intelligence and English literacy, explained 48% of individual differences in mathematics examination scores. In a follow-up study, we took a componential approach and examined the extent to which WM was associated with success in generating model representation and computing solutions from such representations (Lee, Ng, & Ng, 2009). We also obtained separate measures of children's competencies in different subcomponent skills believed to be important for solving algebraic word problems (understanding of problem goal, understanding of quantitative relationships specified in questions as demonstrated by the selection of corresponding arithmetic operators, and arithmetic computation). The results showed that WM alone explained around 25% of variance in the model representation and solution computation tasks. Overall accuracy on the algebraic word problems was particularly dependent on students' abilities to make sense of the quantitative relationships specified in questions and assign the correct arithmetic operators to describe those relations. Performances on the latter task were also associated strongly with WM capacity.

Fuchs et al. (2012) focused on prealgebra and found WM measured in 2nd grade was not a significant direct predictor of performance, but had a small indirect effect on calculation skill at 3rd grade. More recently, focusing on relations between WM, mathematical anxiety, and algebra performance, Trezise and Reeve (2014) found that algebra scores varied among adolescents with different WM and worry profiles. Adolescents with high WM and low worry had the best algebra scores. Working with 15 year olds, Geary, Hoard, Nugent, and Rouder (2015) found WM explained individual differences in algebra achievement, but its explanatory power was mediated fully by measures of specific components of algebraic understanding. These findings suggest that WM and updating predict performance in algebraic tasks, but these relations will likely be partially or fully mediated by domain-specific skills.

THE PRESENT STUDY

All previous studies examined algebraic problem solving ability at one or a limited number of time points in the curriculum. For this reason, questions regarding the development of algebraic skills could not be answered. Here, we describe a four-year longitudinal study in which the contributions of executive functioning and earlier mathematical skills to more complex mathematical skills were examined. The development of general mathematical skills was assessed using a standardized measure: the Numerical Operations test (Wechsler Individual Achievement Test, 2nd edition, Wechsler, 2001). The development of algebraic skills was examined using a battery of relational and algebraic word problems.

In addition to relations between WM, updating, and mathematical achievement, we were particularly interested in whether associations between relational and algebraic skills varied with grade. On the one hand, being practiced at detecting underlying mathematical relations may give children an advantage right from the start and remain so when they are introduced to more complex algebraic questions. On the other hand, as changes in the algebra curriculum involve largely changes in problem solving strategies, being able to detect underlying relations may play a secondary role to abilities to learn and adapt to the new strategies.

Study Design

The study was based on a cross-sequential design, with children (N = 673) recruited from four cohorts: Kindergarten II (M_{age} = 5.72, SD = 0.34), 2nd grade (M_{age} = 7.85, SD = 0.32), 4th grade (M_{age} = 10.05, SD = 0.30), and 6th grade (M_{age} = 12.32, SD = 0.29). Collapsing across cohorts, this design yielded data from ten grades. Each year, children were administered tests of general mathematics achievement, mathematical relational skills, algebraic skills, and domain-general skills. Not all of the mathematical tasks were administered at each grade (Table 9.1). We used the Numerical Operations test (Wechsler, 2001) to assess children's general mathematical skills because it has a broad coverage of domains covered in the mathematics curriculum and contains items that

TABLE 9.1 Math Instruments Administered at Each Grade

Grades	K2	1	2	3	4	5	6	7	8	9
Algebra word problems					XX	XX	XX	X	X	X
Arithmetic word problems		X	XX	XX	XX	XX	XX			
Function machine	X	X	XX	XX	XX	XX	XX	X	X	X
Number sequences	X	X	XX	XX	XX	XX	XX	X	X	X
Functions			XX	XX	XX	XX	XX	X	X	X
Repeating geometric patterns	X									
Repeating number sequences	X									

increases gradually in difficulty and complexity. Critically, it offers a common metric on which we can measure children's general mathematical achievement.

To assess mathematical relational skills, we developed tasks based on items from the Navigating through Algebra series (National Council of Teachers, 2001a,b,c) and the Singapore mathematics curriculum. Three sets of tasks were developed: Number Sequences, Functions, and Function Machines. In the Number Sequences task, younger children were presented with simple number sequences (e.g., 9, 10, 11, _, 13) and were asked to identify the missing number. The underlying rule for the sequences became considerably more difficult across grades. In 8th grade, for example, children were presented with 2, 6, 18, 54, _, _, _ and a, $2(a + d)$, $3a + 4d$, $2(2a + 3d)$, _, _, _. For children in kindergarten, we designed an additional number sequence task using repeating sequences of digits (e.g., 3, 6, 3, 6, 3, _), because some of the children in kindergarten would not be fluent with numbers and counting, we also designed a version using geometric patterns. Each set of patterns was based on a changing characteristic of the constituent geometric figures or a combination of characteristics, such as shape, color, size, and orientation.

In the Functions task (Figs. 9.1 and 9.2), children were first asked to identify which visual stimulus was coupled with various depicted numbers. They were then asked to extrapolate and identify which stimuli were coupled with numbers that were not explicitly depicted. Children in the more senior grades were asked to use algebra to express the functional rule relating the input and the output. In the Function Machine task (Fig. 9.3), children were asked to fill in the missing numbers and identify the rule underpinning the input–output numbers. Similar to the Functions task, children in the more senior grades were expected to use symbolic algebra to express the predictive rule. Algebraic skills were assessed using algebraic word problems; there were described earlier (Figs. 9.4 and 9.5).

WM span was measured using two tasks from the Automated Working Memory Test (Alloway, 2007). We focused on the executive component of WM. In the Listening Recall task, we asked children to verify whether a series of simple statements (e.g., fish live in the sky) were "true" or "false." The children were also told to remember the last word in each statement. At the end of each trial, which contained one to six statements, children were asked to remember the last words in order; capacity was determined by the number of words recalled correctly. In the Mister X task, we presented two figures on-screen. Each figure held a ball in one hand, in one of several fixed positions. Children were asked to decide whether the two figures held their balls on the same side. They were also asked to remember the position at which the target figure held its ball. These positions had to be recalled, in order, at the end of each trial in which children were shown one to seven pairs of figures; capacity was determined by the total number of positions correctly recalled.

We also used a pictorial updating task, in which children were asked to remember the identity of animals that were shown to them one at a time.

The animals were depicted as black and white line drawings. The task started off with children just remembering the last two animals. This increased to the last four. The task was made the more difficult as the number of animals shown in each trial was randomized and children were not told how many animals would be shown. Capacity was determined by the number of animals recalled correctly.

General Mathematical Achievement, Domain-Specific and Domain-General Precursors

As reported in Lee and Bull (2016) cross-sectional correlations between WM, updating, and performance on the Numerical Operations task were moderate at kindergarten, but strong for all other grades, especially 1st and 2nd grade. Furthermore, the strength of these relations was little changed when intelligence was included in the analysis.

To examine the extent to which performances on the Numerical Operations tasks were predicted by prior performance versus WM and updating capacities, the data were also analyzed longitudinally. Using an autoregressive cross-lagged model, performance on the Numerical Operations task was found to be increasingly dependent on prior performance. In fact, from 1st to 4th grade, performances on Numerical Operations were better predicted by WM and updating than by prior performance. From 5th grade onwards, prior performance on the Numerical Operations task predicted future performance better than did WM and updating. Of note was that despite changes in its relative importance, the absolute predictive power of WM and updating remained unchanged across the 10 grades.

Mathematical Relational Skills and Arithmetic Word Problems

Orton and Orton (1999) found children relied on arithmetic strategies for solving number sequences. On a cross-sectional basis, we found performances on the Number Sequences and arithmetic word problems to be correlated strongly and positively. A similar pattern of correlations was found between the Number Sequences and Numerical Operations tasks. These findings are consistent with the hypothesis that competency in arithmetic may be important for finding solutions to number sequences.

With few exceptions, we also found performance on Number Sequences predicted accuracy on the arithmetic word problems in the following year. These predictive relations were somewhat stronger in the earlier grades. Notably, relations in the reverse direction were neither strong nor consistent: arithmetic performance did not consistently predict performance in Number Sequences. The same analyses conducted with the Function Machine task revealed a similar pattern of findings, but with weaker and less consistent associations.

One interpretation of the cross-sectional and longitudinal findings is that arithmetic skills are used in solving the relational problems, but they matter little in children's learning of more complex relational skills. This interpretation is consistent with the demands of the relational tasks. Faced with a number sequence, children need to figure out the structural relations specific in the sequence before they can usefully carry out any arithmetic operations. Arithmetic competency will matter little if children cannot figure out which arithmetic operation to apply. Some children will rely on a guess-and-check heuristic to try to work out the solution. Equivalent to a brute-force approach, the use of such heuristics to solve relational tasks is critically dependent on the efficiency with which arithmetic operations are conducted. However, even with guess-and-check, correct estimation of the structural relations will help narrow the range of operations and functions to be attempted. Conversely, children good at the relational tasks are more likely to have a better sense of relations between numbers, and in applying and reversing operations; both of which will aid in the learning of more complex arithmetic operations.

Algebraic Problems, Earlier Mathematical Skills, and Domain-General Capacities

Children were first introduced to algebraic word problems in 4th grade. As mentioned earlier, children at this grade were not expected to use symbolic algebra for solution, but were taught the model method for depicting the problem (Fig. 9.4). The notion of variable was introduced in 6th grade, but symbolic algebra is not used for problem solving until 7th grade onwards. For this reason, our analyses of relations between algebra and earlier mathematical skills were focused on data from 3rd to 9th grade, inclusive.

Focusing first on the relational tasks, we formed a latent measure with data from the Number Sequence, Function Machine, and Functions tasks and examined whether they correlated with performance on the algebraic word problems. The findings showed strong correlations at each grade. Longitudinal analyses using autoregressive cross-lagged models showed significant predictive relations throughout, but the strength of prediction decreased with age. Nonetheless, even for 9th grade, the relation remained sizeable.

From a cross-sectional perspective, performances on the arithmetic and algebraic word problems were correlated strongly and positively. From a longitudinal perspective, competencies in arithmetic word problems predicted consistently performance in their algebraic counterparts one year later. The strongest relation was seen in 4th grade, the first year in which algebraic word problems were introduced.

WM and updating capacities were correlated strongly and positively with performances on both the relational and algebraic word problems. Correlations with arithmetic word problems were moderate to strong. From a longitudinal perspective, performance on the relational tasks consistently predicted performance on the algebraic word problems, even when variances attributable to arithmetic word problems and WM and updating were controlled. The strongest predictive relations were observed in the primary years, when children were first introduced to algebraic thinking. The predictive relations remained strong, but were relatively weaker in the secondary years. Although analyses on relations between the arithmetic and algebraic problems suggest that laying a good foundation in arithmetic problems is important for learning algebraic word problems, the present finding shows that this relation is largely mediated by relational skills. With the exception of a weak predictive relation in the transition from primary to secondary school, predictive relations at other grades were not significant.

Of note was that despite strong cross-sectional correlations, WM and updating did not consistently predict performances on any of the three mathematical tasks. Interestingly, children with smaller WM and updating capacities tended to do better in 8th grade algebra than those with larger capacities. This relation returned to the canonical direction (i.e., higher capacity, better performance) one year later. We argue that these findings may be due to differences in the way in which children with different WM and updating capacities deploy problem solving heuristics at the two grades. Children with higher WM capacities might have been more willing to use letter-symbolic algebra than those with lower capacities. However, because they were still new to letter-symbolic algebra, they made more errors. With practice, they became more familiar with the method and again exhibited a performance advantage over those with lower WM when they were evaluated one year later at 9th grade.

CONCLUSIONS AND FUTURE DIRECTIONS

Several aspects of these findings are of note. First, in terms of general mathematical achievement, the findings show that WM and updating capacities are important. However, there are grade-related differences when their predictive power is weighed against that of prior mathematical performance: note that the absolute predictive power of WM and updating capacities do not vary with grade; it is the predictive power of prior mathematics achievement that varies. Relative to prior mathematics achievement, WM and updating capacities are more important for the younger than for the older children.

These findings are somewhat counter-intuitive if we view the Numerical Operations test as containing items of increasing difficulties and complexities. Should not more complex items impose greater demands on WM and updating? In Lee and Bull (2016), we argued that this dynamic pattern of relations between domain-general and domain-specific influences might have resulted from the way in which increasing difficulty and complexity were realized in the Numerical Operations test. Though more difficult items require knowledge of additional mathematical concepts, both easier and more difficult test items require a similar number of computational steps for solution. This similarity in computational steps ensures that there are few changes in resource demands from year to year. In contrast, as items become more difficult, only children with the requisite knowledge are likely to succeed.

Regarding algebraic skills, relational, and arithmetic word problem solving skills were hypothesized to contribute to competencies in algebraic word problems. Our findings show that this is indeed the case for the relational tasks, which consistently predicted performances in the algebraic word problems. Though strong associations between arithmetic and algebraic word problems were found when they were considered in isolation, inclusion of relational competencies in the predictive model shows that the predictive power of arithmetic is secondary to and mediated by relational competencies. Despite strong cross-sectional correlations, WM and updating did not consistently predict performances in the relational, arithmetic, or algebraic tasks. Performances in the relational tasks provided the best predictor of performance in the algebraic word problems. These findings suggest that though solving algebra problems is demanding of WM and updating resources, success in learning algebra is largely determined by domain-specific mathematical underpinnings.

Considered from the perspective of learning and solving more complex mathematical problems, these findings show that the importance of WM and updating capacities vary depending on the domain of mathematics under consideration. The findings suggest that WM and updating play a much smaller and less consistent role in predicting future performance in algebra as compared to their role in predicting performance in general mathematics achievement. One explanation for this difference in findings is that relative to the Numerical Operations test, across grades, there is a larger degree of conceptual homogeneity in the relational, arithmetic, and algebraic tests. The algebraic word problems varied the most across 4th–6th grade, with the higher grades containing more complex questions involving different number systems (e.g., fractions) and scenarios (e.g., comparisons involving speed). From 6th grade onwards, the questions remained largely unchanged. Given their relative homogeneity across grades, one explanation for the smaller role of WM and updating is that most of the variance in the algebraic task was explained by prior performance in the preceding grade. Reasonable though this explanation may be, it is not supported by the data. In all grades, performance in the relational task predicted subsequent performance in the algebraic task better than did prior performance in the algebraic task. Furthermore, though large portions of variance in the algebraic tasks are explained, variance explained did not approach ceiling.

From a design perspective, our measure of general mathematics achievement—the Numerical Operations test—provides a different degree of control over prior mathematics achievement than do the relational and arithmetic tasks for algebraic performance. Children were administered the Numerical Operations test and were asked to complete as many questions as they could. The test stopped only when the children showed clear sign of difficulties, and thus provides a measure of children's maximal performance, at each grade, over a range of mathematical domains. As argued by Friso-van den Bos, van der Ven, Kroesbergen, and van Luit (2013), one reason for stronger relations between WM, updating, and performances on standardized mathematics achievement tests is that children have to handle the demands imposed by having to answer questions from a range of mathematical domains. Having different types of questions stop children from getting settled into the procedures for one specific type of mathematical problem, and instead forces them to consider new, albeit related sources of knowledge across questions.

Future Directions

Although we have identified some variables that contribute to individual differences in performances on more complex mathematics problems, the mathematics education literature provides a wealth of additional factors that require consideration. One of the foci of our empirical study was to examine the influences of domain-general versus domain-specific factors in learning and solving algebra word problems. We end this chapter by providing a few ideas for future studies.

First, consider that in the following problem, there are three variables: the number of men, women, and children.

There are 900 people at a parade. There are 40 more men than women. There are twice as many children as there are men. How many children are there?

To solve for the number of children, any of the three variables can be chosen to be the generator for constructing the equation. In solving such questions, the choice of generator is a strategic decision. Designating the number of women, w, as the generator produces $w + (w + 40) + 2(w + 40) = 900$. This is often the generator of choice, perhaps because the appropriate arithmetic operation can be more directly translated from the text-based description. The drawback is that this approach presents an indirect solution, because it is still necessary to work out the number of children represented by $2(w + 40)$. Using children, c, as the generator produces $c + c/2 + (c/2 - 40) = 900$, and leads to a direct solution, but the construction of the algebraic expression requires inversion of operations. This approach is more effortful and the initial formulation of the equation is likely to be more prone to error. Why do some students take the direct route and others the indirect? It is not clear how and on what bases children select which protagonist to be designated as the generator.

Second, algebraic representations do not adhere to the same set of conventions underpinning the use of numbers. With letter-symbolic algebra, for example, $x + x/2$ cannot be simplified to x and $x/2$. In learning algebra, students need to inhibit conventions acceptable in arithmetic, but not algebra (see also Van Hoof et al., Chapter 5). Further work is needed on the extent to which success in making this transition is dependent upon domain-specific processes, such as inhibitory efficiencies.

Finally, because algebraic equations are necessary for solving algebraic word problems, the shift from arithmetic to algebra requires knowing that an algebraic equation is (1) a structure which links two different algebraic expressions and (2) two or more different algebraic expressions can be constructed to represent the same situation. Further work is needed on whether and how mental flexibility interacts with domain-specific knowledge to influence the acquisition of such knowledge.

REFERENCES

Alloway, T. P. (2007). *The automated working memory assessment*. London: Psychological Corporation.

Baddeley, A. D. D. (2000). The episodic buffer: a new component of working memory? *Trends in Cognitive Sciences*, *4*(11), 417–423.

Baddeley, A. D. D., & Hitch, G. J. (1974). Working memory. In G. A. Bower (Ed.), *Recent advances in learning and motivation* (pp. 47–90). (Vol. 8). New York: Academic Press.

Bailey, D. H., Watts, T. W., Littlefield, A. K., & Geary, D. C. (2014). State and trait effects on individual differences in children's mathematical development. *Psychological Science*, *25*(11), 2017–2026.

Bull, R., & Lee, K. (2014). Executive functioning and mathematics achievement. *Child Development Perspectives*, *8*(1), 36–41.

Bush, S. B., & Karp, K. S. (2013). Prerequisite algebra skills and associated misconceptions of middle grade students: a review. *The Journal of Mathematical Behavior*, *32*(3), 613–632.

Cai, J., Lew, H. C., Morris, A., Moyer, J. C., Ng, F. F., & Schmittau, J. (2005). The development of students' algebraic thinking in earlier grades: a cross-cultural comparative perspective. *Zentralblatt für Didaktik der Mathematik*, *37*(1), 5–15.

Carraher, D. W., Schliemann, A. D., & Schwartz, J. L. (2007). Early algebra is not the same as algebra early. In J. J. Kaput, D. W. Carraher, & M. L. Blanton (Eds.), *Algebra in the early grades* (pp. 235–272). Mahwah, NJ: Erlbaum.

Collis, K. F. (1975). *A study of concrete and formal operations in school mathematics: A Piagetian viewpoint*. Hawthorn, Victoria: ACER.

Comprehensive School Mathematics Program. (1975). *CSMP overview*. St Louis, MI: CEMREL.

Cowan, R., & Powell, D. (2014). The contributions of domain-general and numerical factors to third-grade arithmetic skills and mathematical learning disability. *Journal of Educational Psychology*, *106*(1), 214–229.

Davydov, V. V. (1962). An experiment in introducing elements of algebra in elementary school. *Soviet Education*, *5*(1), 27–37.

Delazer, M. (2003). Learning complex arithmetic—an fMRI study. *Cognitive Brain Research.*, *18*(1.), 76–88.

Duncan, G. J., Dowsett, C. J., Claessens, A., Magnuson, K., Huston, A. C., Klebanov, P., et al. (2007). School readiness and later achievement. *Developmental Psychology*, *43*(6), 1428.

English, L. D., & Warren, E. A. (1998). Introducing the variable through pattern exploration. *Mathematics Teacher*, *91*(2), 166–170.

Freudenthal, H. (1974). Soviet research on teaching algebra at the lower grades of the elementary school. *Educational Studies in Mathematics*, *5*(4), 391–412.

Friso-van den Bos, I., van der Ven, S. H. G., Kroesbergen, E. H., & van Luit, J. E. H. (2013). Working memory and mathematics in primary school children: a meta-analysis. *Educational Research Review*, *10*, 29–44.

Fuchs, L. S., Geary, D. C., Compton, D. L., Fuchs, D., Hamlett, C. L., & Bryant, J. D. (2010). The contributions of numerosity and domain-general abilities to school readiness. *Child Development, 81*(5), 1520–1533.

Fuchs, L. S., Compton, D. L., Fuchs, D., Powell, S. R., Schumacher, R. F., Hamlett, C. L., et al. (2012). Contributions of domain-general cognitive resources and different forms of arithmetic development to pre-algebraic knowledge. *Developmental Psychology, 48*(5), 1315–1326.

Fuchs, L. S., Geary, D. C., Fuchs, D., Compton, D. L., & Hamlett, C. L. (2014). Sources of individual differences in emerging competence with numeration understanding versus multidigit calculation skill. *Journal of Educational Psychology, 106*(2), 482–498.

Fuchs, L. S., Geary, D. C., Fuchs, D., Compton, D. L., & Hamlett, C. L. (2016). Pathways to third-grade calculation versus word-reading competence: are they more alike or different? *Child Development, 87*(2), 558–567.

Geary, D. C. (2004). Mathematics and learning disabilities. *Journal of Learning Disabilities, 37*(1), 4–15.

Geary, D. C. (2011). Cognitive predictors of achievement growth in mathematics: a 5-year longitudinal study. *Developmental Psychology, 47*(6), 1539–1552.

Geary, D. C., Hoard, M. K., & Nugent, L. (2012b). Independent contributions of the central executive, intelligence, and in-class attentive behavior to developmental change in the strategies used to solve addition problems. *Journal of Experimental Child Psychology, 113*(1), 49–65.

Geary, D. C., Hoard, M. K., Nugent, L., & Bailey, D. H. (2012a). Mathematical cognition deficits in children with learning disabilities and persistent low achievement: a five-year prospective study. *Journal of Educational Psychology, 104*(1), 206–223.

Geary, D. C., Hoard, M. K., Nugent, L., & Rouder, J. N. (2015). Individual differences in algebraic cognition: relation to the approximate number and semantic memory systems. *Journal of Experimental Child Psychology, 140*, 211–227.

Harper, E. (1987). Ghosts of Diophantus. *Educational Studies in Mathematics, 18*(1), 75–90.

Kaput, J. J., Carraher, D. W., & Blanton, M. L. (2008). *Algebra in the early grades*. New York, NY: Lawrence Erlbaum Associates/National Council of Teachers of Mathematics.

Khng, K. H., & Lee, K. (2009). Inhibiting interference from prior knowledge: arithmetic intrusions in algebra word problem solving. *Learning and Individual Differences, 19*(2), 262–268.

Kieran, C. (1996). The changing face of school algebra. In C. Alsina, J. Alvarez, B. Hodgson, C. Laborde, & A. Pérez (Eds.), *8th International Congress on Mathematical Education: Selected lectures* (pp. 271–290). Seville, Spain: S. A. E. M. Thales.

Kieran, C. (1997). Mathematical concepts at the secondary school level: the learning of algebra and functions. In T. Nunes, & P. Bryant (Eds.), *Learning and teaching mathematics: An international perspective* (pp. 133–158). Hove, England: Psychology Press.

Kieran, C. (2004). Algebraic thinking in the early grades: what is it? *The Mathematics Educator, 8*(1), 139–151.

Kolkman, M. E., Hoijtink, H. J. A., Kroesbergen, E. H., & Leseman, P. P. M. (2013). The role of executive functions in numerical magnitude skills. *Learning and Individual Differences, 24*, 145–151.

Küchemann, D. (1981). Algebra. In K. M. Hart (Ed.), *Children's understanding of mathematics: 11–16* (pp. 102–119). London: John Murray.

Lannin, J. K., Barker, D. D., & Townsend, B. E. (2006). Recursive and explicit rules: how can we build student algebraic understanding? *The Journal of Mathematical Behavior, 25*(4), 299–317.

Lee, K., & Bull, R. (2016). Developmental Changes in Working Memory, Updating, and Math Achievement. *Journal of Educational Psychology, 108*(6), 869–882.

Lee, K., Ng, S. -F., Ng, E. -L., & Lim, Z. -Y. (2004). Working memory and literacy as predictors of performance on algebraic word problems. *Journal of Experimental Child Psychology, 89*(2), 140–158.

Lee, K., Ng, E. L., & Ng, S. F. (2009). The contributions of working memory and executive functioning to problem representation and solution generation in algebraic word problems. *Journal of Educational Psychology, 101*(2), 373–387.

Lee, K., Ng, S. F., Pe, M. L., Ang, S. Y., Hasshim, M. N. A. M., & Bull, R. (2012). The cognitive underpinnings of emerging mathematical skills: executive functioning, patterns, numeracy, and arithmetic. *British Journal of Educational Psychology, 82*(1), 82–99.

Lee, K., Bull, R., & Ho, R. M. H. (2013a). Developmental changes in executive functioning. *Child Development, 84*(6), 1933–1953.

Lee, K., Khng, K. H., Ng, S. F., & Ng Lan Kong, J. (2013b). Longer bars for bigger numbers? Children's usage and understanding of graphical representations of algebraic problems. *Frontline Learning Research, 1*(1), 81–96.

Lesh, R., Post, T., & Behr, M. (1987). Dienes revisited: Multiple embodiments in computer environments. In I. Wirsup, & R. Streit (Eds.), *Development in school mathematics education around the world* (pp. 647–680). Reston, VA: National Council of Teachers of Mathematics.

Li, Y., & Geary, D. C. (2013). Developmental gains in visuospatial memory predict gains in mathematics achievement. *PloS One, 8*(7), 1–9.

Mason, J. (1996). Expressing generality and roots of algebra. In N. Bednarz, C. Kieran, & C. Lee (Eds.), *Approaches to algebra: Perspectives for research and teaching* (pp. 65–86). Dordrecht, Netherlands: Kluwer Academic Press.

Miyake, A., Friedman, N. P., Emerson, M. J., Witzki, A. H., Howerter, A., & Wager, T. D. (2000). The unity and diversity of executive functions and their contributions to complex "frontal lobe" tasks: a latent variable analysis. *Cognitive Psychology, 41*(1), 49–100.

Monette, S., Bigras, M., & Guay, M. -C. (2011). The role of the executive functions in school achievement at the end of Grade 1. *Journal of Experimental Child Psychology, 109*(2), 158–173.

Nathan, M. J., & Kim, S. (2007). Pattern generalization with graphs and words: a cross-sectional and longitudinal analysis of middle school students' representational fluency. *Mathematical Thinking and Learning, 9*(3), 193–219.

National Mathematics Advisory Panel. (2008). *Foundations for success: The final report of the National Mathematics Advisory Panel*. Washington, DC: U.S. Department of Education.

National Council of Teachers of Mathematics. (2000). *Principles and standards for school mathematics*. Reston, VA: NCTM.

National Council of Teachers. (2001a). *Navigating through algebra in grades 3–5*. Reston, VA: NCTM.

National Council of Teachers. (2001b). *Navigating through algebra in grades 6–8*. Reston, VA: NCTM.

National Council of Teachers. (2001c). *Navigating through algebra in prekindergarten*. Reston, VA: NCTM.

Ng, S. F., & Lee, K. (2008). As long as the drawing is Logical, size does not matter. *The Korean Journal of Thinking & Problem Solving, 18*(1), 67–82.

Ng, S. F., & Lee, K. (2009). Model method: Singapore children's tool for representing and solving algebra word problems. *Journal for Research in Mathematics Education, 40*(3), 282–313.

Noël, M. -P. (2009). Counting on working memory when learning to count and to add: a preschool study. *Developmental Psychology, 45*(6), 1630–1643.

Orton, A., & Orton, J. (1999). Pattern and the approach to algebra. In A. Orton (Ed.), *Pattern in the teaching and learning of mathematics* (pp. 104–120). London: Cassell.

Qin, S., Cho, S., Chen, T., Rosenberg-Lee, M., Geary, D. C., & Menon, V. (2014). Hippocampal-neocortical functional reorganization underlies children's cognitive development. *Nature Neuroscience*, *17*(9), 1263–1269.

Raghubar, K. P., Barnes, M. A., & Hecht, S. A. (2010). Working memory and mathematics: a review of developmental, individual difference, and cognitive approaches. *Learning and Individual Differences*, *20*(2), 110–122.

Rittle-Johnson, B., Fyfe, E. R., McLean, L. E., & McEldoon, K. L. (2013). Emerging understanding of patterning in 4-year-olds. *Journal of Cognition and Development*, *14*(3), 376–396.

Rivera, S. M., Reiss, A. L., Eckert, M. A., & Menon, V. (2005). Developmental changes in mental arithmetic: evidence for increased functional specialization in the left inferior parietal cortex. *Cerebral Cortex*, *15*(11), 1779–1790.

Sfard, A. (1995). The development of algebra: confronting historical and psychological perspectives. *The Journal of Mathematical Behavior*, *14*(1), 15–39.

Stacey, K., & MacGregor, M. (1999). Learning the algebraic method of solving problems. *The Journal of Mathematical Behavior*, *18*(2), 149–167.

Stacey, K., & Chick, H. (2004). Solving the problem with algebra. In K. Stacey, H. Chick, & M. Kendal (Eds.), *The future of teaching and learning of algebra. The 12th ICMI study* (pp. 1–20). Boston, MA: Kluwer.

Toll, S. W. M., van der Ven, S. H. G., Kroesbergen, E. H., & Van Luit, J. E. H. (2011). Executive functions as predictors of math learning disabilities. *Journal of Learning Disabilities*, *44*(6), 521–532.

Träff, U. (2013). The contribution of general cognitive abilities and number abilities to different aspects of mathematics in children. *Journal of Experimental Child Psychology*, *116*(2), 139–156.

Trezise, K., & Reeve, R. A. (2014). Working memory, worry, and algebraic ability. *Journal of Experimental Child Psychology*, *121*, 120–136.

Usiskin, Z. (1988). Conceptions of school algebra and uses of variables. In A. F. Coxford, & A. P. Schulte (Eds.), *The ideas of algebra* (pp. 8–19). Reston, VA: National Council of Teachers of Mathematics.

van der Ven, S. H. G., Boom, J., Kroesbergen, E. H., & Leseman, P. P. M. (2012). Microgenetic patterns of children's multiplication learning: confirming the overlapping waves model by latent growth modeling. *Journal of Experimental Child Psychology*, *113*(1), 1–19.

Walkowiak, T. A. (2014). Elementary and middle school students' analyses of pictorial growth patterns. *The Journal of Mathematical Behavior*, *33*, 56–71.

Wechsler, D. (2001). *Wechsler individual achievement test* (2nd ed.). London, UK: The Psychological Corporation.

Zheng, X., Swanson, H. L., & Marcoulides, G. A. (2011). Working memory components as predictors of children's mathematical word problem solving. *Journal of Experimental Child Psychology*, *110*(4), 481–498.

Chapter 10

Learning Geometry: the Development of Geometrical Concepts and the Role of Cognitive Processes

Irene C. Mammarella*, David Giofrè**, Sara Caviola[†]

*Department of Developmental and Social Psychology, University of Padova, Padova, Italy; **Liverpool John Moores University, Natural Sciences and Psychology, Liverpool, United Kingdom; [†]Department of Psychology, University of Cambridge, Cambridge, United Kingdom

Geometry is one of humanity's oldest sciences and concerns the study of the size, shape, length, relative position of figures, and their spatial properties. It has long been considered one of the most important forms of mathematical knowledge, so much so that people who did not know their geometry were not allowed to enter the arcade of the famous school in Athens where Plato taught. Though geometry is one of the main areas of mathematical learning, the cognitive processes underlying the learning of geometry and achievement in geometry-dependent domains are not well understood. The psychological aspects of geometry have been studied by both developmental psychologists (e.g., Piaget, Inhelder, & Szeminska, 1960; Piaget & Inhelder, 1967) and educational psychologists (Clements & Battista, 1992; Clements, 2003, 2004; Crowley, 1987; Owens & Outhred, 2006; van Hiele, 1986), as has the contributions of spatial abilities, imagery (Bishop, 1980; Brown & Presmeg, 1993) and working memory (WM) (Giofrè, Mammarella, & Cornoldi, 2014; Giofrè, Mammarella, Ronconi, & Cornoldi, 2013).

In the present chapter, we first provide a brief review of the principal theories and models relating to learning geometry, focusing on how it is now secondary to arithmetic and algebra in many school curricula. We then turn to the development of geometrical concepts as revealed by classic research in developmental and educational psychology. Results from these classic studies are then integrated with more recent findings focusing on the relationship between cognitive processes (such as WM and intelligence) and geometry learning in primary-school children. We close with discussion of the educational implications.

Acquisition of Complex Arithmetic Skills and Higher-Order Mathematics Concepts
http://dx.doi.org/10.1016/B978-0-12-805086-6.00010-2

CLASSICAL STUDIES ON GEOMETRY

Piaget's theory is full of references to "geometry." For instance, he suggested that children are born with no knowledge of space, or even of the permanence of objects (Piaget & Inhelder, 1967). He claimed that children's notions of space are first topological (i.e., understanding closure and connectedness), then projective (i.e., relationships between the child and surrounding objects), and ultimately Euclidean, or coordinated.

Among the associated experiments conducted by Piaget and Inhelder (1967), we should mention those in which they tried to ascertain when children become able to recognize geometric figures. They concluded that children can construct *perceptual space* already in infancy (e.g., being able to develop their own idea of space through the influence of motor and perceptual actions), but it is only much later—when they are able to coordinate their actions systematically—that they develop ideas about space and figures, or what the authors called *representational space* (e.g., a space reconstruction resulting from the most primitive notations in which the topological representations are rebuilt).

During the 1980s, van Hiele (Fuys, 1984; van Hiele, 1986) described a model of how geometrical thinking develops. The model consists of five levels of understanding; other authors have since suggested adding another, earlier, prerecognition level that would indicate failure to attain the van Hieles' first level (e.g., Clements & Sarama, 2003). On level 0 (prerecognition), during their first years of life, children cannot reliably distinguish circles, triangles, and squares from nonexamples of those classes (i.e., a shape similar to a triangle but with a curved line). Children at this level are just starting to form visual schemes for canonical geometric shapes. On level 1 (the visual level), students can recognize a shape as a whole but cannot form a mental image of it. They can identify and name common geometrical figures (Prevost, 1985), but these figures are recognized by their shape, without considering their parts or attributes, such as the number of sides or angles in the shape (Crowley, 1987). At this level, children's descriptions of shapes are visual; for example, "it looks like a door, so it is a rectangle". On level 2 (the descriptive/analytical level), children recognize and characterize shapes by their properties. For instance, a child might think of a rectangle as a figure that has two pairs of equal sides. Properties are established experimentally by observing, measuring, drawing, and model making. In order to improve awareness about the relationships between the properties of geometric figures, when describing an object, children might be helped by listing all the properties they are able to name, and discerning which properties are necessary and which are sufficient to describe the object. Children on level 3 (informal deduction) understand the hierarchies of geometrical figures. They understand class inclusion; for example, that a square is also a rectangle and a parallelogram, and that they are all uadrilaterals (Crowley, 1987). Students can also deduce properties of a figure, for example, children are able to deduce an inverse property, also if not explicitly explained. On level 4 (deduction), the role of axioms and theorems is comprehensible. Students understand the significance of proof and can construct their own proofs. Finally, on level 5 (rigor), students can learn non-Euclidean geometries.

Crowley (1987) also elaborated some properties of the previously outlined model. He suggested, for example, that students must proceed through the levels one after the other, exactly in the sequence described by van Hiele (1986). This is because learning associated with higher levels is dependent on knowledge associated with preceding levels. However, the rate with which students move from one level to the next depends more on the content and method of the instruction they receive than on their chronological age. Moreover, the instruction must be appropriate for the level students have reached in order for them to progress to the next level, so it is important for teachers to be aware of the level a given student has attained.

CORE INTUITIVE PRINCIPLES OF GEOMETRY

Mathematical abilities seem to emerge from different representational core systems: inherent attentional and representation systems for processing and responding to key features of the world (Spelke, Breinlinger, Macomber, & Jacobson, 1992). Within this perspective, the core knowledge hypothesis provides a somewhat different account of the origin of geometric knowledge. A first system-based on intuition and shared by adults, children, and nonhuman animals relies on nonsymbolic representations, such as clusters of objects or area. A second is specific to humans who have received some formal education, and relies on symbolic representations (Halberda, Mazzocco, & Feigenson, 2008): humans can build on these core systems to create evolutionarily novel abilities, including symbolic mathematics (Geary, 1995). At this time, it is not clear whether there is a single, integrated system of core knowledge that covers space and number or whether these are distinct systems that can interact in some contexts, although some researchers claim that geometry systems are distinct from number systems (Spelke, Lee, & Izard, 2010), and that the mathematical domain is therefore more complex than other core systems. Concerning geometry, a distinction has been suggested between intuitive and academic geometry (Giofrè, Mammarella, & Cornoldi, 2013), similarly to the distinction between evolved and nonevolved concepts (Lourenco, Bonny, Fernandez, & Rao, 2012): the former would include abilities believed to be independent of a person's cultural background and education, while the latter would depend on learning and education.

As mentioned previously, intuitive concepts are mainly implicit and not necessarily associated with schooling, but they are used to interpret a variety of situations and may support academic learning. Fischbein (Fischbein & Nachlieli, 1998; Fischbein, 1999) described the properties of intuitive cognition, saying that it is: (1) a direct and self-evident knowledge; (2) associated with a strong intrinsic certainty; (3) associated with a coercive effect (i.e., a tendency to reject alternative interpretations); (4) able to extrapolate beyond any empirical

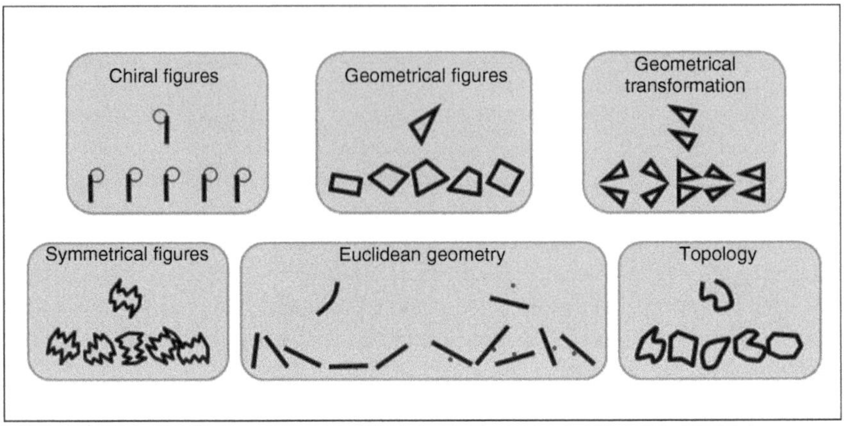

FIGURE 10.1 Examples of items represented the geometrical concepts of topology, symmetry, transformation, and Euclidean geometry used in the Giofrè et al. (2013b) study. Subjects were asked to select the odd one out from each of the six sets of images. For easy reference, the odd one out is shown at the top of each item. *(Adapted and modified from Dehaene, 2006.)*

support; and finally, (5) associated with a globally based general estimation. With respect to geometry, Rosch (1975) wrote that, when unschooled people in a "Stone Age" culture were asked to choose the "best example of a shape" among a set of several shapes (e.g., quadrilaterals and near-quadrilaterals, including square-like shapes that were open, had curved sides, or contained non-90 degree angles), they usually chose a square even when the set contained some close variants. It is likely that people have a preference for closed symmetrical shapes (Bornstein, Ferdinandsen, & Gross, 1981).

More recent studies have explicitly focused on the core knowledge that defines intuitive geometry (Dehaene, Izard, Pica, & Spelke, 2006; Spelke et al., 2010). Specifically, Dehaene et al. (2006) devised a procedure for assessing intuitive geometry and examining to what extent it may be culture-independent. Their test was based on a number of sets of six images, each representing an intuitive concept of geometry (i.e., topology, Euclidean geometry, geometrical figures, and so on). Five images fitted the target concept, while one contradicted it, and participants were asked to point to the odd one out (Fig. 10.1). They tested Amazonian Indian children and adults with no formal geometry training, comparing them with North American children and adults. They found that: (1) the Amazonian Indians were able to correctly identify concepts of topology (e.g., connectedness), Euclidean geometry (e.g., lines, points, parallelism, and right angles) and geometrical figures (e.g., squares, triangles, and circles); and Dehaene et al. (2006) interpreted this as proof of the existence of *core principles* of geometry; and (2) Amazonian Indian adults with no formal geometry training showed much the same competence in the core principles of geometry as young children in the Amazonian Indian and US groups; and the performance of the Amazonian Indian children did not differ from that of the North American children. The North American adults performed significantly better in the whole test, however, revealing how formal education influences the ability to detect the noncore principles of geometry. In particular, the Amazonian Indian adults' performance was much the same as that of the children in both groups when it came to assessing geometrical transformations (e.g., the third item of Fig. 10.1), for which participants needed to use abilities, such as translations and rotations of shapes, and the manipulation of symmetrical figures (e.g. figures showing horizontal or vertical symmetrical axes).

Several years later, Spelke et al. (2010) suggested that natural geometry is founded on two core systems, the first of which is useful for navigating 3D spatial layouts, and the second for analyzing 2D visual forms. Together, the two systems capture all of the fundamental properties of Euclidean geometry (distance, angle, and directional relationships). According to Spelke et al. (2010), evidence for the first core system is provided by studies showing that both humans and nonhuman animals represent distances and directional relationships during navigation (Doeller & Burgess, 2008; Lever, Wills, Cacucci, Burgess, & O'Keefe, 2002). Most of the data available derive from research on reorientation, studying the ability to reestablish a sense of direction after being disoriented (Cheng & Gallistel, 1984; Cheng, 1986), and have been collected in both humans and nonhuman animals (Hermer & Spelke, 1994; Lourenco & Huttenlocher, 2006).

Evidence of the second core system emerges from experiments carried out by Izard and Spelke (2009) that identified changes in the development of sensitivity to length, angle and direction. These experiments, conducted on adults and 4- to 10-year-old children, revealed that children of all ages, as well as adults can easily detect angle and length relationships, but they failed to detect directional relationships until the end of childhood. Izard and Spelke (2009) used a modified version of the detection paradigm used by Dehaene et al. (2006), and participants had to find the odd one out in sets of six L-shaped figures. The authors found that participants of all ages were most sensitive to relationships of angle and length, and least sensitive to differences between a figure and its mirror image. They also showed that for participants of all ages, the sense of the figure (e.g., mirror images) was difficult to detect and easily ignored. The authors concluded that the proposed two core systems (one for navigating 3D spatial layouts, and the other for analyzing 2D visual forms) have some limits in that they capture only a subset of the properties covered by symbolic Euclidean geometry. Children can construct a new, more complete, and general system of geometrical representation, however, by productively combining the representations afforded by the two systems. This process relies on culturally variable artifacts (such as formal instruction) and children learn to use it as they engage with the symbol systems of their culture.

ACADEMIC ACHIEVEMENT IN GEOMETRY

As mentioned earlier, the main difference between intuitive geometry and academic geometry is that the latter only emerges with formal education and is based on a symbolic system, while the former does not. Academic geometry demands an explicit knowledge of mathematics, and particularly of geometrical principles, such as diagonals, parallel lines, and right angles, as well as the ability to solve geometrical problems, such as calculating the area or perimeter of a figure; it therefore relies heavily on formal education.

Achievement in academic geometry, especially at secondary school level, is considered one of the most important areas of mathematical learning and associated with future academic and professional success (Verstijnen, van Leeuwen, Goldschmidt, Hamel, & Hennessey, 1998). Students attending secondary schools must learn basic geometric concepts, definitions, theorems, etc., and apply their knowledge to solving problems that are typically presented in the form of verbal or written words. Without a proficient geometrical knowledge, students are not prepared for advanced study in the fields of science, technology, engineering and mathematics (STEM; Zhang, Ding, Stegall, & Mo, 2012). Given this, it is not surprising that geometry is included in the mathematical curriculum all over the world, in international assessments including the Programme for International Student Assessment (PISA) (e.g., OECD—Organization for Economic Co-operation and Development, 2014), the Trends In International Mathematics and Science Study (TIMSS) (e.g., Mullis, Martin, Foy, & Aror, 2012) and the INVALSI in Italy (INVALSI, 2011). Scores on the PISA predict educational outcomes (such as the likelihood of gaining access to postsecondary education), as well as important outcomes outside school, such as performance in vocational-education and at universities, job performance, and future salary (Fischbach, Keller, Preckel, & Brunner, 2013). Focusing on geometry, in particular, it has been argued that teaching geometry has a positive effect in helping to improve spatial intelligence (Gittler & Glück, 1998). Importantly, geometry and geometrical problem-solving are also needed in the complex society in which we live, and a nation's well-being depends on its capacity to excel in STEM fields, which are strongly related to mathematics and geometry (Hunt, 2012). It can therefore be argued that academic achievement in geometry may contribute to growth in other abilities, including higher-order cognition and mathematical reasoning (Clements & Sarama, 2011b; Tatsuoka, Corter, & Tatsuoka, 2004).

Despite its importance, the results of international assessments have shown that many students struggle to learn geometry. This can be illustrated by recent results from the PISA international assessment, which is administered to large samples of 15 year olds throughout the world every three years. The latest assessment for which data were available was conducted in 2012 and focused on mathematical literacy. The OECD (2013, p. 25) definition for mathematical literacy is:

> *Mathematical literacy is an individual's capacity to formulate, employ, and interpret mathematics in a variety of contexts. It includes reasoning mathematically and using mathematical concepts, procedures, facts, and tools to describe, explain, and predict phenomena. It assists individuals to recognize the role that mathematics plays in the world and to make the well-founded judgments and decisions needed by constructive, engaged and reflective citizens.*

The PISA mathematics section includes four content categories: (1) Quantity; (2) Uncertainty and Data; (3) Change and Relationship; and (4) Space and Shape. This last category was defined by the OECD (2013, p. 34) as:

> *Space and shape encompasses a wide range of phenomena that are encountered everywhere in our visual and physical world: patterns, properties of objects, positions and orientations, representations of objects, decoding and encoding of visual information, navigation and dynamic interaction with real shapes as well as with representations. Geometry serves as an essential foundation for space and shape, but the category extends beyond traditional geometry in content, meaning and method, drawing on elements of other mathematical areas such as spatial visualization, measurement and algebra. For instance, shapes can change, and a point can move along a locus, thus requiring function concepts. Measurement formulas are central in this area.*

From this definition, it is clear that the Space and Shape category includes what we have called academic geometry. PISA scores are standardized with the mean at about 500 with a standard deviation of 100 points. In the last PISA study (OECD, 2014), of the four aforementioned content categories, Space and Shape had the lowest mean value, which was 490, that is, 4 points lower than the mean for mathematical literacy (the mean score was 495 for Quantity; 493 for Uncertainty and Data; and 493 for Change and Relationship). These results do not necessarily indicate strong differences between the four categories, but they do suggest that geometry is not as strongly emphasized as other areas of mathematics in many countries, as shown in Table 10.1. This is true, for instance, of many advanced, "Western" countries where performance across the mathematics domains assessed by the PISA varies considerably, revealing an extremely poor performance in geometry (as measured by the Space and Shape category).

As an example, performance in geometry in the United Kingdom and USA is undeniably poor, and worse than in the other categories, with a difference of about **18** points vis-à-vis the average score for mathematical abilities in both countries. The difference is even more pronounced if we consider the difference between geometry and the best performance in the other mathematical domains; here the differences are **27** points for the United Kingdom and **25** for the USA. These gaps are not found in all countries, however. Germany and Italy, for example, have much smaller gaps between the four mathematical categories: 517, 509, 516, and **507** in Germany (range 10 points); and 491, 482, 477, and **487** (range 14) in Italy, for Quantity, Uncertainty and Data, Change and Relationship, and Space and Shape, respectively. Elsewhere, performance in Space and Shape is quite strong, and no lower than in the other categories: in Hong Kong, for instance, performance in the four categories ranges between 553 and 567 being 567 for Space and Shape. There is an intriguing situation in Shanghai, China (which is a top performer in PISA studies), where children have a mean score of **649** in Space and Shape, and lower scores (though still excellent) in the other mathematical categories (591, 592, and 624 for, Quantity, Uncertainty and Data, and Change and Relationship, respectively).

TABLE 10.1 Performance on the Latest International Assessments (PISA & TIMSS) for Selected Countries

Country	PISA 2012				TIMSS 2011, 8th grade			
	Quantity	Uncertainty and data	Change and relationship	Space and shape	Numbers	Algebra	Geometry	Data and chance
Shanghai	491	592	624	**649**	—	—	—	—
Singapore	569	559	580	**580**	611	614	**609**	607
Hong Kong	566	553	564	**567**	588	583	**597**	581
Germany	517	509	516	**507**	—	—	—	—
Italy	491	482	477	**487**	496	491	**512**	499
United Kingdom[a]	494	502	496	**475**	512	489	**498**	543
USA	478	488	488	**463**	514	512	**485**	527

Country	TIMSS 2011, 4th grade		
	Numbers	Geo. Sha. & Meas.[b]	Data display
Singapore	619	**589**	588
Hong Kong	604	**605**	593
Italy	510	**513**	495
England	539	**545**	549
USA	543	**535**	545

Note: Figures for Geometry or related topics are given in bold. The latest assessment was conducted in 2012 for the PISA and in 2011 for the TIMSS.
[a]*England alone in the case of the TIMSS.*
[b]*Geometrical shapes and measures.*

TABLE 10.2 Percent of Students Reaching Different Proficiency Levels for Selected Countries on the PISA Space and Shape Subscale

Country	<Level 1	Level 1	Level 2	Level 3	Level 4	Level 5	Level 6
Shanghai	**0.7**	**2.4**	5.5	9.8	14.9	20.8	**45.9**
Singapore	**3.2**	**6.4**	11.2	16.7	19.7	19.4	**23.4**
Hong Kong	**3.2**	**6.4**	12.2	18.1	22.6	20.3	**17.1**
Germany	**6.5**	**12.6**	20.8	24.2	20.1	12.2	**4.7**
Italy	**10.7**	**15.9**	22.4	21.7	15.7	9.0	**4.6**
United Kingdom[a]	**12.0**	**17.5**	23.8	22.5	14.5	7.0	**2.7**
USA	**13.5**	**20.9**	25.0	20.6	12.4	5.4	**2.2**

Note: Levels <1, 1 and 6 are shown in bold.
[a]*England alone in the case of the TIMSS.*

In addition to mean values, the PISA study also provides level of proficiency benchmarks; specifically, student's performance is classified from level 1 (a very poor performance, corresponding to scores around 358–419 points) to level 6 (an extremely good performance, higher than 669 points). Performance can also be measured by considering the percentages of children whose performance is below level 2 (see Table 10.2 for further details), which is around 420–481 points and considered a sort of cutoff. Students below level 2 are considered low performers and are at risk for long-term under employment and frequent unemployment (Bynner, 1997). Overall, the average proportion of children below level 2 in math is about 23%, but the proportion is even lower at 26% for Space and Shape. To give some examples, the proportion of Italian students below level 2 in Space and Shape comes close to the OECD average (27%), whereas the proportion of German students is lower than average (19%). Top performing countries naturally have a very small proportion of children below level 2, such as 3% in Shanghai, and around 10% in Hong Kong. As expected, the United Kingdom and USA both have larger than OECD-average proportions of children below level 2 in Space and Shape, that is, about 30 and 34% of students, respectively—meaning that about one in three students fare very poorly in this mathematical area (and presumably in geometry as a whole). There are even quite large proportions of children in countries, such as Italy (11%), the United Kingdom (12%), and the USA (14%), that fall below level 1 (corresponding to about 1.5 SD below the mean), so about 1 in every 10 students do not achieve even the lowest proficiency level.

These patterns are not restricted to the 15 year olds assessed with the PISA. The International Association for the Evaluation of Educational Achievement (IEA) conducts international surveys in the math domain on children in grades 4 and 8, the TIMSS survey. The TIMSS also divides mathematical subjects into four content categories, that is, Numbers, Algebra, Data and Chance, and Geometry (Mullis et al., 2012). When we look at performance in Geometry, Asian countries are again the top performers. Overall, math performance is high—about 1 SD above the average—in Singapore (606 in 4th grade and 611 in 8th grade) and Hong Kong, China (602 in 4th grade and 586 in 8th grade). These countries also perform very well in geometry, as shown in Table 10.1. As clearly reported in Table 10.1, the findings of two independent international surveys (TIMSS and PISA) indicated that performance in geometry deteriorated with age in the United Kingdom, and especially in the USA, being worse in secondary, than in primary-school children (Martin et al., 2008).

In short, international studies of mathematics achievement indicate that students from many countries perform more poorly in geometry than other mathematical domains, and we may wonder why. A glance at the books primary-school teachers use to teach math in the USA reveals that only about 8–15% of the pages are dedicated to geometry, with percentages lower than 1% in the case of special-needs teachers (Cawley, Foley, & Hayes, 2009). To make the situation worse, many teachers are inadequately prepared to teach geometry, and a large proportion of those in the USA and United Kingdom are unable to score beyond levels 1 or 2 of the van Hiele model of geometrical thinking (Clements & Sarama, 2011b; Fujita & Jones, 2006a; van Hiele, 1986). It is therefore hardly surprising that a sizable proportion of the children in these countries struggle with geometry, given that the van Hiele (1986) model of geometrical thinking postulates that levels of proficiency in geometrical thinking are achieved sequentially, meaning that geometry-teaching inadequacies in primary school would have severe consequences for subsequent achievement.

THE DEVELOPMENT OF GEOMETRICAL KNOWLEDGE

Based on the earlier described van Hiele (1986) model, the development of a child's formal geometrical skills depends largely on their country's educational system, since higher proficiency levels can only be built on consolidated lower-level foundations. Given the sequence of learning geometric concepts, children's early foundational skills (e.g., van Hiele, levels 1 and 2) are critical, but little is known about how children develop these skills. In this section, we review our research on the early stages of geometrical learning, referring particularly to preschool children.

Our studies focused on the geometric competences of Italian children from 4 to 6 years old, testing their geometrical knowledge in kindergarten and in the first year of primary school (Giofrè, Mammarella, & Lucangeli, 2009). In a first study, we assessed 269 preschoolers and 1st-graders (from 4 to 6 years old) in terms of their geometrical knowledge and spatial abilities. With the former, we focused on their familiarity with four basic figures (testing their ability to recognize and name a square, a rectangle, a triangle, and a rhombus), and on their ability to produce geometrical explanations (e.g., to explain why two figures, such

TABLE 10.3 Percentages of Correctly Recognized And Named Geometrical Shapes in Kindergarten

Shapes	4-years-old (%)	5-yars-old (%)	6-yars-old (%)
	74	76	89
	47	63	77
	15	27	60
	3	11	32

as a square and a rectangle were different). As for the children's spatial abilities, we examined whether they were able: to identify figures that were the same, but rotated or of a different size; to mentally merge two figures into a different one (e.g., two right-angled triangles to make a rectangle); and, to identify a given type of shape (all the squares or all the triangles) in a complex drawing consisting of various shapes, such as a house. At all ages, the children easily recognized the square, rectangle, and triangle, but many of them had difficulty identifying the rhombus, as shown by data reported in Table 10.3. These data were fairly consistent in revealing developmental changes in the children's capacity to recognize shapes, but also showed that many 1st-graders were still unable to name such basic shapes as squares, triangles and rectangles. These competencies coincide with level 0 or 1 (prerecognition) of the Van Hiele (1986) model.

We also tested the children's ability to compare two different shapes by asking them, for instance, why a square was different from a triangle (the shapes were always presented and remained available to the children). Compared with the previous task, this involved mentioning some basic geometrical properties (e.g., number of sides, concepts of straight lines and parallel lines) and typically demands a more advanced geometrical knowledge. We found that the geometrical skills needed for this task developed and improved with schooling. For example, the proportions of children who mentioned specific geometrical properties when comparing a triangle with a square were 23, 39, and 52% for the 4-, 5-, and 6-year-olds, respectively. When the children were asked to compare a triangle or a square with a rhombus, their performance was very poor, in the range of 2% for the 4-year-olds to 17% for the 6-year-olds.

Taken together, these results showed that a sizable majority of the children had limited geometrical knowledge, meaning that they had probably received little geometrical education. The results were very different when we considered spatial skills: the children's developmental trajectory (which was expected to be at least partially independent of their culture and education) rose steadily as they became increasingly able to identify, rotate, and distinguish shapes among distractors. This is consistent with the claim that very young children are already able to solve basic geometrical and spatial problems even if their geometrical education is poor (the Italian school system, for one, seems to be scarcely focused on improving their competences in this area). This brings us to the question of whether children's performance in geometry can be improved and, if so, what is the evidence to support this? We try to answer in the section dedicated to the educational implications of learning geometry.

COGNITIVE PROCESSES INVOLVED IN GEOMETRY

Academic geometry is a complex domain because there are many different skills associated with it (Geary, 1996). Academic geometry demands an explicit knowledge of mathematics, for example, and particularly of geometrical principles, such as diagonals, parallel lines, and right angles, as well as the ability to solve geometrical problems, such as calculating the area or the perimeter of a figure—all skills that rely heavily on formal education (van Hiele, 1986). At the same time, geometric problem solving also involves maintaining and manipulating visual and verbal material—skills that require WM (Baddeley, 2000). Indeed, WM seems to be very important and strongly related to success in geometry, although WM alone cannot explain success in academic geometry. This is because academic geometry also involves problem-solving, which requires WM but also involves higher-order skills related to higher-order cognitive processes, such as intelligence or reasoning. In this section, we briefly summarize the cognitive mechanisms related to success in academic geometry, with a particular emphasis on WM and intelligence.

WM is a limited-capacity system that enables information to be stored temporarily and manipulated (Baddeley, 2000). In the tripartite model of WM initially proposed by Baddeley and Hitch (1974), the central executive is the component responsible for controlling attentional resources and monitoring information processing across informational domains, while the storage of information is mediated by two domain-specific representational systems, that is, the phonological loop (which handles the temporary storage of verbal information), and the visuospatial sketchpad (which specializes in retaining and manipulating visual and spatial representations) (Baddeley, 1986). This model has been very successful in explaining the WM competencies of both children (Gathercole, Pickering, Ambridge, & Wearing, 2004; Giofrè et al., 2013a) and adults (Baddeley, 2012).

A large body of research has shown that WM predicts achievement across academic domains. WM is involved in complex cognitive tasks, such as reading comprehension (Borella, Carretti, & Pelegrina, 2010; Borella & de Ribaupierre, 2014; Carretti, Borella, Cornoldi, & De Beni, 2009), arithmetical problem-solving (Passolunghi & Mammarella, 2012), mental addition (Caviola, Mammarella, Cornoldi, & Lucangeli, 2012; Mammarella, Caviola, Cornoldi, & Lucangeli, 2013), mental subtraction (Caviola, Mammarella, Lucangeli, & Cornoldi, 2014), nonverbal problem-solving (Rasmussen & Bisanz, 2005), magnitude representation (Pelegrina, Capodieci, Carretti, & Cornoldi, 2014), and mathematical achievement more generally (Bull, Espy, & Wiebe, 2008; Passolunghi, Mammarella, & Altoè, 2008).

Intelligence can be defined as the ability to reason, plan, solve problems, think abstractly, understand complex ideas, learn quickly, and learn from experience (Gottfredson, 1997, p. 13). Intelligence has been variously defined and measured, but the "g-factor" derived from the common variance in a representative sample of achievement and cognitive tests seem to offer a useful approximation of the basic aspects of intelligence (e.g., Hunt, 2011; Spearman, 1904). The g-factor and academic achievement, including geometrical learning, are seen as closely related but distinct constructs (Kaufman, Reynolds, Liu, Kaufman, & McGrew, 2012).

Intelligence is a broad construct with many facets, one of which is WM. There is evidence, obtained in children and adults, to suggest that intelligence and WM are highly correlated (Gignac, 2014; Giofrè et al., 2013a). Nevertheless, the two constructs do not appear to overlap completely, so they can be seen as distinguishable (Conway, Kane, & Engle, 2003). Academic geometry is related to both intelligence and WM but, since they have a large portion of variance in common, it would be interesting to disentangle these processes and their specific influence on geometrical learning

Academic achievement in geometry can be supported by WM in various ways. Academic geometry typically involves retaining and manipulating a variety of information. Giofrè et al. (2013b) found WM contributed significantly to the geometry skills of 12th- and 13th-graders. They demonstrated, for example, that visuospatial WM (VSWM) explained a significant portion of the variance in geometrical performance and mediated the relationship between the intuitive geometry task proposed by Dehaene et al. (2006) and academic achievement in geometry, assessed by using a standardized test specifically devised for secondary school students (Cornoldi, Friso, & Pra Baldi, 2010). The importance of VSWM and of spatial skills generally is not limited to 12th- and

13th-graders; there is evidence to suggest that these skills are also crucial for younger and older age groups. In fact, the success of engineering students in technical drawing correlated closely with their VSWM and other spatial skills (Prieto & Velasco, 2012); and, among children in the early stages of learning geometry, success in academic geometry was found to be associated with their spatial and visuospatial skills (Clements & Battista, 1992). A role for VSWM in the successful completion of geometrical tasks is also indirectly supported by a study conducted by Mammarella et al. (2013b), who found that children with symptoms of nonverbal learning disabilities, whose impaired VSWM has been amply documented (e.g., Mammarella & Cornoldi, 2014), fared poorly in the intuitive geometry task (Dehaene et al., 2006). The study suffers from the limitation that the authors did not consider other variables potentially correlated with the intuitive geometry task, however, thus, it is possible that children with symptoms of nonverbal learning disabilities performed poorly in the intuitive geometry task because it was similar to some nonverbal reasoning tests in which these children likewise fail.

Although spatial abilities and VSWM seem to be closely related to success in academic geometry, verbal WM may be important too. For a start, geometrical problems can also be solved using verbal or logical skills, and students who apply these skills may outperform other students who rely on visuospatial manipulation alone (Lean & Clements, 1981). The variance in geometrical achievement explained in studies that only considered visuospatial WM was also not particularly high, meaning that other factors are likely to be involved (e.g., Giofrè et al., 2013b), such as verbal WM—which could be needed to retain the verbal information used to solve a geometrical problem (the text describing a problem, procedures, formulas, and so on). In addition to WM, intelligence also seems to relate to success in academic geometry.

To shed more light on the complex relationships between intelligence, WM and academic geometry, all of these variables need to be included in the same study, but such studies are rare because of the difficulty and costs of conducting them. In an important exception, Kyttälä and Lehto (2008) conducted a study on 15- and 16-year-olds that included measures of fluid intelligence (Raven's Progressive Matrices), various VSWM tasks, mental rotation and other mathematical tasks that drew on several areas of learning, including geometry. The authors found that geometrical ability correlated with VSWM (0.35), mental imagery (0.45), and fluid intelligence (0.60). These findings were then included in a path model, that showed that the relationship between geometrical ability and VSWM was mediated by fluid intelligence, with the model explaining about 40% of the variance in geometrical performance.

In another study, Giofrè et al. (2014) focused on 4th- and 5th-graders, and administered a complete battery of WM tasks—distinguishing between three components: verbal short-term memory (STM-V), visuospatial short-term memory (STM-VS), and WM—as well as several measures of intelligence (including Raven, Vocabulary, and Reasoning tasks), a measure of performance in

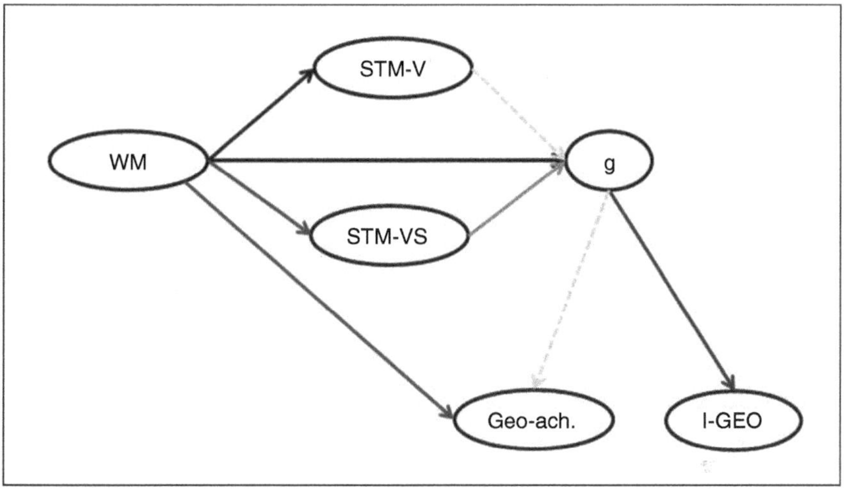

FIGURE 10.2 Structural model, adapted and modified from the Giofrè et al. (2014) study. *Dotted lines* represent not significant relationship and the color of line paths represents the magnitude of the relationship, with *darker color* corresponding to higher relationship. WM, Working memory; STM-V, verbal short term memory; STM-VS, visuospatial short term memory; g, *g* factor; Geo-ach, geometrical achievement; I-GEO, intuitive geometry.

intuitive geometry (the task used by Dehaene et al., 2006), and two measures of performance in academic geometry (geometrical questions and problems; Mammarella et al., 2012). The results showed that: (1) all components of WM correlated strongly with achievement in geometry (measured by geometrical questions and problems); (2) WM explained achievement in geometry over and above the effect of intelligence; and (3) after controlling for intelligence, performance in intuitive geometry did not correlate statistically with achievement in geometry. Further studies, however, should examine whether intuitive geometry is really only important for the initial learning of formal geometry, such as basic shapes and their properties. Fig. 10.2 shows the pattern of relationships between the variables.

At first glance, the results of the two previously described studies (Giofrè et al., 2014; Kyttälä & Lehto, 2008) may appear to diverge. But, taken together, they demonstrate that performance in geometry is related to both WM and intelligence, but the contribution of these two factors may differ in different levels of development depending on the types of geometry assessed and students' level of expertise in geometry (Demetriou, Spanoudis, & Shayer, 2013; Demetriou et al., 2014; Demetriou et al., 2013).

The strong relationship between WM and performance in geometry is largely related to complex-span tasks used to test WM (i.e., the active components of WM), which means tasks that primarily require attentional control, and not only the mere passive recall of information (Cornoldi & Vecchi, 2003; Giofrè et al., 2014). Intelligence also seems to be related to achievement in geometry, but this varies with grade and content of the geometry tests. Unlike high school students (Kyttälä & Lehto, 2008), for younger children WM seems to be relatively more important than intelligence, in agreement with the theory proposed by Demetriou et al. (2013). These results are also consistent with other reports that WM is more important in the early grades, but intelligence is the best predictor of achievement in older children (Giofrè, Borella, & Mammarella, 2015). This picture is also consistent with theories of geometrical development (e.g., van Hiele, 1986). In fact, high-school children are asked to solve complex geometrical problems (e.g., theorems, proofs, trigonometry) that are usually highly complex and abstract, and can be placed within the higher proficiency levels of the van Hiele model (1986). It is therefore not surprising that WM should become less important than intelligence—given that the latter, by definition, concerns the ability to solve complex problems. Finally, if we consider other executive functions too, VSWM is still the best predictor of success on an engineering course, even controlling for other executive functions (Prieto & Velasco, 2012). In fact, among the various executive functions, only updating—which is considered a measure of WM (Miyake et al., 2000)—was found to be related to achievement in geometry, while inhibition and shifting accounted for only a small portion of the variance, which became insignificant after controlling for updating (Lee et al., 2012).

EDUCATIONAL IMPLICATIONS

In the previous sections, we have amply underscored the relevance of learning geometry at school, and it has been demonstrated that a strong foundation in geometry is crucial to success on many standardized tests and in a variety of careers.

School curricula vary widely across countries in terms of their mathematical content and the emphasis placed on geometry as opposed to arithmetic and algebra. In the Netherlands and the USA, geometry teaching is strongly linked to the real world, whereas less emphasis is given to approaching geometry through real-word contexts in countries, such as France, Japan, and Singapore (French, 2004). The main difference concerns how procedural and formal the approach may be, and how precisely from a mathematical perspective the geometrical content required is described. Another aspect concerns the way in which content standards are implemented in the curricula, and the extent to which use of the computer is encouraged or expected.

The complexity of geometry (which involves mathematical reasoning, spatial abilities, verbal knowledge of geometrical concepts, and more) is evident from the difficulties that students encounter when learning this subject. It is noteworthy that the results of international assessments also suggest that these difficulties increase across grades, as the material becomes more abstract and complex (Clements, 2001). Teachers also report finding geometry a demanding topic to teach well: to do so, they need to know how to recognize geometrical problems and theorems, examine the history and cultural background of geometry, and understand the various different uses of geometry in the real world. According to Clements and Sarama (2011a), at all grade levels many teachers sometimes lack a sufficient grounding in geometry, and do not know how to best teach it (Ginsburg et al., 2006). This shortcoming stems from an inadequate learning of geometry in the early stages of their own careers as students, compounded by a poor preparation on this topic during their teacher training (Fujita & Jones, 2006a, 2006b). Combining these considerations with the recommendation of the US National Mathematics Advisory Panel (2008) that teachers often lack sufficient knowledge and understanding of mathematics for the age/grade they teach, it becomes clear that a better teacher preparation programs is a fundamental prerequisite for mathematics teachers at all grade levels.

In the light of the above, several intervention programs have been designed and tested with a view to improving teachers' professional expertise in teaching geometry, but not all of them improved the quality of their teaching (van der Sandt, & Nieuwoudt, 2004), or of the children's geometrical reasoning (Campbell & Rowan, 1995). There is nonetheless a persistent conviction that certain programs can make a difference to both teachers and their students (Jacobson & Lehrer, 2000; Swafford, Jones, & Thornton, 1997). Clements and Sarama (2011a) suggested some guidelines for effective professional development programs in early mathematics/geometry education:

1. They should include direct measures of what the teachers know about the subject in general, and about how it should be learned and taught (National Mathematics Advisory Panel, 2008).
2. It should focus on geometry and spatial reasoning, and be sufficiently intense and extensive, for example, a course of 4 weeks, 4 days a week, and 3 h a day.
3. It should be designed to enhance teachers' knowledge both of geometry and of research on student cognition regarding geometric content and on the qualities required of geometry lessons (what is taught, and how) (Swafford et al., 1997).

Based on these principles, the authors developed an early childhood mathematics curriculum in cooperation with preschool teachers, called *Building Blocks* (Clements & Sarama, 2007), which focused equally on the domains of numbers, geometry and spatial reasoning. The Building Blocks approach involves identifying the mathematical elements in children's activities, and developing their mathematical learning from there. The curriculum was designed to help children extend the mathematization of their everyday activities, from building blocks to art and stories, to puzzles and games, including activities designed to build their capacity for mental actions on objects. The educational goals are to develop competence in the two domains consistently identified as fundamental: (1) number concepts (counting, subitizing, or identifying the numerosity of small quantities) and arithmetical operations; and (2) spatial and geometric concepts and processes. Each of these domains was structured along research-based learning trajectories, which included three main components: a goal (to achieve a confident geometrical ability); a developmental progression (sequential progression from one van Hiele level to the next); and teaching activities (tasks specifically designed to implement geometrical concepts and spatial reasoning in everyday activities) (Sarama & Clements, 2009). Several studies have confirmed that the strengths of this curriculum are in its use of a blend of teaching methods (including explicit instructions) to promote a series of developmentally sequential activities (proposed using different modalities), while providing guidance to help teachers understand, assess, and facilitate student learning on the same sequential learning processes.

The results obtained in a first study with the Building Blocks method seemed very promising. In terms of effect sizes (Cohen, 1988), which measure the "magnitude" of a change from small to large, the improvements were large both for mathematics and for geometry (Clements & Sarama, 2008). More recently, the same authors used the program on a large sample of preschool and, after controlling for various confounding and moderator variables, they confirmed the utility of the program, but the effect sizes were only moderate (Clements & Sarama, 2011a).

In much the same spirit, a paper-and-pencil program was developed in Italy for 4- to 6-year-old children, called "I Know the Shapes" (Conosco le Forme; Lucangeli, Mammarella, Todeschini, Miele, & Cornoldi, 2009). Based on the van Hieles' (1986) theory and the intervention principles outlined by Clements and Sarama (2011a), this program includes both assessment tools and material to use in preparing practical activities to help children improve their recognition and understanding of the properties of shapes. The activities are presented with a clear explanation of the process that they are meant to enhance and associated with different strategy instructions. With much the same goals (i.e., to help teachers and students cope better with geometry), a training program has recently been published for 3rd- to 8th-graders that focuses on geometrical problem solving (Todeschini, Mammarella, Lucangeli, & Pellizzari, 2015).

Another way to help students discover and understand the most useful geometry learning strategy is to use software. The use made of computers to teach geometry varies widely across countries, as mentioned earlier, and the involvement of computers in mathematics teaching is extremely heterogeneous. Their integration in this area of learning is progressing very slowly and there is little consensus on the precise part they should play. There is nonetheless a widespread recognition of the potential value of software, such as Logo and dynamic geometrical software, such as Cabri (Bishop, Clements, Keitel, Kilpatrick, & Leung, 2003). Dynamic software is designed to foster geometrical reflection by means of nonstandard situations because the software requires the students communicating a construction, such as the properties of shapes, rather than just drawing it. The geometrical proprieties of Cabri constructions (e.g., the shape of a triangle) do not vary when basic elements are moved, and this helps students to attribute a meaning to the preservation of geometrical relationships (Pratt & Ainley, 1997).

TABLE 10.4 Examples of Scientifically Based Intervention Programs That are Available Online

Websites	Description
Math Forum	The Math Forum, a service of the Drexel University School of Education that provides an abundance of materials for educators, teachers and students of different ages. The Math Forum is a very extensive online resource for the mathematics education community, and has become a part of the National Council of Teachers of Mathematics (NCTM). Retrieved from: http://mathforum.org/#
Math Landing	The Math Landing, which provides resources and tools for elementary math specialists and teachers. The Math Landing has established a partnership with the Math Forum at Drexel University and with the International Society for Technology in Education. Retrieved from: http://www.mathlanding.org/search/catalog/geometry
NRICH Maths	The NRICH Maths, a project of the University of Cambridge involving a team of qualified teachers who work in and study a broad range of competences, preparing professional development programs for teachers that involve embedding mathematical tasks in everyday classroom activities. The main aim of the project is thus to enrich the mathematical experience of all learners, both students and teachers, providing effective activities and practical tutorials. Retrieved from: https://nrich.maths.org

Within this framework, and with the aim of facilitating students' progression from basic intuitions to higher levels of geometrical reasoning, Clements, Battista, and Sarama (2001) attempted in their Logo Geometry project to develop a research-based geometry curriculum for children from kindergarten to grade 6. Students interact with the Logo software, giving mathematical commands to guide a turtle's movements to induce the turtle to make the component parts of a given geometrical shape, and they see how the parts have to be joined together (Clements & Battista, 1992). The activities in the Logo curriculum help students to externalize their intuitive expectations by writing a sequence of commands (or a procedure) to draw a triangle, for instance. The Logo project prompted a number of empirical and experimental studies that showed, with slight differences, that the program had a positive effect in fostering mathematically correct and coherent angle measurements, reducing confusion regarding the measure of rotation and the measure of the angle, and nurturing the development of concepts, such as symmetry. Other important, scientifically based intervention programs are available online. Table 10.4 mentioned the sources of some good examples.

In addition to the above programs, there is also plenty of other educational material available. There are programs designed specifically for teachers (e.g., Improving Instruction in Geometry and Measurement; Smith, Silver, Stein,

Boston, & Henningsen, 2005) or for students (e.g., Master Math: Pre-Calculus and Geometry; Ross, 1996). The former is intended for middle-grade teachers and contains curricular materials that provide instructors with carefully sequenced, intellectually challenging teaching tasks intended to move away from classroom instruction based on routine skills alone. The material was designed to help teachers improve their knowledge of mathematical content, math teaching, and through these students' math learning, by providing continuing support. The latter is addressed to high-school or college students and is a thorough course that students complete on their own to develop their geometry skills. The content becomes gradually more difficult, building on previously presented material, and the aim of the program is not to replace the use of textbooks, but to serve as a reference book capable of providing clear explanations from a novel perspective.

CONCLUSIONS AND FUTURE DIRECTIONS

In our review of the literature on learning geometry, we have made an effort to consider studies deriving from different approaches: developmental psychology, cognitive psychology, educational psychology, and education. The case of geometry is intriguing because all of these approaches can contribute to our understanding of the teaching and learning of this complex subject. To thoroughly understand how best to go about this issue, we need to consider not only the cognitive processes involved in geometry, not only how geometrical knowledge is developed (are some concepts of geometry innate?), and not only how geometry is taught, but all of them together.

Unfortunately, though we already have several pieces of the puzzle, it remains to be completed. For example, theories of the development of geometrical knowledge have made virtually no progress in the last 30 years. To fill the gaps and complete the puzzle, we need further research in this area. As for the cognitive processes involved in geometry, future research should consider different cognitive measures that might be related to geometrical achievement, such as mental rotation tasks, since geometry problems can also involve the ability to mentally rotate stimuli (Zhang et al., 2012), and it has been argued that the capacity to manipulate and rotate objects is a component of intelligence (Johnson & Bouchard, 2005). Research is also needed on many other factors that presumably affect the acquisition of geometrical knowledge, such as metacognitive skills, motivation, and spatial anxiety (Ramirez, Gunderson, Levine, & Beilock, 2012).

Analyzing school curricula and how geometry is taught seems to suggest that both curriculum developers and teachers need to enrich students' learning of geometry not only by going beyond the traditional teaching methods, but first and foremost by improving their own knowledge of this subject. Finally, as regards educational studies in this area, the Building Blocks project (Clements & Sarama, 2007) and several other programs have been developed, and found valid, although—to the best of our knowledge—we still have no evidence of the long-term benefits of such programs. Hence, more research and more development of curriculum programs in this domain is needed—and this is another important piece of the puzzle that is still missing.

REFERENCES

Baddeley, A. D. (1986). *Working memory (vol. 11)*. Oxford, England: Oxford University Press.

Baddeley, A. D. (2000). The episodic buffer: a new component of working memory? *Trends in Cognitive Sciences*, *4*, 417–423.

Baddeley, A. D. (2012). Working memory: theories, models, and controversies. *Annual Review of Psychology*, *63*, 1–29.

Baddeley, A. D., & Hitch, G. J. (1974). Working Memory. In G. A. Bower (Ed.), *The psychology of learning and motivation: Advances in research and theory* (pp. 47–90). (Vol. 8). New York: Academic Press.

Bishop, A. (1980). Visual abilities and mathematics education. A review. *Educational Studies in Mathematics*, *11*, 257–269.

Bishop, A., Clements, M. K., Keitel, C., Kilpatrick, J., & Leung, F. K. S. (2003). *Second international handbook of mathematics education (vol. 10)*. New York, NY: Springer Science & Business Media.

Borella, E., & de Ribaupierre, A. (2014). The role of working memory, inhibition, and processing speed in text comprehension in children. *Learning and Individual Differences*, *34*, 86–92.

Borella, E., Carretti, B., & Pelegrina, S. (2010). The specific role of inhibition in reading comprehension in good and poor comprehenders. *Journal of Learning Disabilities*, *43*, 541–552.

Bornstein, M. H., Ferdinandsen, K., & Gross, C. G. (1981). Perception of symmetry in infancy. *Developmental Psychology*, *17*, 82–86.

Brown, D.L., & Presmeg, N. (1993). Types of imagery used by elementary and secondary school students in mathematical reasoning. In I. Hirabayashi, N. Nhoda, K. Shigematsu, F. L. Lin, (Eds.). Proceeding of the XVII PME International Conference.

Bull, R., Espy, K. A., & Wiebe, S. A. (2008). Short-term memory, working memory, and executive functioning in preschoolers: longitudinal predictors of mathematical achievement at age 7 years. *Developmental Neuropsychology*, *33*, 205–228.

Bynner, J. (1997). Basic skills in adolescents' occupational preparation. *Career Development Quarterly*, *45*, 305–321.

Campbell, P. F., & Rowan, T. E. (1995). *Project IMPACT: increasing the mathematical power of all children and teachers (Phase One Implementation Final Report)*. Baltimore, MD: Center for Mathematics Education, University of Maryland.

Carretti, B., Borella, E., Cornoldi, C., & De Beni, R. (2009). Role of working memory in explaining the performance of individuals with specific reading comprehension difficulties: a meta-analysis. *Learning and Individual Differences*, *19*, 246–251.

Caviola, S., Mammarella, I. C., Cornoldi, C., & Lucangeli, D. (2012). The involvement of working memory in children's exact and approximate mental addition. *Journal of Experimental Child Psychology*, *112*, 141–160.

Caviola, S., Mammarella, I. C., Lucangeli, D., & Cornoldi, C. (2014). Working memory and domain-specific precursors predicting success in learning written subtraction problems. *Learning and Individual Differences*, *36*, 92–100.

Cawley, J. F., Foley, T. E., & Hayes, A. M. (2009). Geometry and measurement: a discussion of status and content options for elementary school students with learning disabilities. *Learning Disabilities: A Contemporary Journal*, *7*, 21–42.

Cheng, K. (1986). A purely geometric module in the rat's spatial representation. *Cognition*, *23*, 149–178.

Cheng, K., & Gallistel, C. R. (1984). Testing the geometric power of an animal's spatial representation. In H. L. Roitblat, T. G. Bever, & H. S. Terrrace (Eds.), *Animal Cognition* (pp. 409–423). Hillsdale, NJ: Lawrence Erlbaum Associates Publishers.

Clements, D. H. (2001). Teaching and Learning Geometry. In *Research companion to the NCTM standards for mathematics*. Reston, VA: NCTM.

Clements, D. H. (2003). Teaching and learning geometry. In J. Kilpatrick, W. G. Martin, & D. Schifter (Eds.), *A research companion to principles and standards for school mathematics* (pp. 151–178). Reston, VA: National Council of Teachers of Mathematics.

Clements, D. H. (2004). Curriculum research: toward a framework for "research-based curricula." *Educational Research*.

Clements, D. H., & Battista, M. T. (1992). Geometry and spatial reasoning. In D. A. Grouws (Ed.), *Handbook of research on mathematics teaching and learning* (pp. 420–464). New York, NY: MacMillan.

Clements, D. H., & Sarama, J. (2003). Young children and technology: what does the research say? *YC: Young Children*.

Clements, D. H., & Sarama, J. (2007). Effects of a preschool mathematics curriculum: summative research on the Building Blocks project. *Journal for Research in Mathematics Education*, *38*, 136.

Clements, D. H., & Sarama, J. (2008). Experimental evaluation of the effects of a research-based preschool mathematics curriculum. *American Educational Research Journal*, *45*, 443–494.

Clements, D. H., & Sarama, J. (2011a). Early childhood mathematics intervention. *Science*, *333*, 968–970.

Clements, D. H., & Sarama, J. (2011b). Early childhood teacher education: the case of geometry. *Journal of Mathematics Teacher Education*, *14*, 133–148.

Clements, D. H., Battista, M. T., & Sarama, J. (2001). Logo and geometry. *Journal for Research in Mathematics Education. Monograph*, 1–177.

Cohen, J. (1988). *Statistical power analysis for the behavioral sciences* (2nd ed.). Hillsdale, NJ: Lawrence Erlbaum Associates.

Conway, A. R. A., Kane, M. J., & Engle, R. W. (2003). Working memory capacity and its relation to general intelligence. *Trends in Cognitive Sciences*, *7*, 547–552.

Cornoldi, C., & Vecchi, T. (2003). *Visuo-spatial working memory and individual differences*. Hove: Psychology Pr.

Cornoldi, C., Friso, G., & Pra Baldi, A. (2010). Prove MT avanzate-2. Prove MT avanzate di lettura e matematica 2 per il biennio della scuola secondaria di II grado. [Advanced MT 2 test—Advanced MT test of reading and mathematics for 9th and 10th grades]. Florence, Italy: Organizzazioni Speciali.

Crowley, M. L. (1987). The van Hiele model of the development of geometric though. In *Learning and Teaching Geometry, K-12*, Yearbook of the national council of teachers of Mathematics, Mary Montgomery Lindquist, Reston, Va.: National, 1–16.

Dehaene, S., Izard, V., Pica, P., & Spelke, E. S. (2006). Core knowledge of geometry in an Amazonian indigene group. *Science*, *311*, 381–384.

Demetriou, A., Spanoudis, G., & Shayer, M. (2013a). Developmental intelligence: from empirical to hidden constructs. *Intelligence*, *41*, 744–749.

Demetriou, A., Spanoudis, G., Shayer, M., Mouyi, A., Kazi, S., & Platsidou, M. (2013b). Cycles in speed-working memory-G relations: towards a developmental–differential theory of the mind. *Intelligence*, *41*, 34–50.

Demetriou, A., Spanoudis, G., Shayer, M., van der Ven, S., Brydges, C. R., Kroesbergen, E., … Swanson, H. L. (2014). Relations between speed, working memory, and intelligence from preschool to adulthood: Structural equation modeling of 14 studies. *Intelligence*, *46*, 107–121.

Doeller, C. F., & Burgess, N. (2008). Distinct error-correcting and incidental learning of location relative to landmarks and boundaries. *Proceedings of the National Academy of Sciences, 105*, 5909–5914.

Fischbach, A., Keller, U., Preckel, F., & Brunner, M. (2013). PISA proficiency scores predict educational outcomes. *Learning and Individual Differences, 24*, 63–72.

Fischbein, E. (1999). Intuitions and schemata in mathematical reasoning. *Educational Studies in Mathematics*, 11–50.

Fischbein, E., & Nachlieli, T. (1998). Concepts and figures in geometrical reasoning. *International Journal of Science Education, 20*, 1193–1211.

French, D. (2004). *Teaching and learning geometry*. London, UK: Continuum International Publishing Group.

Fujita, T., & Jones, K. (2006a). Primary trainee teachers' knowledge of parallelograms. *Proceedings of the British Society for Research into Learning Mathematics, 26*, 25–30.

Fujita, T., & Jones, K. (2006b). Primary trainee teachers' understanding of basic geometrical figures in Scotland. In J., Novotná, H., Moraová, M., Krátká, N., Stehlíková (Eds.), *Proceedings 30th conference of the international group for the psychology of mathematics education (PME30)* (pp. 129–136). Prague, Czech Republic.

Fuys, D. (1984). English Translation of Selected Writings of Dina van Hiele-Geldof and Pierre M. van Hiele.

Gathercole, S. E., Pickering, S. J., Ambridge, B., & Wearing, H. (2004). The structure of working memory from 4 to 15 years of age. *Developmental Psychology, 40*, 177–190.

Geary, D. C. (1995). Reflections of evolution and culture in children's cognition: implications for mathematical development and instruction. *American Psychologist, 50*(1), 24–37.

Geary, D. C. (1996). Sexual selection and sex differences in mathematical abilities. *Behavioral and Brain Sciences, 19*, 229.

Gignac, G. E. (2014). Fluid intelligence shares closer to 60% of its variance with working memory capacity and is a better indicator of general intelligence. *Intelligence, 47*, 122–133.

Ginsburg, H. P., Kaplan, R. G., Cannon, J., Cordero, M. I., Eisenband, J. G., & Galanter, M. (2006). Helping early childhood educators to teach mathematics. In *Critical issues in early childhood professional development* (pp. 171–202). Baltimore, MD: Brookes Publishing.

Giofrè, D., Mammarella, I. C., & Lucangeli, D. (2009). L'apprendimento della geometria in bambini da 4 a 6 anni. *Difficoltà in Matematica, 5*, 21–38.

Giofrè, D., Mammarella, I. C., & Cornoldi, C. (2013a). The structure of working memory and how it relates to intelligence in children. *Intelligence, 41*, 396–406.

Giofrè, D., Mammarella, I. C., Ronconi, L., & Cornoldi, C. (2013b). Visuospatial working memory in intuitive geometry, and in academic achievement in geometry. *Learning and Individual Differences, 23*, 114–122.

Giofrè, D., Mammarella, I. C., & Cornoldi, C. (2014). The relationship between geometry, working memory and intelligence in children. *Journal of Experimental Child Psychology, 123*, 112–128.

Giofrè, D., Borella, E., & Mammarella, I. C. (2015). The relationship between intelligence, working memory, self-esteem and academic achievement. *Manuscript Submitted for Publication.*

Gittler, G., & Glück, J. (1998). Differential transfer of learning: effects of instruction in descriptive geometry on spatial test performance. *Journal of Geometry and Graphics, 2*, 71–84.

Gottfredson, L. (1997). Mainstream science on intelligence: an editorial with 52 signatories, history, and bibliography. *Intelligence, 24*, 13–23.

Halberda, J., Mazzocco, M. M. M., & Feigenson, L. (2008). Individual differences in non-verbal number acuity correlate with maths achievement. *Nature, 455*, 665–668.

Hermer, L., & Spelke, E. S. (1994). A geometric process for spatial reorientation in young children. *Nature, 370*, 57–59.

Hunt, E. (2011). *Human Intelligence*. New York, NY: Cambridge University Press.

Hunt, E. (2012). What makes nations intelligent? *Perspectives on Psychological Science*, *7*, 284–306.

INVALSI. (2011). *Servizio Nazionale Di Valutazione 2010–11 [National evaluation service 2010–11]*. Rome, Italy: INVALSI.

Izard, V., & Spelke, E. S. (2009). Development of sensitivity to geometry in visual forms. *Human Evolution*, *23*, 213–248.

Jacobson, C., & Lehrer, R. (2000). Teacher appropriation and student learning of geometry through design. *Journal for Research in Mathematics Education*, *31*, 71.

Johnson, W., & Bouchard, T. J. (2005). The structure of human intelligence: it is verbal, perceptual, and image rotation (VPR), not fluid and crystallized. *Intelligence*, *33*, 393–416.

Jones, K. (2000). Teacher knowledge and professional development in geometry. *Proceedings of the British Society for Research into Learning Mathematics*, *20*, 109–114.

Kaufman, S. B., Reynolds, M. R., Liu, X., Kaufman, A. S., & McGrew, K. S. (2012). Are cognitive g and academic achievement g one and the same g? An exploration on the Woodcock–Johnson and Kaufman tests. *Intelligence*, *40*, 123–138.

Kyttälä, M., & Lehto, J. E. (2008). Some factors underlying mathematical performance: the role of visuospatial working memory and non-verbal intelligence. *European Journal of Psychology of Education*, *23*, 77–94.

Lean, G., & Clements, M. A. (Ken) (1981). Spatial ability, visual imagery, and mathematical performance. *Educational Studies in Mathematics*, *12*, 267–299.

Lee, K., Ng, S. F., Pe, M. L., Ang, S. Y., Hasshim, M. N. A. M., & Bull, R. (2012). The cognitive underpinnings of emerging mathematical skills: executive functioning, patterns, numeracy, and arithmetic. *British Journal of Educational Psychology*, *82*, 82–99.

Lever, C., Wills, T., Cacucci, F., Burgess, N., & O'Keefe, J. (2002). Long-term plasticity in hippocampal place-cell representation of environmental geometry. *Nature*, *416*, 90–94.

Lourenco, S. F., & Huttenlocher, J. (2006). How do young children determine location? Evidence from disorientation tasks. *Cognition*, *100*, 511–529.

Lourenco, S. F., Bonny, J. W., Fernandez, E. P., & Rao, S. (2012). Nonsymbolic number and cumulative area representations contribute shared and unique variance to symbolic math competence. *Proceedings of the National Academy of Sciences USA*, *109*, 18737–18742.

Lucangeli, D., Mammarella, I. C., Todeschini, M., Miele, G., & Cornoldi, C. (2009). *Know the geometrical shapes: assessment and intervention for children from 4 to 6 years old. [Conoscere le forme geometriche: Valutazione delle competenze e schede di intervento per bambini dai 4 ai 6 anni]*. Trento, Italy: Erickson.

Mammarella, I. C., & Cornoldi, C. (2014). An analysis of the criteria used to diagnose children with Nonverbal Learning Disability (NLD). *Child Neuropsychology*, *20*, 255–280.

Mammarella, I. C., Caviola, S., Cornoldi, C., & Lucangeli, D. (2013a). Mental additions and verbal-domain interference in children with developmental dyscalculia. *Research in Developmental Disabilities*, *34*, 2845–2855.

Mammarella, I. C., Giofrè, D., Ferrara, R., & Cornoldi, C. (2013b). Intuitive geometry and visuospatial working memory in children showing symptoms of nonverbal learning disabilities. *Child Neuropsychology*, *19*, 235–249.

Mammarella, I. C., Todeschini, M., Englaro, G., Lucangeli, D., & Cornoldi, C. (2012). *Geometria test [Geometry assessment]*. Trento: Erickson.

Martin, M. O., Mullis, I. V. S., & Foy, P. (with Olson, J. F., Erberber, E., Preuschoff, C., & Galia, J.) (2008). TIMSS 2007 international science report. Chestnut Hill, MA: TIMSS & PIRLS Study Center 6. Giofrè, D., Mammarella, I. C., & Cornoldi, C. (2014). The relationship among geometry, working memory, and intelligence in children. *Journal of Experimental Child Psychology, 123*, 112–128.

Miyake, A., Friedman, N. P., Emerson, M. J., Witzki, A. H., Howerter, A., & Wager, T. D. (2000). The unity and diversity of executive functions and their contributions to complex "Frontal Lobe" tasks: a latent variable analysis. *Cognitive Psychology, 41*, 49–100.

Mullis, I. V. S., Martin, M. O., Foy, P., & Aror, A. (2012). *TIMSS 2011 international results in mathematics*. Boston, MA: IMSS & PIRLS International Study Center.

National Mathematics Advisory Panel. (2008). *Foundations for success: the final report of the National Mathematics Advisory Panel*. Washington, DC: Department of Education, Office of Planning, Evaluation and Policy Development.

OECD. (2013). *PISA 2012 Assessment and Analytical Framework*. OECD Publishing.

OECD. (2014). *Results: What students know and can do—student performance in mathematics, reading and science (Vol. I, Revised edition, February 2014)*. Pisa: OECD Publishing.

Owens, K., & Outhred, L. (2006). The complexity of learning geometry and measurement. In A. Gutierrez, & P. Boero (Eds.), *Handbook of research on the psychology of mathematics education* (pp. 83–115). Rotterdam, The Netherlands: Sense publishers.

Passolunghi, M. C., & Mammarella, I. C. (2012). Selective spatial working memory impairment in a group of children with mathematics learning disabilities and poor problem-solving skills. *Journal of Learning Disabilities, 45*, 341–350.

Passolunghi, M. C., Mammarella, I. C., & Altoè, G. (2008). Cognitive abilities as precursors of the early acquisition of mathematical skills during first through second grades. *Developmental Neuropsychology, 33*, 229–250.

Pelegrina, S., Capodieci, A., Carretti, B., & Cornoldi, C. (2014). Magnitude representation and working memory updating in children with arithmetic and reading comprehension disabilities. *Journal of Learning Disabilities, 48*(6), 658–668.

Piaget, J., & Inhelder, B. (1967). *The child's conception of space*. New York, NY: W. W. Norton.

Piaget, J., Inhelder, B., & Szeminska, A. (1960). *The child's conception of geometry*. London, UK: Routledge.

Pratt, D., & Ainley, J. (1997). The construction of meanings for geometric construction: two contrasting cases. *International Journal of Computers for Mathematical Learning, 1*, 293–322.

Prevost, F. J. (1985). Geometry in the junior high school. *Mathematics Teacher, 78*, 411–418.

Prieto, G., & Velasco, A. D. (2012). Predicting academic success of engineering students in technical drawing from working memory. *Journal for Geometry and Graphics, 16*, 111–120.

Ramirez, G., Gunderson, E. A., Levine, S. C., & Beilock, S. L. (2012). Spatial anxiety relates to spatial abilities as a function of working memory in children. *The Quarterly Journal of Experimental Psychology, 65*, 474–487.

Rasmussen, C., & Bisanz, J. (2005). Representation and working memory in early arithmetic. *Journal of Experimental Child Psychology, 91*, 137–157.

Rosch, E. (1975). Cognitive representations of semantic categories. *Journal of Experimental Psychology: General, 104*, 192–233.

Ross, D. A. (1996). *Master math: Pre-calculus and geometry*. Franklin Lakes, NJ: Career Press.

Sarama, J., & Clements, D. H. (2009). *Early childhood mathematics education research: Learning trajectories for young children*. New York, NY: Routledge.

Smith, M. S., Silver, E. A., Stein, M. K., Boston, M., & Henningsen, M. A. (2005). *Improving instruction in geometry and measurement: Using cases to transform mathematics teaching and learning (Volume 3)*. New York, NY: Teachers College Press.

Spearman, C. (1904). "General Intelligence," objectively determined and measured. *The American Journal of Psychology, 15*, 201.

Spelke, E. S., Breinlinger, K., Macomber, J., & Jacobson, K. (1992). Origins of knowledge. *Psychological Review*, *99*, 605–632.

Spelke, E. S., Lee, S. A., & Izard, V. (2010). Beyond core knowledge: natural geometry. *Cognitive Science*, *34*, 863–884.

Swafford, J. O., Jones, G. A., & Thornton, C. A. (1997). Increased knowledge in geometry and instructional practice. *Journal for Research in Mathematics Education*, *28*, 467–483.

Tatsuoka, K. K., Corter, J. E., & Tatsuoka, C. (2004). Patterns of diagnosed mathematical content and process skills in TIMSS-R across a sample of 20 countries. *American Educational Research Journal*, *41*, 901–926.

Todeschini, M., Mammarella, I. C., Lucangeli, D., & Pellizzari, E. (2015). *Learn to solve geometrical problems. [Imparo a risolvere i problemi di geometria]*. Trento, Italy: Erickson.

Van der Sandt, S., & Nieuwoudt, H. D. (2004). Prospective mathematics teachers' geometry content knowledge: a typical case? In S. Nieuwoudt, S. Froneman, & P. Nkhoma (Eds.), *Proceedings of tenth national congress of the association for mathematics education of South Africa* (pp. 250–262). Potchefstroom, South Africa: AMESA.

van Hiele, P. M. (1986). *Structure and insight: A theory of mathematics education*. Orlando, FL: Academic Press.

Verstijnen, I., van Leeuwen, C., Goldschmidt, G., Hamel, R., & Hennessey, J. (1998). Creative discovery in imagery and perception: combining is relatively easy, restructuring takes a sketch. *Acta Psychologica*, *99*, 177–200.

Zhang, D., Ding, Y., Stegall, J., & Mo, L. (2012). The effect of visual-chunking-representation accommodation on geometry testing for students with math disabilities. *Learning Disabilities Research & Practice*, *27*, 167–177.

Chapter 11

The Unit Circle as a Grounded Conceptual Structure in Precalculus Trigonometry

Kevin W. Mickey, James L. McClelland
Department of Psychology, Stanford University, Stanford, CA, United States

The United States remains behind other nations in mathematics and science education. According to results from the 2012 assessment by the Program on International Student Assessment (PISA; Kelly, Nord, Jenkins, Chan, & Kastberg, 2013), only 25% of US 15-year-olds reached the level where they could understand and complete higher order tasks, such as "solv[ing] problems that involve visual or spatial reasoning…in unfamiliar contexts" (OECD, 2004, p. 55), less than the average level (31%) of students across the 34 participating OECD-member countries. Like the National Mathematics Advisory Panel (2008), we believe that our nation must find ways to enhance the training of its high-school students in mathematics and science, so they can have the opportunity to compete for positions in highly STEM-dependent technology fields in industry, government, research, and education.

Our research explores the role of a visuospatial representation that grounds the mathematical concepts covered in precalculus trigonometry. A difficult subject for many, trigonometry sits at the gateway to entry into university-level science, mathematics, and technology coursework and ultimately into careers in STEM disciplines. Our research explores whether it may be possible to enhance reliance on a spatially grounded conceptual model, which we refer to as the unit circle, and thereby increase high school students' success in this difficult subject. In preliminary studies reviewed later, we found that individuals who report visualizing trigonometric quantities as measurements within the unit circle perform better than those who do not report using this model. We have also found that a brief lesson in the use of the unit circle produces postlesson benefits relative to no lesson or a rule-based lesson for some participants, further supporting the view that the unit circle provides a grounded conceptual framework that supports acquiring an understanding of trigonometry.

Acquisition of Complex Arithmetic Skills and Higher-Order Mathematics Concepts
http://dx.doi.org/10.1016/B978-0-12-805086-6.00011-4

These promising results are limited, however, in that the participants were Stanford undergraduates who had prior exposure to trigonometry, and those who performed poorly in the pretest showed little or no benefit. Our proposed research seeks to adapt and extend the lesson to make it more suitable for high school and community college students who are just learning trigonometry for the first time, and to examine whether such a lesson might help promote mastery of essential trigonometry concepts among such students. At the end of this chapter we consider several specific challenges we face in addressing this goal. These include the possibility that students may lack relevant knowledge of mathematical ideas that could be considered to be prerequisites for understanding the unit circle; the possibility that their knowledge of these elements may not be strong enough to allow them to manipulate the elements of the representation internally (mentally) rather than relying on external supports; and finally, the possibility that many students may hold beliefs about the nature of mathematics and/or their own abilities that make it difficult for them to treat trigonometry as a coherent conceptual system to be learned.

GROUNDED CONCEPTUAL STRUCTURES IN MATHEMATICAL COGNITION

Historically, multiple points of view have emerged regarding the nature of mathematical reasoning and of factors that allow mathematical reasoning to be successful (Dantzig, 2007). On one hand, it is common to view mathematics as an essentially formal system, in which structured arrangements of abstract symbols are manipulated according to structure sensitive rules (Marcus, 2003; Newell & Simon, 1961; Russell, 1903). On the other hand, many have suggested that mathematical and scientific reasoning often operates on idealized objects humans can manipulate in their minds, mirroring external manipulations of real objects (Barwise & Etchemendy, 1996; Harnad, 1990; Lakoff & Núñez, 2000). As one example, Shepard (2008), (Fig. 11.1) described an ancient proof of the Pythagorean theorem that is carried out through the mental manipulation of four congruent right triangles with sides of lengths a and b and hypotenuse of length c, within a bounding square with sides of length $a + b$. In panel C, the region within the bounding square not covered by the four triangles has area $a^2 + b^2$. In D, which can be constructed by translating three of the triangles in C (without changing their shape), the region not covered by the four triangles has area c^2. To individuals who have an intuitive appreciation for conservation of area under translation and who appreciate how area depends on the lengths of sides of right-angled figures, the proof seems intuitively compelling, though care is required to make the proof rigorous.

Shepard's (2008) article makes bold claims about the role of spatial reasoning in mathematics and even in scientific discovery, and there is now a body of evidence to support these claims. Wai, Lubinski, and Benbow (2009) present evidence supporting the claim that spatial ability represents a third psychological

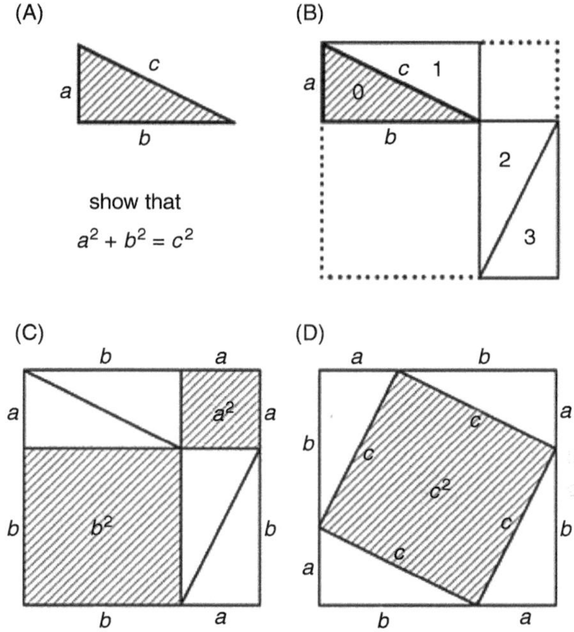

FIGURE 11.1 **A proof of the Pythagorean theorem.** *(From Shepard, R. N. (2008). The step to rationality: the efficacy of thought experiments in science, ethics, and free will.* Cognitive Science, *32(1), 6. Copyright 2008 Cognitive Science Society, Inc. Reprinted with permission.)*

attribute over and above verbal and mathematical ability, and is strongly associated with success in the physical sciences, mathematics, and engineering. Adolescents with greater spatial ability are more likely to major in a STEM field, to enter a STEM occupation, and to produce STEM publications and patents (Kell, Lubinski, Benbow, and Steiger, 2013). A broad range of other work, reviewed in a meta-analysis by Uttal et al. (2013), supports the view that spatial ability can be taught. This work suggests that individuals across the spectrum of spatial ability can benefit to a comparable extent, and that the typical extent of benefit from the interventions considered was equal to a shift of nearly half of one standard deviation of spatial ability scores in the whole population.

The work just described has been construed as supporting a perspective in which there is some generalized *visuospatial faculty*, perhaps malleable but quite general—or, if not fully general, its partitioning is still based on a broad distinction, such as a distinction between within-object and between-object processes (Uttal et al., 2013). An alternative perspective draws on insights from seminal work in cognitive science, developmental psychology, and the learning sciences on the role of an acquired understanding of a set of integrated conceptual relationships that, in many domains of mathematics, can be represented by externalized visual depictions like those in Shepard's proof. This perspective contrasts with the symbolic perspective (i.e., that we manipulate expressions using structure sensitive rules) and with the visuospatial faculty idea. We emphasize this third perspective, and focus on helping students acquire an integrated understanding, while acknowledging that both mastery of the rules of symbol manipulation and generalized visuospatial ability may also have roles to play in mathematical cognition.

For us, the essential idea in what we will call the *grounded conceptual structure* perspective is that symbolic expressions—be they sentences or equations—need not be manipulated simply as such. Instead, we argue that they can (and should) be used to construct internal representations in which tokens standing for objects referred to in the expressions are placed in particular relationships with each other (Bransford, Barklay, & Franks, 1972; Glenberg, Gutierrez, Levin, Japuntich, & Kaschak, 2004), within a conceptual framework that the comprehender understands (Bartlett, 1932; Bransford & Johnson, 1972). Consider the passage "Ben needs to feed the animals. He pushes the hay down the hole. The goat eats the hay." (from Glenberg et al., 2004). Comprehension of this vignette is thought to involve mentally acting out Ben's and the goat's activities. This process can be supported by manipulation of toy objects corresponding to those mentioned in a play farm set, or it can occur entirely in the imagination. Either way, implied relationships (Ben is above the goat) may be captured in the constructed result. A similar visuospatial construction can occur when a person understands the expression $a^2 + b^2 = c^2$. The symbol manipulation approach to conceptualizing mathematical ability fails to make contact with these ideas, and leaves the student performing calculations according to arbitrary memorized rules rather than constructing representations of meaningful quantities that can be connected to the properties of referenced or imagined objects in the world (Thompson, Philipp, Thompson, & Boyd, 1994).

When it comes to mathematical expressions, they may in some instances be constructed with reference to very specific situations. Thus we may convert "Together Ann and Ben have 7 apples. If Ben has 3 apples, how many apples does Ann have?" into expressions, such as $a + b = 7$ and $b = 3$. Purely symbolic actions could then be carried out, but mathematics educators have argued persuasively that such actions are highly error prone. Instead, maintaining contact with the referenced context deepens engagement with the underlying relationships (Thompson et al., 1994), helps prevent errors (Mayer & Hegarty, 1996), and facilitates transfer (Lewis, 1989) by providing a specific conceptual grounding from which a more generalizable representation of the meanings of the mathematical expressions may ultimately emerge, or at least, to which a more generalizable representation may be linked, as discussed more fully later.

In the framework just described, a visuospatially grounded representation might provide a coherent conceptual framework within which statements in a particular mathematical domain "make sense." In this view, articulated by Case (1996), the visuospatial representation participates (along with other elements) in providing an abstract, generalizable, framework that can be applied to a wide range of specific instantiations—constituting what he calls a *core conceptual structure*. Case (1996) builds up these ideas for natural number (his "mental number line") and extends this to encompass fractions (Moss & Case, 1999). The representation is abstract and generalizable, because (in the case of the number line) it can be applied to sets of enumerable items of any kind, be they coins, toy frogs, or even items of a less tangible nature (days of the week, even ideas).

The visuospatially grounded conceptual representations we propose are related in some ways to other proposals in the psychological literature that are posed as alternatives to symbol manipulation, but are different in some important ways. Shepard (2008; Shepard & Chipman, 1970) and Dehaene (1992) discuss visuospatial representations in mathematics but do not emphasize the role of learning or of culturally defined conventions in structuring such representations. The mental structures we describe may also be like Johnson-Laird's (1983) mental models in some ways, in that they involve internal representations that capture features of and relations among entities that can have external referents. One difference is that in our case we are specifically interested in representations with visuospatial structure, whereas Johnson-Laird focuses primarily on representations that do not have an explicitly spatial character. Also, we emphasize the role of extensive exposure to a culturally constructed system that may adhere to a set of partially arbitrary conventions, usually acquired in a formal educational setting. These different perspectives may be mutually compatible, but the emphasis is certainly different.

Several issues for learning and learners arise in our framework. One of these is *gradual integration and elaboration*. Acquisition of a fully elaborated core conceptual structure is a process that occurs gradually over time, in part because of a gradual associative process that interlinks relevant parts and in part because the full structure may depend on previous consolidation of constituent structures. For example, an initial simple conceptual structure for numbers from 1 to 10 may be elaborated to provide the basis for a more complex conceptual structure for the numbers from 1 to 100 by using one copy of the structure for the 10 decades and embedding 10 additional copies to represent the numbers within each decade. The ability to create the linkages is thought to depend on the gradual consolidation and integration of the representation of the constituent structures, groups of 10 in this case. Furthermore, a range of factors may play a role in fostering the acquisition and construction of a particular conceptual structure or the engagement of relevant background knowledge that could contribute to supporting performance while the structure is developing.

One such issue that we will consider is *externalization then internalization*. Based on the findings of Glenberg et al. (2004) in their research described previously, requiring learners to engage in the actual manipulation of a set of physical toys to act out the content of a symbolic expression (Ben needs to feed the animals. He pushes the hay down the hole), then gradually removing this requirement while instructing the learner to imagine the externalized representation while reading other similar texts could help establish the practice of mapping symbolic expressions to their meaning in the conceptual system represented by the diagram, rather than relying on surface representations.

Another potentially important issue is *generic versus specific content.* The unit circle representation that we introduce later is generic in that it can be applied to a wide range of specific contexts; relations that hold in this generic model also hold in a range of specific instantiations to which the model can then be applied. As Kaminski, Sloutsky, and Heckler (2008) have argued, this may have advantages over models of specific systems that can bring additional irrelevant content that may obscure the relationships that the model is intended to convey. While there may be benefits of starting with a specific familiar instance of a content domain and then shifting to the more abstract/generic unit circle (Goldstone & Son, 2005), this could complicate our effort to compare lessons based on the unit circle to rule-based lessons. We have, therefore chosen to stay with the abstract unit-circle framework in our initial investigations.

The Unit Circle as a Grounded Conceptual Structure for Trigonometry

The aforementioned ideas provide the setting for our exploration of the role of a particular visuospatial representation in trigonometry. Specifically, we hypothesize that a particular visuospatial representation—the unit circle—links trigonometric expressions to measurable properties of a physically realizable visuospatial model that learners can internalize and then use as a basis for reasoning about relationships between trigonometric expressions that they encounter in precalculus trigonometry. The topics and ideas encompassed under this heading build on prior exposure to the concepts of sine, cosine, and tangent as these are encountered in right triangles (sometimes called "right-triangle trigonometry"), extending their definitions and introducing additional trigonometric functions. The topics include relationships among these functions, graphs of these functions, rules for converting among expressions for such functions, and applications of such functions to problems throughout the physical sciences and engineering.

A reader might be forgiven for thinking that the domain of trigonometry would be one in which, if any, spatial thinking and visualization would be emphasized. Trigonometry appears at first glance to be predominantly visuospatial in nature: Standard definitions of the three elementary trigonometric functions are usually given in terms of ratios of lengths of sides of right triangles, which can be drawn on paper or even constructed out of wood. Yet, an examination of textbooks currently in use in precalculus trigonometry courses and of the articles we have found on the understanding and teaching of trigonometry presents a fragmentary and in some ways inconsistent picture. Many texts (e.g., Foerster, 1990; Hornsby, Lial, & Rockswold, 2011, recently-used texts at Palo Alto High School and Foothill College, respectively) emphasize graphing and the use of the unit circle as a starting place for generalizing the sine and cosine functions beyond the right triangle, but move quickly away from this representation.

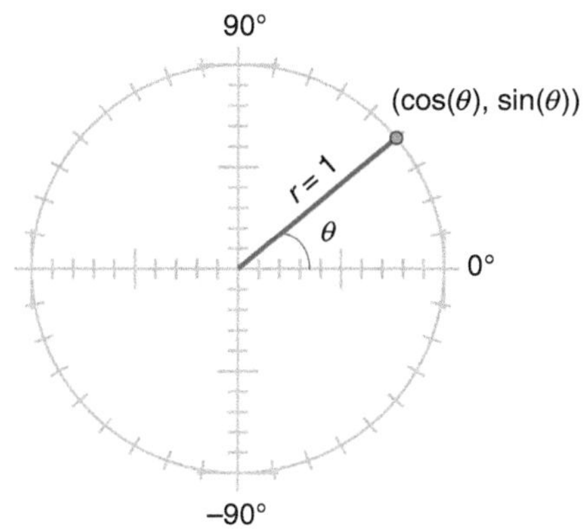

FIGURE 11.2 **The unit circle.**

To illustrate this inconsistent treatment, consider the initial definition of the sine and cosine functions. The presentation of these concepts may begin with the unit circle (as shown in Fig. 11.2). The diagram treats angles as arising by rotating a radial line of length 1, emanating from the center of a circle centered at (0,0) on a (x,y) coordinate plane. Angles are measured in degrees or radians, with 360 degrees or 2π radians corresponding to a full rotation. The rotation is either counter-clockwise (for positive angles) or clockwise (for negative angles) from the starting position [where the radial line points horizontally from the origin to the right and corresponds to the line from (0,0) to (1,0)]. The cosine of the angle is defined as the x-coordinate of the endpoint of the line, and the sine as corresponding to the y-coordinate. From this clearly visuospatial starting place, it is possible to shift almost entirely into algebra. For example, a fundamental trigonometric identity can be derived by applying the formula for the distance between two points (here 0,0 and the x,y coordinates of the endpoint of the radial line) and noting that, by the definition of the unit circle, this distance (the radius of the circle) is equal to 1. Applying the Pythagorian theorem, $x^2 + y^2 = 1$. Replacing x and y with $\cos\theta$ and $\sin\theta$ respectively, we obtain the identity $\cos^2\theta + \sin^2\theta = 1$. Furthermore, all of the other trigonometric functions can be defined algebraically (e.g., $\tan\theta = \sin\theta/\cos\theta$; $\sec\theta = 1/\cos\theta$). Additional key identities, such as $1 + \tan^2\theta = \sec^2\theta$ may then be derived purely by algebraic manipulation or simply stated as rules to be learned, and the wave graphs of such functions (i.e., graphs of the form $y = \cos\theta$ or $y = \sec\theta$) may then be introduced, with little or no engagement with the underlying construct that gives rise to the values of these functions. Thereby some of the features of these graphs [e.g., the fact that $\cos(0) = 1$ and $\sin(0) = 0$] are stripped of their relation to the underlying meaning of the functions and become arbitrary facts to be memorized by the student. Our experience with teachers in four precalculus classrooms at two high schools suggests considerable variability among teachers in their tendency to engage students with this or other visuospatial representations.

While in some classrooms, whole 2-h time blocks may be dedicated to teaching the unit circle, in others there is a tendency to rely on the teaching of rules or procedures that, if mastered, would allow students to calculate answers to particular problems without any comprehension of the expressions whatsoever.

What, indeed, are the best practices here? One can find passionately written articles, such as the one by Shear (1985), a dedicated teacher of trigonometry, decrying the algebraic approach. Shear quotes Poincaré (1968) and Einstein (1979) for proclaiming their understanding of mathematical relationships in visuospatial terms, lays out a visuospatial framework in which the identities aforementioned are explicitly made visually apparent (or so he claims), and then asserts that use of his framework may allow students to gain "some sense of that elegance of thought and expression which, in the words of Poincaré, 'all true mathematicians know'." Unfortunately, Shear presents no evidence other than his own intuition and personal observations of his own students, and many textbook writers and classroom teachers continue to rely very heavily on the algebraic approach, especially when producing derivations. It is true that there is a textbook by Gelfand and Saul (2001) that develops a visuospatially grounded approach, but it is only one among many. It is also true that all teachers and all textbooks make use of some visuospatial examples (often, based on wave functions), but these uses vary widely in their extent and stated purpose and are not based on a scientific foundation.

Our effort to find scientific evidence has unearthed rather little beyond the above. We found one study based on a master's thesis (Kendal & Stacey, 1996) comparing two alternative visuospatial approaches to teaching right-triangle trigonometry problems of the kind covered in 9th or 10th grade geometry, as well as some more recent studies by Ninness et al. (2006, 2009) studying methods for teaching formula-to-graph relationships involving trigonometric functions and the ability to distinguish different types of relationships between formulas. These studies point to the difficulties many students have mastering the relationships between trigonometric functions and their graphs, and show that training using interactive graphing software can be effective, but the small sample size and lack of comparison conditions makes it difficult to draw inferences about the relative advantage of the approach. We also found a recent study by Moore (2013) applying the conceptual approach favored by Thompson et al. (1994) to helping students construct a cogent representation of the meanings of angular and radian measure in terms of arc length relative to the radius of a circle. Again only a small sample was used and there was no comparison of different approaches, but the approach is consistent with our perspective and informs our own studies.

PRELIMINARY INVESTIGATIONS

In accordance with the ideas that the use of a conceptual model within which symbolic expressions have meaning can contribute to success in mathematics and that reliance on such skills and models can be enhanced, we have undertaken

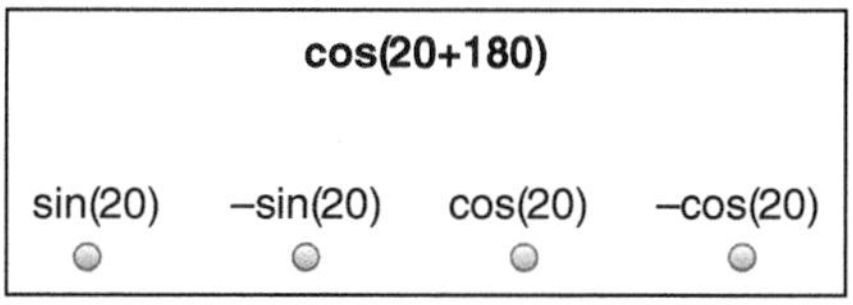

FIGURE 11.3 **An example problem.** Participants saw a probe expression at the top of an on-screen display and were instructed to choose the equivalent expression from the alternatives below it.

a series of preliminary studies using undergraduate students at Stanford University. Our focus has been on a specific topic nearly all of these students were exposed to in high school, namely identities among trigonometric expressions involving the sine and cosine functions. We focused on a subset of such identities, involving the sine and cosine functions with positive or negative arguments between 10 and 80 degrees and subject to offsets of [–180, –90, 0, +90, or +180] degrees. While Stanford is a highly selective institution, not all Stanford undergraduates are especially strong in mathematics, allowing us to sample a range of prior experience and ability levels with respect to knowledge of these relationships. We asked three questions in our studies: (1) to what extent do students rely on visuospatial representations, as compared with rules or mnemonics, and among visuospatial representations, which do they use in solving the identity problems we used? (2) To what extent is the use of a particular representation associated with success in solving the class of problems we posed to our students? (3) For selected representations, how much can a student's performance be improved by a brief lesson based on a particular representation?

An example of the type of problem we used is shown in Fig. 11.3. These problems all involved a probe expression, created by choosing either the sine or cosine function, followed by an expression in parentheses consisting of a signed two-digit base value and an offset (participants were told these values corresponded to degrees). There were 20 probe types, defined by all combinations of the function (sin or cos), the sign of the base value (positive or negative), and the value of the offset {–180, –90, 0, 90, 180}. Specific problems within each type further varied in the magnitude of the base value [equal to 20 in the example shown in the figure; always a multiple of 10 in the range (10, 80)], and order (base value first or offset first; when the first element was positive, its sign was omitted, as in this example). The four choice alternatives were always $\sin(\theta)$, $-\sin(\theta)$, $\cos(\theta)$, and $-\cos(\theta)$, where θ stands for the base value. One and only one alternative is correct for each problem. The participants were instructed to respond "quickly but still accurately" and were asked not to use paper and pen/pencil or a calculator and not to refer to any outside sources.

How might problems of the kind described be solved? One approach would be to rely on rules and formulas. Indeed a very small number of rules, together with a few very general rules of algebra, are sufficient to solve all of these problems. For example the rule $\cos(\theta + 180) = -\cos(\theta)$ yields the correct answer in this case. A set of rules sufficient to answer all of the problems is given in

Spatial representations

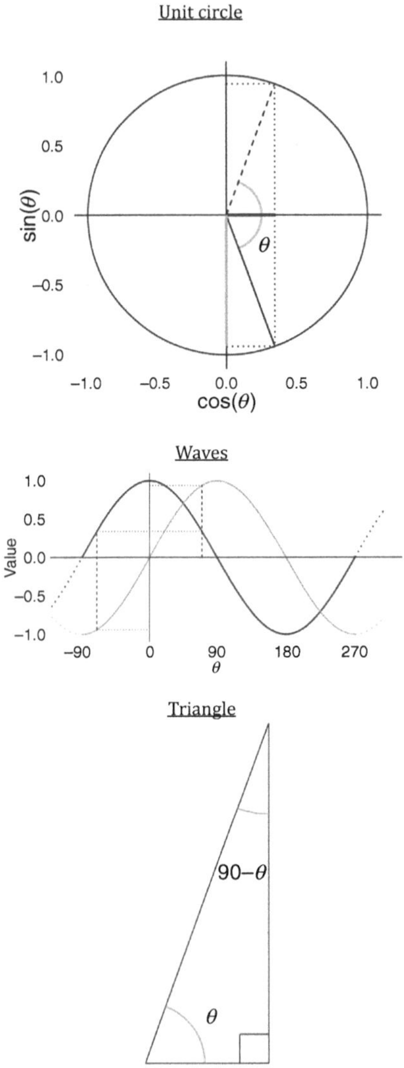

Symbolic rules

Trigonometric identities

$\sin(-\theta) = -\sin(\theta)$
$\cos(-\theta) = \cos(\theta)$
$\text{func}(\theta \pm 180) = -\text{func}(\theta)$
$\text{func}(90-\theta) = \text{opp}(\theta)$

Principles of algebra

$x \pm 0 = x$
$-(-x) = x$
$-x-y = -(x+y)$
$x+y = y+x$

A mnemonic

Symbolic rule component

Mnemonic	Positive functions	In quadrant:
All	**All**	I
Students	**S**ine	II
Take	**T**angent	III
Calculus	**C**osine	IV

Visuospatial component

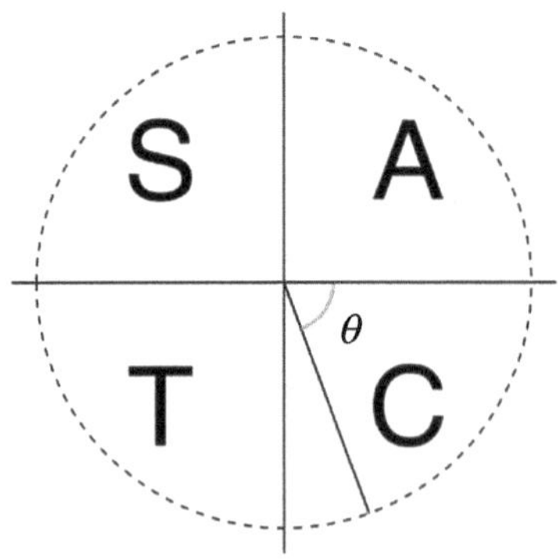

FIGURE 11.4 Some representations used in the domain of trigonometry. In the rules shown, func could be sin or cos; when func refers to one of these, opp refers to the other.

Fig. 11.4 (where func appears in Fig. 11.4, it can be replaced with sin or cos; opp refers to sin if func refers to cos and vice versa).

In line with our hypotheses, we considered three visuospatial representations: (1) the unit circle; (2) the wave representation of the sine and cosine functions; and (3) the right triangle. These representations are all shown in Fig. 11.4. As previously discussed, the unit circle treats angles as arising by rotating a radial line emanating from the center of a circle of radius 1 centered at (0,0) on a (x,y) coordinate system. Compound angles can be viewed as a sequence of rotations (e.g., 20 + 180 corresponds to rotating counterclockwise by 20 degrees and then continuing for another 180 degrees). The cosine of the overall angle can be visualized as a line segment extending either to the right (positive) or the left (negative) of the origin of the coordinate system. The sine can be visualized as a line segment extending either upward (positive) or downward (negative) from the origin. With this representation, the problems can be solved by comparing the directions and lengths of the line segments corresponding to the value of the probe and each of the choice alternatives. As shown on the unit circle diagram in Fig. 11.4, for $\cos(-\theta)$, the line segment corresponding to its value is the same as the line segment corresponding to the value of $\cos(\theta)$, indicating that $\cos(-\theta) = \cos(\theta)$.

In the wave representation, sine and cosine are represented as waves, each with a period of 360 degrees, with sin (lighter curve in Fig. 11.4) shifted 90 degrees to the right relative to cos. The arguments of these functions in this representation specify positions along the x axis and compound arguments can be viewed as a sequence of translations from the 0 position; the value of the function can then be read off by examining the height of the wave at the appropriate point on the x axis [again, it can be seen that $\cos(-\theta) = \cos(\theta)$].

We also considered a third visuospatial representation, the right triangle. In such a triangle, the two acute angles must equal 90 degrees, the cosine of an angle is defined as the length of the adjacent side over the length of the hypotenuse, and the sine of an angle is defined as the length of the opposite side over the hypotenuse. Although this representation is not useful for all of our problems, it is useful for the $\cos(90-\theta)$ or $\sin(90-\theta)$ problems. For example, in the right triangle shown in Fig. 11.4, the side adjacent to the angle θ is the side opposite the angle $(90-\theta)$ so that $\cos(\theta)$ and $\sin(90-\theta)$ both refer to the ratio of the length of the same leg of the right triangle (the horizontal leg) to the hypotenuse; in other words, $\sin(90-\theta) = \cos(\theta)$. In addition, we explored the possibility that participants might rely on mnemonics, such as SOHCAHTOA (sine equals opposite over hypotenuse, cosine equals adjacent over hypotenuse, tangent equals opposite over adjacent) and All Students Take Calculus. These mnemonics combine rules and visual representations as illustrated for All Students Take Calculus in the lower right of Fig. 11.4.

PRELIMINARY STUDY: OBSERVING USE AND SUCCESS OF THE UNIT CIRCLE

In our preliminary study, students encountered 2 examples of each of the 20 problem types in each of 2 blocks of trials. At the end of the experiment, we asked participants to rate how frequently they used each of the five possible representations on a scale from 1 to 5 (1 = never, 5 = always; a sixth "other representation" category was included but rarely used). Ratings were given first for

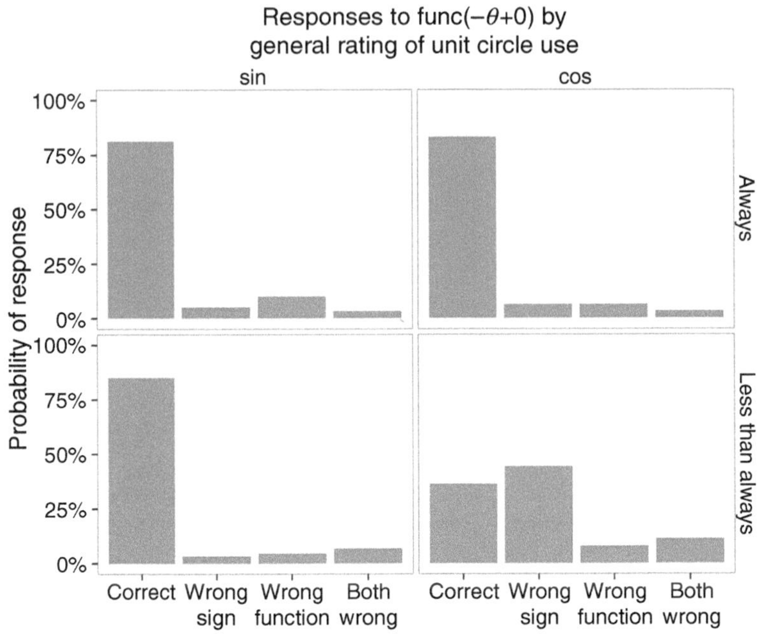

FIGURE 11.5 The distribution of responses to func($-\theta + 0$), split by function and by general rating of unit circle use.

block 1 and then for block 2 of the experiment. We then asked them to report the number and recency of courses in which they had used trigonometry. One group of participants ($n = 12$) had a brief break between blocks, while the remaining two groups had a brief lesson, either involving rules ($n = 12$) or waves ($n = 13$). Only ratings of block 1 were used in analyses.

Overall performance varied widely, with the median score falling at 66.3% correct and several participants performing at near-chance levels (all participants appeared to be trying to respond correctly and nearly all did well on the very easy problems of the form [$\cos(\theta + 0)$ and $\sin(\theta + 0)$]. Order of arguments in parentheses (base or offset first) did not affect responding. Neither the rule lesson nor the wave lesson led to greater improvement in the second block compared to the no-lesson control condition.

The unit circle representation was generally favored by our participants. Overall, participants reported using the unit circle more frequently (mean 3.5) than any other representation (largest other mean 2.4). Furthermore, we found a positive correlation between extent of reported unit circle use and performance score (proportion of problems correct) that remained a significant independent predictor of performance after taking into account both number of courses and recency.

Exploratory analysis revealed a striking pattern on one subset of problems (Fig. 11.5), namely problems involving the sine or cosine of a negative argument and an offset of 0, that is, $\sin(-\theta + 0)$ and $\cos(-\theta + 0)$. Most participants answered $\sin(-\theta + 0)$ problems correctly [choosing $-\sin(\theta)$], regardless of use of the unit circle. For $\cos(-\theta + 0)$, however, the correct answer is not $-\cos(\theta)$ but simply $\cos(\theta)$ (as noted previously, Fig. 11.4 shows this on the unit circle). In this case, participants who reported "always" using the unit circle ($n = 15$) answered this class of problems correctly 83% of the time, whereas other participants ($n = 22$) chose the correct answer only 36% of the time. The predominant error for these participants was $-\cos(\theta)$, a response these participants made more frequently than the correct response.

These results provide promising evidence for a positive role of visualizing the unit circle in solving trigonometric identity problems of the kinds that we used, but are limited in several ways. The ratings were global and retrospective, and the presence of a lesson may have altered the representations participants used or their recollections of these representations (though this should, if anything, have reduced reports of unit circle use relative to rules or waves). More importantly, the groups were too small to meaningfully assess the effect of the lessons, and the lessons did not include a circle lesson, preventing a causal inference about its usefulness.

STUDY 2: COMPARING A UNIT CIRCLE LESSON TO A RULES LESSON AND BASELINE KNOWLEDGE

Two further studies addressed these limitations. The first (Study 2.1) followed up on the relationship between unit circle use and performance in a larger group of participants ($N = 50$) without any lesson. The experiment followed the same protocol as in the no-lesson condition of the preliminary study, avoiding influence of a lesson on retrospective ratings (we did not solicit ratings during performance of either of the two main blocks of trials to avoid suggesting strategies to participants). To supplement the retrospective ratings, an additional block of 20 problems was added. After selecting a response for each of these problems, the participants rated their use of each representation on that problem on a three-point scale (1 = not at all, 2 = a little, 3 = a lot). Participants then answered further questions about the extent of their prior exposure to and use of each of the five representations, and finally participated in a short videotaped session in which the experimenter interviewed each participant on their strategy in solving $\cos(-50 + 0)$, $\cos(20 + 180)$, and $\sin(90 - 70)$. In the second further Study 2.2, participants also completed the same protocol, with the modification that they received either a lesson on the unit circle ($n = 35$) or in the use of rules and formulas ($n = 35$) between the first and second block. Lessons covered the inputs (signed argument plus offset) and outputs (resulting values) of the sine and cosine functions, and specifically described how either the unit circle or a simple rule could be used to correctly solve problems of the forms $\cos(\theta + 180)$, $\sin(\theta + 180)$, $\cos(-\theta + 0)$, $\sin(-\theta + 0)$, $\sin(\theta + 90)$, and $\cos(\theta + 90)$.

Study 2.1 replicated and extended all of the findings from the preliminary study. Average retrospective ratings of representations used in both the first and second block once again showed greater rated use of the unit circle compared to any of the other representations, and once again circle use rating strongly co-varied with overall performance, even after controlling not only for number of courses and recency, but also for prior unit circle exposure and prior unit circle use. That is, controlling for prior unit circle exposure and use, higher reported use in our experiment predicted higher overall performance. We also observed that reported circle use was associated with longer response times, consistent with the idea that those reporting circle use explicitly visualized and manipulated an internal visuospatial representation.

In a now planned comparison, we replicated the pattern of findings on the relation between average retrospective rated circle use and performance on $\cos(-\theta + 0)$ during blocks 1 and 2. The pattern was largely the same as before, except that among those who said they always used the unit circle in their retrospective ratings, there were fewer correct responses than in the preliminary study, and the predominant error was $-\cos(\theta)$, suggesting that some participants who stated that they relied on the unit circle either lacked sufficient mastery or imagery of the unit circle construct, or perhaps did not use the unit circle on these specific problems. Our problem-specific ratings allowed us to follow up on this pattern. Here we found that those who reported using the unit circle "a lot" on problems of this type averaged 75% correct and did not tend to make the $-\cos(\theta)$ error on such problems during blocks 1 and 2.

In summary, Study 2.1 supports the conclusion that unit circle use is generally associated with more accurate performance but requires more time than some other strategies. Some who say they always use the unit circle in retrospective ratings do less well than others, but both response times and problem-specific circle ratings suggest that most of these participants do not rely on the unit circle for $\cos(-\theta + 0)$ problems. Instead, we hypothesize that these and many other participants may rely on a faulty rule or heuristic corresponding to "pulling out the minus sign" from both the sin and cos functions. This error reflects a lack of engagement with the underlying meaning of these functions. It takes less time than visualizing the functions on the circle, and gives the correct answer for $\sin(-\theta + 0)$, but fails for $\cos(-\theta + 0)$. A small number of relatively expert users may know and understand correct rules that allow them to by-pass constructing a visuospatial representation at least for certain problems. Others may attempt to construct a visuospatial representation but fail to do so with sufficient reliability.

The results of our follow-up training Study 2.2 provide further support for the value of the unit circle. The unit circle lesson led to more improvement from block 1 to block 2 than the rule lesson. Since the participants in both Study 2.1 and 2.2 were drawn from the same pool and since their experiences were identical through the end of the second block except for the presence of a lesson, we used the participants in Study 2.1 as a no-lesson control, comparing the improvement from block 1 to block 2 shown by these participants to the improvement exhibited by each of the two lesson groups from Study 2.2. We also compared problem types directly taught in the lessons with problems that were not directly taught, hereafter called transfer problems. [The trivial $\cos(\theta + 0)$ and $\sin(\theta + 0)$

problems were excluded from this analysis.] This analysis revealed that both lessons helped participants with problems that were directly taught, but participants who received the unit circle lesson benefitted on transfer problems (16% gain) compared to controls (6% gain), while participants who received the rule lesson did not show a benefit on transfer problems (3% gain) compared to controls.

A further analysis considered whether the improvement in performance shown by participants who received the circle or rule lesson was moderated by their pretest performance. We examined the overall improvement (combined over taught and transfer items) by those whose pretest performance was "near chance" (within the 95% confidence interval of 25% correct on nontrivial problems), versus those whose pretest performance was above the near chance level but below a 95% correct ceiling level. Among participants receiving the unit circle lesson, those in the above chance group benefited more overall than those in the near chance group; for those in the rule lesson condition, both groups showed a similar, small overall benefit compared to corresponding controls. (Also, after controlling for pretest performance, there was no significant sex difference in gains after either lesson.) Caution is required in interpreting this pattern since it is based on an unplanned analysis. That said, the pattern strongly resonates with a large body of other work in which improvement based on experience depends on the participant having a foundation (perhaps implicit) on which the experience can build, spanning fields as diverse as development of an intuitive understanding of balance (Siegler, 1976) and the spatial memory formation in rodents (Tse et al., 2007, 2011), as well as our own modeling work addressing these phenomena (McClelland, 1995, 2013).

In summary, we found that participants exhibit a generalizable performance gain after a brief lesson based on the unit circle, relative to those receiving no lesson or those receiving a lesson based on rules. Importantly, we also found that the ability to benefit from the lesson may depend on the participants' being in a state of readiness to learn (Siegler, 1976) or what Vygotsky (1978) called the "zone of proximal development" (Chaiklin, 2003). In the context of the grounded conceptual structures framework described in the introduction, one can potentially understand what readiness means in terms of the availability of sufficient degree of prior establishment of the relevant structures and their constituents, a point to which we will return. Given this issue and other limitations of these studies, it will be important to examine these issues further.

CHALLENGES IN LEARNING THE UNIT CIRCLE

The research we have begun to undertake resonates with calls over many years to enhance the reliance on use of spatial representations that instantiate the meanings of mathematical expressions in visualizable form rather than rote memorization of rules (Polya, 1945; Shear, 1985; Thompson et al., 1994; Wertheimer, 1959). While all trigonometry teachers and texts connect with visuospatial representations to some degree, they do so to strikingly varying extents, and in different, apparently *ad hoc* ways. By conducting studies aimed to provide evidence about which types of representations are most useful for solving problems of different types by students of varying ability levels, our work may pave the way to establishing the evidence needed to begin to construct evidence-based teaching programs.

Our principal goal is to further assess, and potentially establish a broader basis of support for, the proposition that grounding precalculus trigonometry instruction on a meaningful visuospatial representation can contribute substantially to student learning of this material. This entails further assessing the relative merits of teaching students to rely on such a representation relative to a rule-based lesson. In order to do this, we intend to correct two limitations of Study 2.2: (1) the fact that the study was undertaken with a group of students at an elite university, almost all of whom reported having prior experience or exposure to the relevant ideas and concepts and (2) the relatively small size of the effect of the lessons, especially among students who performed relatively poorly in the pretest. In an informal replication of the unit circle lesson condition from Study 2.2, administered to 17 precalculus students at Palo Alto High school, all but one of these students fell into the "near chance" range of our previous study, even though they were just coming to the end of an 8-week instructional block that covered the relevant material. Like our Stanford students, these students systematically erred on the $\cos(-\theta + 0)$ problem, with 70% incorrectly choosing $-\cos(\theta)$. Furthermore, the gain in performance from pretest to posttest was even more modest (6%) than it was for the near-chance pretest subgroup of the Stanford group. It should be noted that there are several faster-paced "lanes" at Palo Alto High School, and those who participated largely came from the middle or bottom thirds of their classroom cohorts; thus their performance relative to Stanford students should not necessarily be surprising. Nevertheless, these results underscore the need to find ways of improving mastery of precalculus concepts for middle-of-the-road students, to enhance their ability to move into STEM careers. These results also point to the challenge we face in developing a lesson that can address the needs of these students.

As we work toward the eventual goal of addressing this challenge, our work will use community college and high-school students who have not already taken mathematics courses beyond Geometry and Algebra 2. Work of other researchers suggests that college students encountering precalculus relevant mathematical content for the first time experience similar difficulties as high-school students (Moore & Carlson, 2012; see also Carlson, Oehrtman, & Engelke, 2010), and ready access to a large population of such students makes this an attractive next step for our efforts.

Unit Circle Instruction for Students Without Prior Precalculus Trigonometry

The lesson used in the studies discussed previously was designed with Stanford undergraduates in mind, and in order to be more effective for students with less prior experience, we hypothesize that it must be adapted in several ways. In keeping with the idea that the emergence of a coherent, grounded conceptual structure is a gradual process, the new versions of our lessons will be longer, allowing for repetition and practice using the material covered in each lesson. We will incorporate active visuospatial responding with the mouse or trackpad (constructing angles and projections in the unit circle lesson) or manipulating symbolic expressions at the keyboard (in the rule-based lesson); this is likely to increase the mental engagement during learning (compared with the requirement to simply read the material as in the existing lesson) and therefore enhance learning (Ninness et al., 2009; Schmidt & Bjork, 1992).

Rather than needing more repetition or enhancement, perhaps some students do not have the appropriate prerequisite knowledge or experience in order to be ready to learn the unit circle. The unit circle provides a conceptual structure that integrates two component structures (the x, y plane and the circle) into one, and those components that are assumed to be familiar already may not be readily accessible to a student. In order to use the unit circle, one must first comprehend an angle and be able to mark the endpoint of a radius representing one side of that angle on a circle. Then, one must be able to add or subtract positive and negative angles, or mentally visualize a compound angle as the appropriate sequence of rotations. Then, to relate the resulting endpoint of a radial line to its sine or cosine, a student must project the endpoint onto the appropriate axis of the coordinate plane to find a number between −1 and 1. These skills should have been acquired with relevant experience in previous mathematical settings, but may not be fully mastered by all students. Students must have a high fluency with these steps as they are embedded in the context of problems that are presented as symbolic expressions such that any reliance on visuospatial representation would require participants to form and manipulate the visuospatial representation in their minds. We also note that this sequence of procedures may recruit general visuospatial abilities that may be malleable (Uttal et al., 2013) like those used in mentally rotating block shapes (Shepard & Metzler, 1971; Vandenberg & Kuse, 1978) or in finding simple shapes or parts in more complex figures (Ekstrom, French, Harman, & Derman, 1976).

We presented a new and considerably extended set of materials intended to address the points aforementioned to a small pilot group of eleventh graders near the beginning of their exposure to precalculus trigonometry, and found that, although some students had initial misconceptions, they progressed easily through the lesson sessions focused on knowledge prerequisites for the unit circle: the number line, the coordinate plane, and angles rotations of a radial line around the circle. They showed (and self-reported) greater difficulty in fully gaining an understanding of how the sine and cosine functions linked the position of the endpoint of a radial line corresponding to an angle to the endpoints' x and y coordinates, and they encountered additional difficulty when required to compare values of expressions involving the sine and cosine functions when these were presented symbolically without the availability of an external unit circle. These difficulties suggest that our lesson may need to be adapted to allow more practice internalizing a representation of the meaning of the sine and cosine functions and also allowing the transition from reliance on an external representation of the unit circle to an internalized representation of it to occur more gradually. We also suspect the materials may need to be adapted to address counterproductive epistemic beliefs and mindsets some students may bring to trigonometry learning. We consider these issues in turn in the next two sections.

Internalizing the Unit Circle

To facilitate the development of a sufficiently internalized and integrated understanding of the unit circle, we will adapt the *Externalization then Internalization* approach of Glenberg et al. (2004). In our unit circle lesson, students will initially be taught their way around the basic conceptual structure of the unit circle by constructing angles initially under carefully guided instruction and with external supports that will gradually be removed. For example, students will be introduced to positive and negative angles, and given opportunities to construct such angles by rotating lines around a unit circle; they will initially have critical landmark angles labeled, but these labels will then be gradually removed. Students will then be introduced to compound angles, and will be given opportunities to construct compound angles by successive rotations corresponding to each of the compound angle's parts. They will be given the opportunity to construct equivalent resulting angles by carrying out the corresponding rotations in different orders, and observing the end result.

Once the concepts of sin and cos are introduced, students will at first be asked to position a radial line to correspond to a specific example angle and then to construct the projection of the endpoint of the radial line onto the appropriate axis to represent the angle's sine or cosine value. We will gradually eliminate reliance on the externalization, replacing it with the instruction to carry out the corresponding operations mentally, since this instruction was found necessary to produce robust comprehension gains in Glenberg et al. (2004). Rather than eliminating external support all at once, we plan to do so in stages. For example, one intermediate step might be to ask students to construct an approximate representation of the sine or cosine of a particular angle (as points or segments on the x–y coordinate plane) without externally positioning the radial line corresponding to the angle. After repeated practice along this progression, students will be better prepared to transition to comparison of the values of two trigonometric expressions. Here again, we will first allow them to explicitly construct the value of each expression externally, one on each of a pair of side-by-side copies of the unit circle, before gradually removing these external aids and asking the students to perform approximations of these comparisons completely "in their heads." One intriguing possibility that we are exploring is that the ability to make these comparisons mentally may rely on internalization of schematic representations that do not represent specific angles exactly but do represent relationships between the projection onto the x axis of the endpoint of a radial line corresponding to positive and negative rotations of the same amount. One may not be able to visualize a particular given angle correctly, but one may be able to "see" that the endpoints of both the positive and negative rotations of the same amount project to the same point on the x axis through the unit circle.

The Role of Epistemic Belief in Acquiring an Integrated Conceptual Representation

We are also exploring the roles of what others have called students' epistemic beliefs and mindsets in enabling their ability to acquire the unit circle framework as an integrated conceptual structure. If students believe the goal of our lesson is to gain an understanding of trigonometry, what do they believe is the nature of such an understanding? The unit circle provides a conceptual framework that supports an integrated problem solving procedure that can be applied across a range of problem types, whereas rules are narrower, with each one applying only to a limited set of problem types. If students think that understanding mathematics is a matter of learning a set of unrelated rules, they may struggle when confronted with new types of problems not covered by existing rules (like the transfer problems in our task). In a study on learning statistics, students who believed mathematics consisted of a set of simple facts tended to show worse comprehension than those who saw mathematics as an integrated system of relationships (Schommer, Crouse, & Rhodes, 1992). These are examples of what the authors called epistemic beliefs.

Other beliefs they call epistemic include beliefs about one's innate ability to learn, the quickness of learning, and the source and nature of knowledge (Schommer, 1990). Such beliefs may be malleable: students who were encouraged to believe that intelligence is a malleable quality (displaying a "growth mindset") showed greater motivation as well as stronger grades in their mathematics classes (Blackwell, Trzesniewski, & Dweck, 2007). Students with a growth mindset also more strongly endorsed the idea that effort is necessary to learn, which may play a role in expectations about the quickness of learning and thus the ability of students to be patient and take the necessary time to learn, rather than rushing through materials for the sake of getting through them. Because these factors seem likely to us to have important influences on a student's ability to master the unit circle as an integrated conceptual structure underlying trigonometry, the next version of our materials will explicitly encourage viewing trigonometry as an integrated system of knowledge that requires time and engagement to learn. The materials will be introduced as an integrated framework as the students start into the program of lessons, and each block of the lesson will described as playing a specific role within an integrative approach to understanding the meaning of trigonometric expressions within the unit circle framework.

CONCLUSIONS AND FUTURE DIRECTIONS

In this chapter, we have argued that a particular visuospatial model called the unit circle acts as a grounded conceptual structure in trigonometry. In preliminary studies, we found that students who reported using the unit circle performed better at solving trigonometry identities than those who did not report visualizing this representation. We also found that a brief lesson in the use of the unit circle produced postlesson benefits relative to no lesson or a rule-based lesson. These benefits were seen in participants who exhibited some partial understanding of the relevant concepts in a pretest, whereas high school students who lacked facility with components of the unit circle had more difficulty learning in a unit circle lesson. The difficulty students without sufficient prior knowledge experienced illustrates the challenges we face in helping students learn this conceptual structure.

We have described some of the steps we are taking to address the challenges we have identified. First, we hope to strengthen students' command of the components that are integrated in the unit circle. Second, we hope to foster their gradual internalization, so that when students confront expressions in trigonometry, they can visualize the component elements and their relationship to each other mentally, rather than relying on an external visuospatial representation. In addition, we plan to measure learners' initial epistemic beliefs and mindsets and to provide explicit encouragement of beliefs that will, we hope, encourage the learner to understand that the goal is to build an integrated conceptual structure and that to do so may require patience and practice. Finally, our future work will also explore how individual differences in visuospatial or other aspects of cognitive abilities affect students' acquisition and use of the unit circle representation. As educators and policymakers examine the structure and content of the trigonometry curriculum, we hope our work will eventually lead to the development of effective practices for allowing learners to acquire an integrated conceptual framework for understanding trigonometry.

REFERENCES

Bartlett, F. C. (1932). *Remembering*. Cambridge: Cambridge University Press.

Barwise, J., & Etchemendy, J. (1996). Visual information and valid reasoning. In G. Allwein, & J. Barwise (Eds.), *Logical reasoning with diagrams* (pp. 3–26). New York: Oxford University Press.

Blackwell, L. S., Trzesniewski, K. H., & Dweck, C. S. (2007). Implicit theories of intelligence predict achievement across an adolescent transition: a longitudinal study and an intervention. *Child Development*, *78*(1), 246–263.

Bransford, J. D., & Johnson, M. K. (1972). Contextual prerequisites for understanding: some investigations of comprehension and recall. *Journal of Verbal Learning and Verbal Behavior*, *11*(6), 717–726.

Bransford, J. D., Barclay, J. R., & Franks, J. J. (1972). Sentence memory: a constructive versus interpretive approach. *Cognitive Psychology*, *3*(2), 193–209.

Carlson, M., Oehrtman, M., & Engelke, N. (2010). The precalculus concept assessment (PCA) instrument: a tool for assessing reasoning patterns, understandings, and knowledge of precalculus level students. *Cognition and Instruction*, *28*(2), 113–145.

Case, R. (1996). Introduction: reconceptualizing the nature of children's conceptual structures and their development in middle childhood. *Monographs of the Society for Research in Child Development*, *61*(1–2), 1–26.

Chaiklin, S. (2003). The zone of proximal development in Vygotsky's analysis of learning and instruction. In A. Kozulin, B. Gindis, V. Ageyev, & S. Miller (Eds.), *Vygotsky's educational theory and practice in cultural context* (pp. 39–64). Cambridge: Cambridge University.

Dantzig, T. (2007). *Number: The language of science* (Masterpiece Science ed.). New York: Penguin Group.

Dehaene, S. (1992). Varieties of numerical abilities. *Cognition*, *44*, 1–42.

Einstein, A. (1979). *Ideas and opinions*. New York: Dell Publishing Co, 1–42.

Ekstrom, R. B., French, J. W., Harman, H. H., & Derman, D. (1976). *Manual for Kit of Factor-Referenced Cognitive Tests*. Princeton, NJ: Educational Testing Service.

Foerster, P. A. (1990). *Trigonometry: Functions and applications* (3rd ed.). Reading, MA: Addison Wesley Publishing.

Gelfand, I. M., & Saul, M. (2001). *Trigonometry*. Boston, MA: Brinkhauser.

Glenberg, A. M., Gutierrez, T., Levin, J. R., Japuntich, S., & Kaschak, M. P. (2004). Activity and imagined activity can enhance young children's reading comprehension. *Journal of Educational Psychology*, *96*(3), 424–436.

Goldstone, R. L., & Son, J. Y. (2005). The transfer of scientific principles using concrete and idealized simulations. *Journal of the Learning Sciences*, *14*(1), 69–110.

Harnad, S. (1990). The symbol grounding problem. *Physica D: Nonlinear Phenomena*, *42*(1), 335–346.

Hornsby, E. J., Lial, M. L., & Rockswold, G. K. (2011). *A graphical approach to precalculus with limits: A unit circle approach* (5th ed.). Boston, MA: Addison-Wesley.

Johnson-Laird, P. N. (1983). *Mental models: Towards a cognitive science of language, inference, and consciousness*. Cambridge, MA: Harvard University Press.

Kaminski, J. A., Sloutsky, V. M., & Heckler, A. F. (2008). The advantage of abstract examples in learning math. *Science*, *320*(5875), 454–455.

Kell, H. J., Lubinski, D., Benbow, C. P., & Steiger, J. H. (2013). Creativity and technical innovation spatial ability's unique role. *Psychological Science*, *24*(9), 1831–1836.

Kelly, D., Nord, C. W., Jenkins, F., Chan, J. Y., & Kastberg, D. (2013). Performance of US 15-Year-Old Students in Mathematics, Science, and Reading Literacy in an International Context. First Look at PISA 2012. NCES 2014-024. National Center for Education Statistics.

Kendal, M., & Stacey, K. (1996). Trigonometry: Comparing ratio and unit circle methods. In P.C. Clarkson, (Ed.), *Technology in mathematics education* (Proceedings of the 19th annual conference of the Mathematics Education Research Group of Australasia). Melbourne: MERGA.

Lakoff, G., & Núñez, R. E. (2000). *Where mathematics comes from: How the embodied mind brings mathematics into being*. New York: Basic books.

Lewis, A. B. (1989). Training students to represent arithmetic word problems. *Journal of Educational Psychology*, *81*(4), 521–531.

Marcus, G. F. (2003). *The algebraic mind: Integrating connectionism and cognitive science*. The MIT Press: Cambridge, MA.

Mayer, R. E., & Hegarty, M. (1996). The process of understanding mathematical problems. In R. J. Sternberg, & T. Ben-Zeev (Eds.), *The nature of mathematical thinking* (pp. 29–53). New York: Routledge.

McClelland, J. L. (1995). A connectionist approach to knowledge and development. In T. J. Simon, & G. S. Halford (Eds.), *Developing cognitive competence: New approaches to process modeling* (pp. 157–204). Mahwah, NJ: LEA.

McClelland, J. L. (2013). Incorporating rapid neocortical learning of new schema-consistent information into complementary learning systems theory. *Journal of Experimental Psychology: General*, *142*(4), 1190–1210.

Moore, K. C. (2013). Making sense by measuring arcs: a teaching experiment in angle measure. *Educational Studies in Mathematics*, *83*(2), 225–245.

Moore, K. C., & Carlson, M. P. (2012). Students' images of problem contexts when solving applied problems. *The Journal of Mathematical Behavior*, *31*(1), 48–59.

Moss, J., & Case, R. (1999). Developing children's understanding of the rational numbers: a new model and an experimental curriculum. *Journal for Research in Mathematics Education*, *30*(2), 122–147.

National Mathematics Advisory Panel. (2008). Foundations for success: The final report of the National Mathematics Advisory Panel. US Department of Education.

Newell, A., & Simon, H. A. (1961). Computer simulation of human thinking. *Science*, *134*(3495), 2011–2017.

Ninness, C., Barnes-Holmes, D., Rumph, R., McCuller, G., Ford, A.M., Payne, R., …, & McGinty, J. (2006). Transformations of mathematical and stimulus functions. *Journal of Applied Behavior Analysis*, *39*(3), 299–321.

Ninness, C., Dixon, M., Barnes-Holmes, D., Rehfeldt, R.A., Rumph, R., McCuller, G., …, & McGinty, J. (2009). Constructing and deriving reciprocal trigonometric relations: a functional analytic approach. *Journal of Applied Behavior Analysis*, *42*(2), 191–208.

Organization for Economic Cooperation and Development (OECD). (2004). *Learning for tomorrow's world—First results from PISA 2003*. Paris: Author.

Poincaré, J. H. (1968). Mathematical Creation. In M. Kline (Ed.), *Mathematics in the modern world: Readings from Scientific American*. San Francisco: W. H. Freeman & Co.

Polya, G. (1945). *How to solve it*. Princeton, NJ: Princeton University Press.

Russell, B. (1903). *The principle of mathematics*. Cambridge: University Press.

Schmidt, R. A., & Bjork, R. A. (1992). New conceptualizations of practice: common principles in three paradigms suggest new concepts for training. *Psychological Science*, *3*(4), 207–217.

Schommer, M. (1990). Effects of beliefs about the nature of knowledge on comprehension. *Journal of Educational Psychology*, *82*(3), 498–504.

Schommer, M., Crouse, A., & Rhodes, N. (1992). Epistemological beliefs and mathematical text comprehension: believing it is simple does not make it so. *Journal of Educational Psychology*, *84*(4), 435–443.

Shear, J. (1985). Visual thinking, algebraic thinking, and a full unit-circle diagram. *The Mathematics Teacher*, *78*(7), 518–522.

Shepard, R. N. (2008). The step to rationality: the efficacy of thought experiments in science, ethics, and free will. *Cognitive Science*, *32*(1), 3–35.

Shepard, R. N., & Chipman, S. (1970). Second-order isomorphism of internal representations: shapes of states. *Cognitive psychology*, *1*(1), 1–17.

Shepard, R. N., & Metzler, J. (1971). Mental rotation of three dimensional objects. *Science*, *171*(3972), 701–703.

Siegler, R. S. (1976). Three aspects of cognitive development. *Cognitive Psychology*, *8*(4), 481–520.

Thompson, A. G., Philipp, R. A., Thompson, P. W., & Boyd, B. A. (1994). Calculational and conceptual orientations in teaching mathematics. In A. Coxford (Ed.), *1994 Yearbook of the NCTM* (pp. 79–92). Reston, VA: NCTM.

Tse, D., Langston, R.F., Kakeyama, M., Bethus, I., Spooner, P.A., Wood, E.R., …, & Morris, R.G. (2007). Schemas and memory consolidation. *Science*, *316*(5821), 76–82.

Tse, D., Takeuchi, T., Kakeyama, M., Kajii, Y., Okuno, H., Tohyama, C., … & Morris, R. G. (2011). Schema-dependent gene activation and memory encoding in neocortex. *Science*, *333*(6044), 891–895.

Uttal, D. H., Meadow, N. G., Tipton, E., Hand, L. L., Alden, A. R., Warren, C., & Newcombe, N. S. (2013). The malleability of spatial skills: a meta-analysis of training studies. *Psychological Bulletin*, *139*(2), 352–402.

Vandenberg, S. G., & Kuse, A. R. (1978). Mental rotations, a group test of three-dimensional spatial visualization. *Perceptual and Motor Skills*, *47*(2), 599–604.

Vygotsky, L. S. (1978). Interaction between learning and development (M. Lopez-Morillas, Trans.). In M. Cole, V. John-Steiner, S. Scribner, & E. Souberman (Eds.), *Mind in society: The development of higher psychological processes* (pp. 79–91). Cambridge, MA: Harvard University Press.

Wai, J., Lubinski, D., & Benbow, C. P. (2009). Spatial ability for STEM domains: aligning over 50 years of cumulative psychological knowledge solidifies its importance. *Journal of Educational Psychology*, *101*(4), 817–835.

Wertheimer, M. (1959). *Productive thinking* (Enlarged Ed.). New York: Harper & Brothers.

Part IV

Instructional Approaches

Chapter 12

The Power of Comparison in Mathematics Instruction: Experimental Evidence From Classrooms

Bethany Rittle-Johnson*, Jon R. Star**, Kelley Durkin†
*Department of Psychology and Human Development, Vanderbilt University, Nashville, TN, United States; **Graduate School of Education, Harvard University Cambridge, MA, United States; †Peabody Research Institute, Vanderbilt University, Nashville, TN, United States

INTRODUCTION

We often learn through comparison. For example, we compare new words, objects, and ideas to ones we already know, and these comparisons help us recognize what features are important and merit more attention. Indeed, comparison aids learning across a broad array of topics, ranging from babies learning the distinction between dogs and cats (Oakes & Ribar, 2005), to preschoolers learning new words (e.g., Namy & Gentner, 2002), to business-school students learning contract negotiation skills (Gentner, Loewenstein, & Thompson, 2003). A recent metaanalysis confirmed that comparison promotes learning across a range of domains (Alfieri, Nokes-Malach, & Schunn, 2013). As Goldstone, Day, and Son (2010) noted: "Comparison is one of the most integral components of human thought…. Furthermore, research has demonstrated that the simple act of comparing two things can produce important changes in our knowledge" (p. 103).

In this chapter, we focus on using comparison to support mathematics learning. Comparison is integral to best practices in mathematics education. Having students share and compare solution methods for solving a particular problem (e.g., discuss the similarities and differences in the methods) lies at the core of reform pedagogy in many countries throughout the world (Australian Education Ministers, 2006; Brophy, 1999; Kultusministerkonferenz, 2004; NCTM, 2014; Singapore Ministry of Education, 2006; Treffers, 1991), including the Common Core State Standards (2010). Its inclusion as a best practice

Acquisition of Complex Arithmetic Skills and Higher-Order Mathematics Concepts
http://dx.doi.org/10.1016/B978-0-12-805086-6.00012-6

was based on observational research that expert teachers in the United States, as well as teachers from high-performing countries, have students compare multiple methods for solving problems during mathematics instruction (Ball, 1993; Lampert, 1990; Richland, Zur, & Holyoak, 2007).

Based on this convergence in the cognitive science and mathematics education literatures, we identified comparison as a promising instructional practice for improving mathematics learning. Our goal was to evaluate how comparison supports learning of school mathematics within a classroom setting. In this chapter, we present results from our classroom-based research on using comparison to improve mathematics learning, supplementing with recent research by others. We begin with short-term, researcher-led classroom research. Then, we describe a year-long study on helping teachers use comparison throughout the Algebra I curriculum.

SHORT-TERM, RESEARCHER-LED CLASSROOM RESEARCH

In our initial research on the effectiveness of comparison, we redesigned 2–3 middle-school math lessons on a particular topic in several different ways and implemented these lessons during students' mathematics classes. This approach ensured that students were accountable for learning the material and that the material was readily usable in a typical classroom setting. We evaluated two types of comparison: comparing multiple methods for solving the same problem (*comparing methods*) and comparing different problems solved with the same method (*comparing problems*).

Typically, students in the control condition studied the same material sequentially, without comparison. This allowed us to isolate the effectiveness of comparison. Students worked with partners within the same classroom, and each pair of students was randomly assigned to a comparison or sequential condition. By randomly assigning pairs of students within the same classroom to a condition, we avoided confounding classroom effects with condition effects.

Instructional Materials

Before reviewing individual studies, we describe the design of the instructional materials, both for the comparison and control conditions. Worked examples along with prompts to explain the examples were the core of our instructional materials across experimental and control conditions. Worked examples present solution methods step-by-step and are a very effective way to help novices learn new procedures and related concepts (Atkinson, Derry, Renkl, & Wortham, 2000; Sweller & Cooper, 1985). They are commonly used in textbooks, so they are also familiar to students. Worked examples are also an effective way to introduce students to alternative, more efficient methods (Star & Rittle-Johnson, 2008). To improve learning from worked examples, students should be prompted to generate explanations while studying the examples (see Atkinson et al., 2000). Generating explanations aids comprehension and transfer by promoting integration of new information with prior knowledge (Chi, 2000) and by guiding attention to structural features over surface features of the to-be-learned content (McEldoon, Durkin, & Rittle-Johnson, 2013; Siegler & Chen, 2008). Asking students to generate explanations is also a recommended instructional practice for mathematics (Common Core State Standards, 2010). Thus, for each condition, we created a packet of worked examples with appropriate explanation prompts. We also included practice problems, in line with findings that worked examples should be mixed with practice problems to solve (Atkinson et al., 2000).

When designing our experimental materials, we extracted design principles from the comparison literature. First, examples to be compared were presented simultaneously to facilitate comparison (Begolli & Richland, 2015; Gentner, 1983). Second, spatial cues (e.g., side-by-side presentation) and common language were used to help students align and map the solution steps, to facilitate noticing of important similarities and differences in the examples (Namy & Gentner, 2002; Richland et al., 2007). Third, the explanation prompts focused on specific aspects of the examples to compare because this is encouraged by expert mathematics teachers (Fraivillig, Murphy, & Fuson, 1999; Huffred-Ackles, Fuson, & Sherin Gamoran, 2004; Lampert, 1990) and improves learning from comparison relative to generic, open-ended prompts "to compare" (Catrambone & Holyoak, 1989; Gentner et al., 2003).

Materials for the control condition included the same worked examples presented one at a time without supports for comparison. All students worked with a partner when studying the worked examples because working with a partner provides a familiar context for students to generate explanations, and students who collaborate with a partner tend to learn more than those who work alone (e.g., Johnson & Johnson, 1994; Webb, 1991).

Students in all conditions received the same direct instruction after studying the worked examples. We provided some direct instruction at the end of the intervention because direct instruction has been found to improve learning from comparison (Gick & Holyoak, 1983; Schwartz & Bransford, 1998; VanderStoep & Seifert, 1993).

The instructional materials often focused on multistep equation solving. Consider the equation $3(x + 2) = 6$. Two possible first steps are to distribute the 3 (i.e., to get $3x + 6$) or to divide both sides by 3. Although the former is almost universally taught as part of the algorithm for solving this type of equation, the latter approach is arguably more efficient because it reduces the number of computations and steps needed to solve the equation. Regrettably, students often memorize rules and do not learn flexible and meaningful methods for solving equations (Kieran, 1992; Robinson & LeFevre, 2011; see also Robinson chapter in this volume). They also struggle to understand key algebraic concepts. For example, only 59% of US eighth graders were able to find an equation that is equivalent to $n + 18 = 23$ (National Assessment of Educational

Progress, 2011). Thus, improving students' knowledge of multistep equation solving is greatly needed.

For student outcomes, we focused on three critical components of mathematical competence: procedural knowledge, procedural flexibility, and conceptual knowledge. Procedural knowledge is the ability to execute action sequences to solve problems, including the ability to adapt known procedures to unfamiliar problems (Rittle-Johnson, Siegler, & Alibali, 2001). Procedural flexibility includes the knowledge of multiple methods as well as the ability to choose the most appropriate method based on specific problem features (Kilpatrick, Swafford, & Findell, 2001; Star, 2005; Verschaffel, Luwel, Torbeyns, & Van Dooren, 2009). Procedural flexibility supports efficient problem solving and is also associated with greater accuracy solving novel problems and greater understanding of domain concepts (e.g., Blöte, Van der Burg, & Klein, 2001; Carpenter, Franke, Jacobs, Fennema, & Empson, 1998; Hiebert et al., 1996). Finally, conceptual knowledge is "an integrated and functional grasp of mathematical ideas" (Kilpatrick et al., 2001). This knowledge is flexible and not tied to specific problem types, and is therefore generalizable. However, the knowledge may be implicit and not easily articulated (Alibali & Nathan, 2012; Prather & Alibali, 2009). Mathematics competence rests on students developing all three types of knowledge (Kilpatrick et al., 2001).

Studies on Comparing Methods

Our initial studies focused on comparing multiple methods for solving the same problem. Expert mathematics teachers in the United States (e.g. Ball, 1993; Lampert, 1990) and teachers in high performing countries, such as Japan (Richland et al., 2007) often use this approach. All of the methods presented were correct, but they varied in terms of which method was most appropriate and efficient for solving a particular problem. Students studied pairs of worked examples and were prompted to compare them (*compare-methods condition*) or studied the same examples one at a time and were prompted to reflect on them individually (*sequential condition*). Our comparison prompts focused student attention on recognizing that both methods adhered to domain principles, but that a particular method was more efficient for solving a particular problem.

We hypothesized that the compare-methods condition would support procedural flexibility better than the sequential condition. Comparing methods should highlight the accuracy and efficiency of multiple solution methods and facilitate knowledge and use of these methods. We also hypothesized that comparing methods would support better conceptual and procedural knowledge, as procedural flexibility is associated with both types of knowledge (e.g., Blöte et al., 2001; Carpenter et al., 1998; Hiebert et al., 1996), and reflection prompts asked students why the methods were valid.

In Rittle-Johnson and Star's (2007) study, seventh-grade students ($N = 70$) in prealgebra classes learned about solving multistep linear equations during three

A. Compare-Methods Condition

Jessica's Solution:		**Mary's Solution:**	
$2(t-1)+3(t-1)=10$		$2(t-1)+3(t-1)=10$	
$2t-2+3t-3=10$	________	$5(t-1)=10$	________
$5t-5=10$	*Combine*	$t-1=2$	*Divide on Both*
$5t=15$	*Add on Both*	$t=3$	*Add on Both*
$t=3$	*Divide on Both*		

Label the first step for each solution in the blank space provided above.

1. Jessica and Mary did different first steps. Is it OK to do either step first in this problem? Explain your reasoning.

2. Why might it be helpful to know two different ways to solve equations like this one?

B. Sequential Condition

Jessica's Solution:	
$2(t-1)+3(t-1)=10$	
$2t-2+3t-3=10$	________
$5t-5=10$	*Combine*
$5t=15$	*Add on Both*
$t=3$	*Divide on Both*

Label the first step in the blank space provided above.
1. Why did Jessica divide as her last step?

-----*NEXT PAGE*-----

Mary's Solution:	
$4(t-5)+3(t-5)=14$	
$7(t-5)=14$	________
$t-5=2$	*Divide on Both*
$t=7$	*Add on Both*

Label the first step in the blank space provided above.
1. Do you think the solution method used on this problem is a good one? Why?

FIGURE 12.1 Sample pages from the compare-methods and sequential conditions from Rittle-Johnson and Star (2007).

class periods. Students completed a packet of worked examples with their partner, explaining the procedures and answering explanation prompts; a sample worked example for each condition is illustrated in Fig. 12.1. Before and after participating in the intervention, students completed an assessment of our three outcome measures. The procedural knowledge measure involved solving algebra equations and the conceptual knowledge measure involved recognizing or explaining algebra concepts, such as like terms indexed. Procedural flexibility was measured in two ways. The first was *use* of more efficient solution methods when solving equations; the second was knowledge of multiple ways to solve equations, including acceptance of nonstandard ways to solve equations. As predicted, those who compared methods gained greater procedural flexibility.

They also acquired greater procedural knowledge. The two groups did not differ in conceptual knowledge.

Students' explanations during the intervention confirmed that those who compared methods often compared the similarities and differences in solution steps across examples and evaluated their efficiency and accuracy. In turn, frequency of making explicit comparisons during the intervention predicted procedural knowledge gain.

We found similar results for 157 fifth- and sixth-grade students learning about estimating answers to multiplication problems (e.g., about how much is 37×29?) (Star & Rittle-Johnson, 2009). Comparing methods supported greater procedural flexibility. For students with above-average knowledge of estimation at pretest, students in the comparison condition were also more likely to maintain their gains in conceptual knowledge over a delay than students in the sequential condition. Thus, comparing methods consistently supported procedural flexibility in two domains that differed in numerous ways, including whether there was a single, correct answer and what features of the methods needed to be considered (e.g., efficiency vs. proximity to the correct answer). Its impact on procedural and conceptual knowledge was less consistent.

Theories of analogical learning help to explain how comparing methods aids learning (Gentner, 1983; Hummel & Holyoak, 1997). In both of our studies, most students were familiar with one of the solution methods at pretest. When students are familiar with one method, they can learn new methods via analogy to the familiar one. Students can make inferences about the new method by identifying its similarities and differences to a known method and making inferences about how the new method works based on its alignment with the known method. For example, students who compared methods identified how the unfamiliar methods were similar to and different from the method that they already knew; in turn these types of comparative explanations predicted learning (Rittle-Johnson & Star, 2007).

Learning via analogy to a known method requires that students are familiar with one of the methods. When one method is not familiar, an alternative learning mechanism is *mutual alignment*. During mutual alignment, people notice potentially relevant features in two unfamiliar examples by identifying their similarities and then focusing attention on and making sense of these similarities (Gentner et al., 2003; Kurtz, Miao, & Gentner, 2001). Thus, the analogy literature pointed to the potential importance of prior knowledge of one of the solution procedures when learning from comparison.

Our subsequent research explored the influence of prior knowledge on learning from comparing methods. In one study, we worked with 236 seventh- and eighth-grade students whose schools did not use a prealgebra curriculum, and thus had had limited experience solving equations (Rittle-Johnson, Star, & Durkin, 2009). Students who did not attempt algebraic methods at pretest (i.e., novices) benefited most from studying examples sequentially, rather than from comparing methods. These students had higher procedural knowledge, conceptual knowledge and flexible use of procedures in the sequential condition than in the compare methods condition. The novices in the compare-methods condition seemed overwhelmed during the intervention—they completed less of the intervention materials and were less successful implementing nonstandard methods when prompted (Rittle-Johnson et al., 2009). In contrast, students who attempted algebraic methods at pretest learned more from comparing methods; they had higher procedural knowledge and flexible use of procedures in the compare methods condition than in the sequential condition. Further, general mathematics achievement did not influence the effectiveness of comparison in this study; rather, it was prior knowledge of one of the demonstrated methods that influenced the effectiveness of comparison. These findings are consistent with cognitive load theory (Sweller, 1988); simultaneous comparison of examples has very high working memory demands, so when both methods are unfamiliar, the working memory demands may be too high and impede learning. Analogical learning can be more effective when comparing a new example to a known example, so instructors must be attentive to students' prior knowledge.

A follow-up study suggested that slowing the pace of instruction allowed novices to learn from comparing methods (Rittle-Johnson, Star, & Durkin, 2012). We worked with 198 eighth-grade students who had little prior instruction on equation solving, so we modified the materials from Rittle-Johnson et al. (2009) to cover less content in more time by focusing on fewer problem types, cutting the number of examples and explanation prompts, and adding 30 min to the intervention time. The impact of the compare-methods condition did not interact with use of algebra at pretest in this study. Regardless of students' prior knowledge, comparing methods supported more flexible use of procedures than sequential study, including on a 1-month retention test. On other outcome measures (i.e., conceptual and procedural knowledge), the compare-methods and sequential conditions resulted in comparable learning (Rittle-Johnson et al., 2012).

In this same study, we also explored the effectiveness of delaying comparison of methods. Students studied one method on the first day, and on the second day, they compared it to alternative methods. The goal was to develop knowledge (although not mastery) of one solution method before comparing it to alternatives. However, students learned the least in this condition, relative to always comparing methods or always studying the examples sequentially. We expected delayed comparison of methods to be effective for novices and suspect that alternative instantiations of this approach could be beneficial. For example, it may be beneficial to delay comparison of multiple methods, but not to delay introducing multiple methods (e.g., initially study multiple methods sequentially, and then compare them).

There were some advantages to immediately comparing methods and no disadvantages in this study, suggesting that novices were able to learn from comparing two unfamiliar methods (i.e., mutual alignment) when the pace of instruction was slowed. Indeed, novices who compared methods often made comparisons between the two examples in their explanations, focusing on comparing problem features, solution steps, answers and the relative efficiency of the methods. Given adequate support, novices seemed able to learn by making analogies between two unfamiliar methods. However, learning via mutual alignment appears to be more difficult, and thus requires more instructional support, than learning an unfamiliar method via analogy to a known method.

Overall, comparing methods supports procedural flexibility, and sometimes conceptual and procedural knowledge as well (see Table 12.1 for a summary of findings from our studies). It can be used early in the learning process, but it must be carefully scaffolded (e.g., focusing on only one or two problem types and allowing ample time for study and explanation) and may result in less benefit than for students with some prior knowledge of a solution method.

Studies on Comparing Problems

Comparing correct methods is only one type of comparison. Although it is the type of comparison most often advocated for in mathematics education reforms, research on comparison in the cognitive science literature focuses on a different type of comparison—comparing different problems solved with the same method. For example, illustrating the same solution method in two stories with different cover stories, and prompting for comparison, greatly increased adults' spontaneous transfer of the solution to a new problem (Catrambone & Holyoak, 1989; Gick & Holyoak, 1983). Comparing problems is thought to support transfer by helping people abstract the key features of the method so that it is not tied to overly narrow problem features, such as the cover story. For example, adults who compared worked examples of different problems solved the same way were more likely to describe the solution to the example problems in general terms, rather than being tied to the specifics of the problem context (Catrambone & Holyoak, 1989; Gentner et al., 2003; Gick & Holyoak, 1983). In addition, explicitly stating the general method *after* learners had compared two example problems improved transfer; stating the general method without use of comparison did little to improve transfer (Gick & Holyoak, 1983). However, this previous research has been conducted with adults solving nonmathematical problems.

To explore the potential of comparing problems for helping children learn math content, we implemented two versions of comparing problems solved with the same method. In our *compare-equivalent-problems* condition, students compared two equivalent equations that varied only in the particular numbers and variables [e.g., $3(x + 2) = 6$ and $5(x + 3) = 15$]; the two were solved using the same method. Prompts focused on the similarities in the solution steps and when a particular solution step could be used. In the context of solving equations, a second variation of comparing problems emerged that may better focus attention on when particular solution steps can be used. In the *compare-problem-types* condition, students compared problems with different features

TABLE 12.1 Design Features and Outcomes From Our Experimental Studies on Using Comparison to Support Mathematics Learning

	Rittle-Johnson and Star (2007)	Star and Rittle-Johnson (2009)	Rittle-Johnson et al. (2009)	Rittle-Johnson et al. (2012)	Rittle-Johnson and Star (2009)
Design features					
Instructional conditions	Compare-methods or sequential	Compare-methods or sequential	Compare-methods, sequential or compare-problem-types	Compare-methods, sequential, or delayed-compare-methods	Compare-methods, compare-equivalent-problems or compare-problem-types
Target task	Linear equations	Computational estimation	Linear equations	Linear equations	Linear equations
Most children familiar with a target method at pretest	Yes	Yes	Mixed	Mixed	Yes
Condition(s) with highest performance					
Conceptual knowledge	Same (poor measure)	Depends on prior knowledge for retention	Depends on prior knowledge	Same	Compare-methods
Procedural knowledge	Compare-methods	Same	Depends on prior knowledge	Similar	Same
Flexibility	Compare-methods	Compare-methods	Depends on prior knowledge	Compare-methods	Compare-methods

solved using similar, although not identical, methods. For example, they compared solutions to $3(x + 2) = 6$ and $3(x + 2) + 5(x + 2) = 16$. Prompts focused on similarities as well as differences in the solution methods due to different problem features.

In Rittle-Johnson and Star (2009), we worked with 162 seventh- and eighth-grade students who had previous experience solving equation and were randomly assigned to the compare-equivalent-problems, compare-problem-types, or compare-methods condition. Students worked on the materials with a partner during three math classes. Frequency of exposure to different solution methods was the same across conditions. We could not predict which condition would be more effective based on prior research. We found that students who compared methods gained greater procedural flexibility and conceptual knowledge than students in either compare-problem condition. Students in all three conditions made similar gains in procedural knowledge, including success transferring the methods to new problem types. These findings suggest that for mathematics learning, comparing methods supports transfer as well as comparing problems, and it supports procedural flexibility and conceptual knowledge better than comparing problems.

We also included a compare-problem-types condition in the Rittle-Johnson et al. (2009) study described in the previous section, with students randomly assigned to a compare-problem-types, compare-methods or sequential condition. Students in the compare-problem-types condition did not differ from students in the sequential condition on any outcome, regardless of prior knowledge (Rittle-Johnson et al., 2009). Differences between the compare-problem-types and compare-methods condition depended on prior knowledge and mirrored the results for the compare-methods versus sequential conditions. For students who attempted to use algebra at pretest, the compare-problem-types condition tended to be less effective than the compare-methods condition for procedural knowledge and flexible use of procedures. However, for students who did not use algebra at pretest, the compare-problem-types condition was more effective than the compare-methods condition for procedural knowledge, conceptual knowledge and flexible use of procedures.

Overall, our comparing problems conditions have not been especially effective in supporting learning relative to comparing methods or sequential study of examples (Table 12.1). These findings corroborate concerns raised by Reed and coworkers that for complex, multistep methods, comparing problems may not effectively support transfer (Reed, 1989; Reed, Stebick, Comey, & Carroll, 2012). Across five experiments with college students, comparing two algebra word problems and their solutions did not support learning or transfer of the solution methods. Comparing two easier arithmetic word problems supported learning of the solution method, but not transfer of the method to algebra word problems (Reed et al., 2012). In part, the lack of benefit for comparison arose because students often generated similarities and differences that were too generic or superficial to help them abstract the key features of the solution method. Complex multistep methods may be more difficult to learn via problem comparison than simpler methods like the one learned in Gick and Holyoak (1983) for the Dunker radiation problem. Our findings also suggest that comparing problems solved with complex methods does not support procedural flexibility or conceptual understanding as well as comparing methods.

Other variations of comparing problems show more promise. First, comparing easily confusable problem types helps learners distinguish between the two problem types and solve more problems correctly (Cummins, 1992; Day, Goldstone, & Hill, 2010; VanderStoep & Seifert, 1993). For example, comparison of algebraic addition and multiplication examples supported better problem-solving accuracy than sequential study of addition examples, followed by multiplication examples (Ziegler & Stern, 2014, 2016). Second, comparing positive and negative examples of key ideas may improve conceptual knowledge. Students who compared problems that were positive and negative examples of each key idea (e.g., a line segment that was vs. was not the altitude of a triangle) gained greater conceptual knowledge than students who studied only positive examples (Guo & Pang, 2011). Note that the control condition was not exposed to negative examples, making it impossible to know whether comparison was critical. Overall, comparing problems may be particularly useful in helping people recognize important problem features that differ between carefully selected problems.

Summary of Researcher-Led Classroom Studies and Proposed Guidelines

Designing studies that could be conducted in classrooms pushed our thinking about different types of comparisons, what learning outcomes each supports, and when learners are prepared to learn from comparison. We have evaluated two types of comparison in mathematics classrooms—comparing methods and comparing problems. Comparing correct methods consistently supported procedural flexibility across studies (Table 12.1). For students who knew one of the solution methods at pretest, comparing methods sometimes supported greater procedural knowledge (Rittle-Johnson & Star, 2007; Rittle-Johnson et al., 2009) or greater conceptual knowledge (Rittle-Johnson & Star, 2009; Rittle-Johnson et al., 2009; Star & Rittle-Johnson, 2009). For novices, who did not know one of the solution methods at pretest, comparing methods was only helpful after we slowed the pace of the lesson. Overall, comparing methods can help a variety of students learn, but its advantages are more substantial if students have sufficient prior knowledge. How best to develop this prior knowledge is an important topic for future research.

Comparing problems solved with the same method has shown less promise for supporting mathematics learning, at least for solving equations and algebraic word problems (Reed, 1989; Reed et al., 2012; Rittle-Johnson & Star, 2009; Rittle-Johnson et al., 2009). It has been generally less effective than comparing methods, especially for students with prior knowledge in the domain. A different version of comparing problems has shown promise for helping novices learn important problem features and concepts that can be hard to learn, such as the altitude of a triangle (Guo & Pang, 2011); and future research is needed to investigate different versions of comparing problems.

Based on this research, we have begun to develop guidelines for effectively using comparison to support mathematics learning. First, *what* is compared matters. Comparing solution methods is particularly promising and can support greater math learning than other forms of comparison (Rittle-Johnson & Star, 2009). In addition, comparing a familiar method to an unfamiliar one is typically more effective than comparing two unfamiliar methods (Rittle-Johnson et al., 2009). Second, it matters *who* is learning from comparison. Students with some prior knowledge in the domain often learn more effectively from comparison than novices in the domain (Rittle-Johnson et al., 2009). At the same time, once students develop strong knowledge in the domain, explicit comparison may not be needed, as students may learn equally well without comparison (Guo & Pang, 2011). Our series of studies contributed significantly to a recent Practice Guide from the US Department of Education (Woodward et al., 2012) that identified comparing multiple solution methods as one of five recommendations for improving mathematical problem solving in the middle grades. Thus, this series of studies has provided important information about *what* to compare and *when* in instruction to best use different types of comparison.

YEAR-LONG STUDY HELPING TEACHERS USE COMPARISON IN ALGEBRA I CLASSROOMS

Given the promise of comparison for supporting mathematics learning, we developed and evaluated a set of supplementary materials for incorporating comparison throughout Algebra I instruction. Mastery of algebra is an important milestone for students, in that algebra serves as a "gatekeeper" for full participation in society and also provides students with the ability to harness new technologies and take advantage of the job opportunities resulting from them (Moses & Cobb, 2001; National Mathematics Advisory Panel, 2008). Regrettably, too few students are graduating from high school with the algebra skills needed for college or the workforce (American Diploma Project, 2004). Thus, improving algebra instruction is of great importance.

Unfortunately, US teachers often struggle to effectively support comparison of multiple methods during mathematics instruction (Stein, Engle, Smith, & Hughes, 2008). Analyses of video records of mathematics instruction indicate that comparison is often not well enacted in US classrooms (Richland, Holyoak, & Stigler, 2004; Richland et al., 2007). Algebra I teachers reported that they introduced multiple solution methods for at least some problem types, but that they often did not explicitly compare the methods (Lynch & Star, 2014b). These findings indicate that comparison is a reasonable adaptation of current teaching practices, but that there is a need for materials and training to help teachers effectively use comparison in mathematics instruction.

Supplemental Curriculum Materials

Based on the promise of our worked example based approach to supporting comparison, we developed a set of supplementary worked example pairs that could be used in conjunction with any Algebra I curriculum. A team of mathematics education experts, including researchers, mathematicians, and Algebra I teachers, developed the materials by going through a typical Algebra I course syllabus, identifying core concepts, common student difficulties, and key misconceptions, and then creating comparison materials to attempt to address them.

This led to a set of 141 worked example pairs (WEPs, available from: scholar.harvard.edu/contrastingcases). A sample WEP is shown in Fig. 12.2. Each WEP showed the mathematical work and dialog of two hypothetical students, Alex and Morgan, as they attempted to solve one or more algebra problems. The curriculum contained four types of WEPs, with the types varying in what was being compared and the instructional goal of the comparison, as outlined in Table 12.2. *Which is better?* WEPs showed the same problem solved using two correct, but different methods, with the goal of understanding when and why one method was more efficient or easier than another method for a given problem, as we did in our short-term research on comparing methods (e.g., Rittle-Johnson & Star, 2007). *Which is correct?* WEPs showed the same problem solved with a correct and incorrect method, with the goal of understanding and avoiding common errors. Comparing correct and incorrect methods supports gains in procedural knowledge, retention of conceptual knowledge, and a reduction in misconceptions (Durkin & Rittle-Johnson, 2012).

Two new comparison types were also included. The new types first emerged during a classroom study by Newton et al. (2010) and were further developed by mathematics educators on the research team. *Why does it work?* WEPs showed the same problem solved with two different correct solution methods with the goal of illuminating the conceptual rationale in one method that is less apparent in the other method. This is in contrast to the *Which is better?* comparisons, where the goal was to learn when and why one method was better for solving particular types of problems. *How do they differ?* WEPs showed two different problems solved in related ways, with an interest in illustrating what the relationship between problems and answers of the two problems revealed about an underlying mathematical concept. This emergence of new comparison types is one clear benefit of bridging between cognitive science and education, and highlights that the benefits are not unidirectional from theory to practice.

The WEPs were designed to maximize their potential impact on student learning based on previous research. As before, the two worked examples were

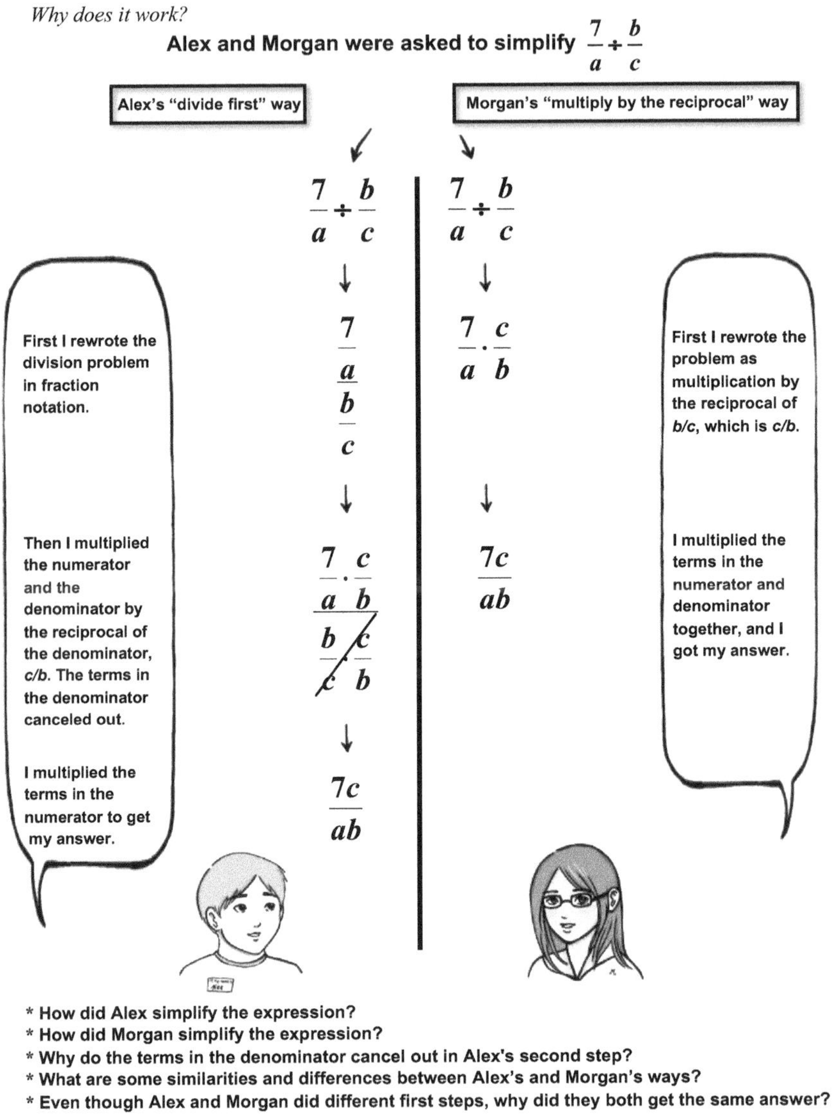

FIGURE 12.2 Sample worked example pair from our Algebra I scale-up project, illustrating a *Why does it work?* Comparison.

presented side-by-side. To facilitate processing of the examples, we included thought bubbles where two students (Alex and Morgan) described their solution methods. We used common language in these descriptions as much as possible to help facilitate alignment of the examples. We also formalized an instructional routine to help improve the effectiveness of using our comparison materials. Each WEP had three types of reflection prompts (understand, compare, and

TABLE 12.2 Four Types of Comparisons Used in Our Year-Long Algebra I Classroom Study

Comparison Type				
Comparison goal	*Which is better?*	*Which is correct?*	*Why does it work?*	*How do they differ?*
Features of examples	Same problem solved with two different, correct methods	Same problem solved with a correct and an incorrect method	Same problem solved with two different, correct methods	Different problems. Focus is not on solution methods
Focus of comparison	When and why a method is more efficient or easier	Why one method works and one does not	Conceptual rationale revealed in one method that is less apparent in other method	What relations between problems and answers reveal about underlying concept
Sample worked example pair	Solving the proportion $4/5 = 24/n$ by finding equivalent fractions or by cross multiplying	Solving $45y + 90 = 60y$ by first subtracting $45y$ from both sides or by incorrectly combining $45y + 60$	Expanding the expression $(x^4)^2$ by applying the power rule ($x^{4 \times 2}$) or by expanding and then squaring	Graphing the equations $y = x^2$ and $y = -x^2$
Sample *Make Connections* prompt	Can you make up a general rule for when Alex's way is better and when Morgan's way is better?	Can you state a general rule about combining like terms?	Even though Alex and Morgan did different first steps, why did they get the same answer?	How does changing the sign of the coefficient of x^2 affect the graph of the quadratic function?

make connections) meant to culminate in a discussion of the learning goal for the pairs of WEPs. First, *Understand* prompts, such as, "How did Morgan solve the equation?" were intended to provide students the opportunity to understand each worked example individually, prior to comparing them. Second, *Compare* prompts, such as "What are some similarities and differences between Alex's and Morgan's ways?" were meant to encourage comparison of the two worked examples.

Understand and *Compare* prompts were very similar across WEP types and were intended to prepare students to engage in productive reflection on the final, *Make Connections* prompts, such as "On a timed test, would you rather use Alex's way or Morgan's way? Why?" and "Even though Alex and Morgan did different first steps, why did they both get the same answer?" Our pilot work revealed that sometimes teachers skipped or inadequately addressed the *Make Connection* prompts, so we supplemented each WEP with an additional "take-away" page. On this page, the fictitious students Alex and Morgan identify the learning goal for that WEP. Our intent was that the teacher would use the take-away page to provide an explicit summary statement of the instructional goal of the WEP. Prior research suggests that direct instruction is needed to supplement student-generated comparisons (Schwartz & Bransford, 1998), and a feature of high-quality instruction is that teachers summarize the instructional goals of a lesson (Brophy, 1999).

Teachers were asked to use our materials once or twice a week as a supplement to their usual curriculum. Teachers were given complete latitude in deciding which WEPs to use, when to use them, and for how long to use them.

In collaboration with Kristie Newton, we also designed a 1-week, 35-h professional development institute to familiarize teachers with the materials and approach (Newton & Star, 2013). Teachers read through and discussed the supplemental curriculum materials and viewed videotaped exemplars of other teachers using the curriculum. In addition, teachers worked in groups to plan and teach sample lessons to their peers using the materials, which were implemented and then debriefed by the group.

Implementation and Evaluation

We first piloted our materials with 13 Algebra I teachers. We then conducted a year-long randomized controlled trial that explored the effectiveness of implementing our Algebra I supplemental curriculum in typical classrooms, that is, its impact on teachers' instruction and students' mathematical knowledge (Star et al., 2015b). Initially, 141 Algebra I teachers from public schools were randomly assigned to either implement the comparison curriculum as a supplement to their regular curriculum (*treatment* condition) or to continue using their existing curriculum and methods ("*business as usual*" *control* condition). However, there was large attrition between the spring when teachers volunteered to participate and the fall when implementation began, due to a range of factors, such as teachers no longer being assigned to teach Algebra 1. Such attrition reflects the difficulties of conducting research in public high schools, but is a limitation of the study. The final sample consisted of 76 teachers and their students (39 treatment teachers with 781 students and 29 control teachers with 586 students).

Before the school year began, treatment teachers completed the 35-h professional development institute. Observations, surveys, and interviews indicated that the professional development was successful in familiarizing teachers with our approach (Lynch & Star, 2014a). During the school year, treatment teachers implemented the materials on their own, without researcher involvement. Each time teachers used our materials, they were asked to call and report on what WEP they had used, how they had used it, and their impressions on the strengths and weaknesses of the materials. Teachers were also asked to videotape their use of our materials every other week, and teachers in both conditions were asked to videotape their instruction (without use of our materials) once a month.

Implementation results. Analysis of the videotapes indicated that treatment teachers were able to implement our materials as intended (Star et al., 2015b). Further, control teachers did not frequently use specific instructional practices that were integral to the intervention; they rarely presented multiple methods side-by-side or explicitly compared examples. However, there was large variation in how frequently treatment teachers used our curriculum, and many treatment teachers used the supplemental materials much less frequently than intended, using our materials an average of 19 class periods during the 180-day school year (range: 0–56 days). In fact, 18% of participating treatment teachers did not report using the materials even once; 30% of them reported using them 5 times or fewer. Overall, treatment teachers used our supplemental materials much less often than intended (i.e., low degree of implementation), but when they chose to use the materials, it was with high fidelity (i.e., high quality of implementation).

Student outcomes. To assess student learning, teachers administered a standardized algebra readiness test, the Acuity Algebra Diagnostic Readiness Exam (CBT/McGraw Hill, 2007), as well as a researcher-designed assessment of algebra knowledge to their students at the beginning and end of the school year. Students in the two conditions did not differ in algebra knowledge at the beginning of the year. At the end of the school year, students' algebra knowledge was not higher in classrooms in which our materials were available (see Star et al., 2015b for full results). This was not surprising given how infrequently our supplemental curriculum was used in many classrooms.

Given the large variability in use of our curriculum, we evaluated whether increased use of our materials was associated with increased algebra knowledge in treatment classrooms. We calculated the *dosage* of curriculum given by each treatment teacher by multiplying the number of reported days each teacher used our materials by how long on average that teacher spent using our materials in a single lesson (based on the videos). Dosage ranged from 0 to 864 min ($M = 140$). We used an instrumental variable estimation (IVE) approach to analyze the data because it can account for a variety of potential factors that might teacher-level decisions about dosage (Murnane & Willett, 2011). Simply including dosage as an additional predictor in a multilevel model would create bias in the estimate of the dose-response relationship, but an IVE approach adjusts for this bias.

Increased dosage was predictive of greater procedural knowledge of algebra at the end of the school year (Star et al., 2015b). This included solving equations, graphing equations, and factoring expressions. Dosage did not have a significant effect on conceptual knowledge, procedural flexibility, or scores on the standardized algebra test. This may be because all comparison materials focused attention on procedural knowledge, whereas attention to flexibility and conceptual knowledge varied by comparison type.

We also explored whether frequency of using particular comparison types was associated with increased algebra knowledge. Frequency of using the *Why does it work?* comparisons was positively correlated with the amount of gain in students' algebra knowledge; correlations for the other types of comparison were positive but not significant.

Discussion of Year-Long Study

Overall, providing our supplemental materials to teachers did not significantly affect students' algebra knowledge. However, many teachers used our materials infrequently. Increased use of our materials had a positive effect on procedural knowledge, although not on other student knowledge outcomes. These findings suggest a potential role for comparison in supporting aspects of algebra learning, but also point to the challenges of supporting teachers' integration of this approach into the curriculum.

The largest limitation was that teachers used our materials much less than intended. Almost half of the sample reported using our materials on 5 or fewer occasions across the entire school year; on average, teachers used our materials for 19 class periods for a total of 140 min across the entire school year. We intentionally chose to give teachers a great deal of choice in which materials to use and when to use them, but this freedom seemed to lead to a substantial number of teachers not using the materials or using them only infrequently. Interest, training, and carefully designed materials were not enough for many teachers to adopt our materials with much regularity. It is time consuming and potentially challenging to select, plan, and integrate supplemental materials with the existing curriculum. Teachers would likely benefit from more clear structure and direction for which materials to use when.

Further, teachers could use additional supports to implement higher-quality instruction. Too frequently in our study, teachers were providing the explanations, with little student explanation or discussion (Star et al., 2015a, 2015b). In addition, our professional development only occurred over the summer, but effective mathematics teacher professional development typically provides teachers with multiple, intensive opportunities to establish meaningful connections between existing instructional practices and the practices and beliefs advocated by the new approach across the school year (Garet, Porter, Desimone, Birman, & Yoon, 2001). More sustained professional development is likely needed to increase high-quality instruction that emphasizes student explanation and discussion when integrating our materials.

CONCLUSIONS AND FUTURE DIRECTIONS

Comparison is a promising instructional approach for supporting mathematics learning. In short-term classroom studies, comparing different solution methods for solving the same problem often promoted procedural flexibility and sometimes promoted conceptual and procedural knowledge. Further, because comparison makes substantial demands on working memory resources, it is often more effective when learners have some prior domain knowledge; novices can become overwhelmed by comparison without adequate support. In addition, comparison can be infused throughout curriculum materials for an entire course, and some teachers are able to implement comparison-based materials in their instruction. However, providing comparison-based materials and professional development in the summer was not sufficient for many teachers to incorporate comparison throughout the school year or to reliably improve student outcomes. A variety of changes to the supplemental materials and professional development may be needed.

Future research is needed to advance both theory and practice. First, research is needed on how different types of comparison support different learning outcomes. For example, the *Why does it work?* comparison type that emerged from our classroom research is particularly promising for supporting knowledge of critical math concepts and procedures. Do *Why does it work?* comparisons support greater student learning than sequential study of the same materials? We predict that they will. More generally, theories of analogical learning need to explicitly consider what is being compared. For example, structure-mapping theory (Gentner, 1983, 2010) should be refined to consider how comparing problems versus methods impact alignment and mapping of elements between two examples. Second, research is needed on the supports teachers need to effectively integrate comparison in their instruction, including the features of the supplemental materials (e.g., use of worked examples vs. student-generated solutions) and the importance of student explanation and discussion. Comparison is an effective practice for improving mathematics learning, and future research should further refine theory and practice to maximize its potential benefits.

ACKNOWLEDGMENTS

Much of the research reported in this article was supported by grants from the National Science Foundation (DRL0814571) and the Institute of Education Sciences (R305H050179); the ideas in this paper are those of the authors and do not represent official positions of NSF or IES.

REFERENCES

Alfieri, L., Nokes-Malach, T. J., & Schunn, C. D. (2013). Learning through case comparison: a meta-analytic review. *Educational Psychologist, 48*(2), 87–113.

Alibali, M. W., & Nathan, M. J. (2012). Embodiment in mathematics teaching and learning: evidence from learners' and teachers' gestures. *Journal of the Learning Sciences, 21*(2), 247–286.

American Diploma Project (2004). Ready or not: Creating a high school diploma that counts. Achieve Inc. Available from: http://www.achieve.org/files/ReadyorNot.pdf.

Atkinson, R. K., Derry, S. J., Renkl, A., & Wortham, D. (2000). Learning from examples: instructional principles from the worked examples research. *Review of Educational Research, 70*(2), 181–214.

Australian Education Ministers (2006). Statements of Learning for Mathematics. Curriculum Corporations. Available from: http://www.curriculum.edu.au/verve/_resources/SOL_Mathematics_2006.pdf.

Ball, D. L. (1993). With an eye on the mathematical horizon: dilemmas of teaching elementary school mathematics. *The Elementary School Journal, 93*, 373–397.

Begolli, K. N., & Richland, L. E. (2015). Teaching mathematics by comparison: analog visibility as a double-edged sword. *Journal of Educational Psychology, 108*(2), 194–213.

Blöte, A. W., Van der Burg, E., & Klein, A. S. (2001). Students' flexibility in solving two-digit addition and subtraction problems: instruction effects. *Journal of Educational Psychology, 93*(3), 627–638.

Brophy, J. (1999). *Teaching education practices series no. 1*. Geneva: International Bureau of Education.

Carpenter, T. P., Franke, M. L., Jacobs, V. R., Fennema, E., & Empson, S. B. (1998). A longitudinal study of invention and understanding in children's multidigit addition and subtraction. *Journal for Research in Mathematics Education, 29*(1), 3–20.

Catrambone, R., & Holyoak, K. J. (1989). Overcoming contextual limitations on problem-solving transfer. *Journal of Experimental Psychology: Learning, Memory, and Cognition, 15*(6), 1147–1156.

CBT/McGraw Hill. (2007). Acuity Algebra. Available from: http://www.acuityforschools.com

Chi, M. T. H. (2000). Self-explaining: the dual processes of generating inference and repairing mental models. In R. Glaser (Ed.), *Advances in instructional psychology: Educational design and cognitive science* (pp. 161–238). (Vol. 5). Mahwah, NH: Lawrence Earlbaum.

Common Core State Standards (2010). National Governors Association Center for Best Practices and Council of Chief State School Officers. Washington DC.

Cummins, D. (1992). Role of analogical reasoning in the induction of problem categories. *Journal of Experimental Psychology: Learning, Memory, and Cognition, 18*(5), 1103–1124.

Day, S., Goldstone, R. L., & Hill, T. (2010). The effects of similarity and individual differences on comparison and transfer. Paper presented at the Cognition in Flux: Proceedings of the 32nd Annual Meeting of the Cognitive Science Society, Austin, TX.

Durkin, K., & Rittle-Johnson, B. (2012). The effectiveness of using incorrect examples to support learning about decimal magnitude. *Learning and Instruction, 22*(3), 206–214.

Fraivillig, J. L., Murphy, L. A., & Fuson, K. (1999). Advancing children's mathematical thinking in everyday mathematics classrooms. *Journal for Research in Mathematics Education, 30*, 148–170.

Garet, M. S., Porter, A. C., Desimone, L., Birman, B. F., & Yoon, K. S. (2001). What makes professional development effective? Results from a national sample of teachers. *American Educational Research Journal, 38*(4), 915–945.

Gentner, D. (1983). Structure-mapping: a theoretical framework for analogy. *Cognitive Science: A Multidisciplinary Journal, 7*(2), 155–170.

Gentner, D. (2010). Bootstrapping the mind: analogical processes and symbol systems. *Cognitive Science, 34*(5), 752–775.

Gentner, D., Loewenstein, J., & Thompson, L. (2003). Learning and transfer: a general role for analogical encoding. *Journal of Educational Psychology, 95*(2), 393–405.

Gick, M. L., & Holyoak, K. J. (1983). Schema induction and analogical transfer. *Cognitive Psychology, 15*(1), 1–38.

Goldstone, R. L., Day, S., & Son, J. (2010). Comparison. In B. Glatzeder, V. Goel, & A. von Müller (Eds.), *On thinking: Towards a theory of thinking* (pp. 103–122). (Vol. II). Heidelberg, Germany: Springer Verlag GmbH.

Guo, J. -p., & Pang, M. F. (2011). Learning a mathematical concept from comparing examples: the importance of variation and prior knowledge. *European Journal of Psychology of Education, 26*(4), 495–525.

Hiebert, J., Carpenter, T. P., Fennema, E., Fuson, K. C., Human, P., Murray, H., ... Wearne, D. (1996). Problem solving as a basis for reform in curriculum and instruction: The case of mathematics. *Educational Researcher, 25*(4), 12–21.

Huffred-Ackles, K., Fuson, K., & Sherin Gamoran, M. (2004). Describing levels and components of a math-talk learning community. *Journal for Research in Mathematics Education, 35*(2), 81–116.

Hummel, J. E., & Holyoak, K. J. (1997). Distributed representations of structure: a theory of analogical access and mapping. *Psychological Review, 104*(3), 427–466.

Johnson, D. W., & Johnson, R. T. (1994). *Learning together and alone: Cooperative, competitive and individualistic learning* (4th ed.). Boston, MA: Allyn and Bacon.

Kieran, C. (1992). The learning and teaching of school algebra. In D. Grouws (Ed.), *Handbook of research on mathematics teaching and learning* (pp. 390–419). New York: Simon & Schuster.

Kilpatrick, J., Swafford, J. O., & Findell, B. (2001). *Adding it up: helping children learn mathematics*. Washington DC: National Academy Press.

Kultusministerkonferenz. (2004). Bildungsstandards im Fach Mathematik für den Primarbereich. [Educational Standards in Mathematics for Primary Schools].

Kurtz, K., Miao, C. -H., & Gentner, D. (2001). Learning by analogical bootstrapping. *The Journal of the Learning Sciences, 10*, 417–446.

Lampert, M. (1990). When the problem is not the question and the solution is not the answer: mathematical knowing and teaching. *American Educational Research Journal, 27*, 29–63.

Lynch, K., & Star, J. R. (2014a). Exploring teachers' implementation of comparison in Algebra I. *Journal of Mathematical Behavior, 35*, 144–163.

Lynch, K., & Star, J. R. (2014b). Teachers' views about multiple strategies in middle and high school mathematics. *Mathematical Thinking and Learning: An International Journal, 16*(2), 85–108.

McEldoon, K. L., Durkin, K. L., & Rittle-Johnson, B. (2013). Is self-explanation worth the time? A comparison to additional practice. *British Journal of Educational Psychology, 83*(4), 615–632.

Moses, R. P., & Cobb, C. E. (2001). *Radical equations: Math literacy and civil rights*. Boston: Beacon Press.

Murnane, R. J., & Willett, J. B. (2011). *Methods matter: Improving causal inference in educational and social science research*. New York: Oxford University Press.

Namy, L. L., & Gentner, D. (2002). Making a silk purse out of two sow's ears: young children's use of comparison in category learning. *Journal of Experimental Psychology: General, 131*(1), 5–15.

National Assessment of Educational Progress. (2011). Mathematics Performance. Available from: http://nces.ed.gov/programs/coe/indicator_cnc.asp

National Mathematics Advisory Panel. (2008). *Foundations of success: The final report of the National Mathematics Advisory Panel*. Washington, DC: US Department of Education.

NCTM. (2014). *Principles to actions: Ensuring mathematical success for all*. Reston, VA: National Council of Teachers of Mathematics, Inc.

Newton, K. J., & Star, J. R. (2013). Exploring the nature and impact of model teaching with worked example pairs. *Mathematics Teacher Educator, 2*, 86–102.

Newton, K. J., Star, J. R., & Lynch, K. (2010). Understanding the development of flexibility in struggling Algebra students. *Mathematical Thinking and Learning: An International Journal, 12*, 282–305.

Oakes, L. M., & Ribar, R. J. (2005). A comparison of infants' categorization in paired and successive presentation familiarization tasks. *Infancy, 7*(1), 85–98.

Prather, R. W., & Alibali, M. W. (2009). The development of arithmetic principle knowledge: how do we know what learners know? *Developmental Review, 29*(4), 221–248.

Reed, S. K. (1989). Constraints on the abstraction of solutions. *Journal of Educational Psychology, 81*(4), 532–540.

Reed, S. K., Stebick, S., Comey, B., & Carroll, D. (2012). Finding similarities and differences in the solutions of word problems. *Journal of Educational Psychology, 104*(3), 636–646.

Richland, L. E., Holyoak, K. J., & Stigler, J. W. (2004). Analogy use in eighth-grade mathematics classrooms. *Cognitiion and Instruction, 22*, 37–60.

Richland, L. E., Zur, O., & Holyoak, K. J. (2007). Cognitive supports for analogies in the mathematics classroom. *Science, 316*(5828), 1128–1129.

Rittle-Johnson, B., & Star, J. R. (2007). Does comparing solution methods facilitate conceptual and procedural knowledge? An experimental study on learning to solve equations. *Journal of Educational Psychology, 99*(3), 561–574.

Rittle-Johnson, B., & Star, J. R. (2009). Compared to what? The effects of different comparisons on conceptual knowledge and procedural flexibility for equation solving. *Journal of Educational Psychology, 101*(3), 529–544.

Rittle-Johnson, B., Siegler, R. S., & Alibali, M. W. (2001). Developing conceptual understanding and procedural skill in mathematics: an iterative process. *Journal of Educational Psychology, 93*(2), 346–362.

Rittle-Johnson, B., Star, J. R., & Durkin, K. (2009). The importance of prior knowledge when comparing examples: influences on conceptual and procedural knowledge of equation solving. *Journal of Educational Psychology, 101*(4), 836–852.

Rittle-Johnson, B., Star, J. R., & Durkin, K. (2012). Developing procedural flexibility: are novices prepared to learn from comparing procedures? *British Journal of Educational Psychology, 82*, 436–455.

Robinson, K. M., & LeFevre, J. -A. (2011). The inverse relation between multiplication and division: concepts, procedures, and a cognitive framework. *Educational Studies in Mathematics, 79*(3), 409–428.

Schwartz, D. L., & Bransford, J. D. (1998). A time for telling. *Cognition and Instruction, 16*(4), 475–522.

Siegler, R. S., & Chen, Z. (2008). Differentiation and integration: Guiding principles for analyzing cognitive change. *Developmental Science, 11*(4), 433–448.

Singapore Ministry of Education. (2006). Secondary Mathematics Syllabuses.

Star, J. R. (2005). Reconceptualizing procedural knowledge. *Journal for Research in Mathematics Education, 36*, 404–411.

Star, J. R., & Rittle-Johnson, B. (2008). Flexibility in problem solving: the case of equation solving. *Learning and Instruction, 18*, 565–579.

Star, J. R., & Rittle-Johnson, B. (2009). It pays to compare: an experimental study on computational estimation. *Journal of Experimental Child Psychology, 101*, 408–426.

Star, J. R., Newton, K., Pollack, C., Kokka, K., Rittle-Johnson, B., & Durkin, K. (2015a). Student, teacher, and instructional characteristics related to students' gains in flexibility. *Contemporary Educational Psychology, 41*, 198–208.

Star, J. R., Pollack, C., Durkin, K., Rittle-Johnson, B., Lynch, K., Newton, K., & Gogolen, C. (2015b). Learning from comparison in algebra. *Contemporary Educational Psychology, 40*, 41–54.

Stein, M. K., Engle, R. A., Smith, M. S., & Hughes, E. K. (2008). Orchestrating productive mathematical discussions: five practices for helping teachers move beyond show and tell. *Mathematical Thinking and Learning, 10*(4), 313–340.

Sweller, J. (1988). Cognitive load during problem solving: effects on learning. *Cognitive Science, 12*(2), 257–285.

Sweller, J., & Cooper, G. A. (1985). The use of worked examples as a substitute for problem solving in learning algebra. *Cognition and Instruction, 2*(1), 59–89.

Treffers, A. (1991). *Realistic mathematics education in the Netherlands 1980-1990*. Realistic Mathematics Education in Primary School. Utrecht, the Netherlands: Freudenthal Institute, Utrecht University.

VanderStoep, S. W., & Seifert, C. M. (1993). Learning "how" versus learning "when": improving transfer of problem-solving principles. *Journal of the Learning Sciences, 3*(1), 93–111.

Verschaffel, L., Luwel, K., Torbeyns, J., & Van Dooren, W. (2009). Conceptualizing, investigating, and enhancing adaptive expertise in elementary mathematics education. *European Journal of Psychology of Education, 24*(3), 335–359.

Webb, N. M. (1991). Task-related verbal interaction and mathematics learning in small groups. *Journal for Research in Mathematics Education, 22*(5), 366–389.

Woodward, J., Beckmann, S., Driscoll, M., Franke, M. L., Herzig, P., Jitendra, A. K., & … Ogbuehi, P. (2012). Improving mathematical problem solving in grades 4 to 8 A practice guide. Washington, DC: National Center for Education Evaluation and Regional Assistance, Institute of Education Sciences.

Ziegler, E., & Stern, E. (2014). Delayed benefits of learning elementary algebraic transformations through contrasted comparisons. *Learning and Instruction, 33*, 131–146.

Ziegler, E., & Stern, E. (2016). Consistent advantages of contrasted comparisons: algebra learning under direct instruction. *Learning and Instruction, 41*, 41–51.

Chapter 13

Evidence for Cognitive Science Principles that Impact Learning in Mathematics

Julie L. Booth*, Kelly M. McGinn*, Christina Barbieri**, Kreshnik N. Begolli*, Briana Chang*, Dana Miller-Cotto*, Laura K. Young*, Jodi L. Davenport†

**Temple University, Philadelphia, PA, United States; **University of Delaware, Newark, DE, United States; †WestEd, San Francisco, CA, United States*

INTRODUCTION

Students in the United States consistently underperform on state tests of mathematical proficiency (e.g., Kim, Schneider, Engec, & Siskind, 2006; Pennsylvania Department of Education, 2011) and in international comparisons on the Trends in International Mathematics and Science Study (TIMSS; e.g., Mullis, Martin, Foy, & Arora, 2012) and Programme for International Student Assessment (PISA; e.g., Fleischman, Hopstock, Pelczar, & Shelley, 2010). Among other issues, aspects of early mathematics instruction can interfere with later learning (McNeil et al., 2006; see also McNeil et al., Chapter 8; Van Hoof et al., Chapter 5) and students are not adequately prepared to tackle difficult gatekeepers, such as fractions (Booth & Newton, 2012) and algebra (Department of Education, 1997), that in turn prevent them from advancing in the fields of Science, Technology, Engineering, and Mathematics (STEM). To address these issues, educators and researchers have repeatedly called for mathematics instruction in the United States to be based on evidence (e.g., CCSSI, 2010; NCLB, 2002; NMAP, 2008).

Over the past few decades, the field of cognitive science has identified factors that have the potential to substantively improve students' learning. These principles have been detailed in several reviews (e.g., Dunlosky, Rawson, Marsh, Nathan, & Willingham, 2013; Koedinger, Booth, & Klahr, 2013; Pashler et al., 2007), and many involve a comparison of different approaches (such as spaced vs. massed practice), showing that one type of instructional technique

Acquisition of Complex Arithmetic Skills and Higher-Order Mathematics Concepts
http://dx.doi.org/10.1016/B978-0-12-805086-6.00013-8

is superior to another type of instructional technique (Koedinger et al., 2013). Many of these cognitive science principles are potentially useful for improving mathematics instruction, perhaps especially those concerning abstract and concrete representations, analogical comparison, feedback, error reflection, scaffolding, distributed practice, interleaved practice, and worked examples.

However, a vast divide persists between research and practice, as many of these principles are not consistently used in US classrooms. For example, mathematics textbooks in the UnitedStates frequently include decorative images alongside mathematically relevant images (Lehman, Schraw, McCrudden, & Hartley, 2007) and lack worked examples for students to study or explain (Mayer, Sims, & Tajika, 1995). Moreover, US teachers tend to shy away from talking about errors that could be useful for targeting students' conceptual misunderstandings (Lannin, Townsend, & Barker, 2006). Many of the recommended practices mimic those common in mathematics instruction in countries with traditionally higher mathematics achievement. For example, teachers in Hong Kong and Japan make more connections between abstract and concrete representations (Richland, Zur, & Holyoak, 2007), worked examples are more common in Japanese textbooks (Mayer et al., 1995), students are taught to use particularly effective concrete models in Singapore (Hoven & Garelick, 2007; see also Lee, Ng, & Bull, Chapter 9), and errors are considered to be critical opportunities for learning in Japanese classrooms (Stigler & Hiebert, 1999).

The fact that these practices are prominent in countries that consistently outperform the United States suggests these practices may also be useful in US classrooms. However, before the wholesale implementation of these practices, it is important to first decompose them into the underlying cognitive principles and then test them within the US educational system. This is important because experiences outside of the classroom, prior instruction, and parental expectations may also differ across countries and in ways that might influence the effectiveness of instructional techniques in different contexts.

In the present chapter, we review the evidence for several principles that are especially promising for improving mathematics instruction (Table 13.1). Using the classification scheme proposed by Koedinger et al. (2013), we begin with principles that focus on improving memory and fluency: scaffolding, distributed practice, and feedback. We then move to worked examples, interleaved practice, and abstract and concrete representations, which primarily promote induction and refinement. Finally, we review principles that are geared towards improving sense making and understanding: Error reflection and analogical comparison. In each section, we first describe the principle and the instructional implications. We then evaluate the evidence for the effectiveness of the principle for promoting mathematics learning, first from laboratory studies and then from classroom studies. Finally, we summarize what is and is not known about how these principles relate to mathematics learning, and identify gaps to be addressed in further translational research. The chapter concludes with an overall evaluation of the current state of evidence regarding how the use of these cognitive principles

TABLE 13.1 Cognitive Principles

Principle	Basic description
Scaffolding	Having instructional support gradually faded enables learners to solve problems fluently
Distributed practice	Spacing out practice leads to better memorization than practicing all at once
Feedback	Receiving feedback about responses improves achievement
Worked examples	Studying (or explaining) worked out examples plus solving problems is better for refining knowledge than problem solving alone
Interleaving	Practicing solving different types of problems in a mixed order is better for problem solving skill than practicing the same type of problems in a row.
Abstract and concrete representations	Linking between abstract and concrete representations yields increased learning and transfer
Error reflection	Thinking about errors improves problem representation and increases conceptual understanding
Analogical comparison	Comparing and contrasting multiple instances leads to better understanding than studying one instance

can influence mathematics instruction, both for simple and for more complex skills and concepts; we also suggest research priorities necessary to maximize the potential impact of cognitive science for mathematics instruction.

SCAFFOLDING PRINCIPLE

Scaffolding derives from Vygotsky's (1978) Zone of Proximal Development (ZPD), or the distance between the individual learner's current development and their potential development if guided by parents, teachers, or peers, within socio-cultural theory. With regard to learning, the theory suggests that adults play a role in children's problem solving activities while taking into account the child's level of understanding (Kupers, van Dijk, & van Geert, 2015). During the initial learning stages, adults offer large amounts of assistance and as the child becomes more competent, the adult removes assistance gradually. Under this assumption, learning begins as a social activity before it becomes an independent one. As an instructional strategy, scaffolding holds promise for promoting improved problem-solving performance (Kim & Hannafin, 2011). Broadly, scaffolding has been defined as any activity that involves modeling a specific behavior for a child or student, and then removing this modeling as the learner becomes more knowledgeable (Wood, Bruner, & Ross, 1976). More specifically, scaffolding involves three components: (1) scaffolding should be contingent on students' level of understanding (Pratt & Savoy-Levine, 1998), (2) there should be a gradual withdrawal of scaffolding (Calder, 2015), and (3) there should be a responsibility transfer from the teacher or adult to the student (van de Pol, Volman, & Beishuizen, 2010). Thus, not only should a teacher monitor the students' learning as they progress, but an initial assessment of the students' knowledge level is necessary to adapt scaffolding strategies for effective learning.

Evidence from Laboratory Studies

With respect to mathematics learning, laboratory studies demonstrate that considering students' prior knowledge about the content is a key component when determining how much or how little scaffolding is needed (Schwonke, Renkl, Salden, & Aleven, 2011; see also Thevenot, Chapter 3). Nevertheless, providing no scaffolding may be more effective than providing predetermined levels of scaffolding to all participants. In a metaanalysis comparing fixed scaffolds to those that were contingent on students' ZPD, Belland, Walker, Olsen, and Leary (2015) found that predetermined, fixed amounts of scaffolding delays the transfer of responsibility process likely because it did not account for the learners' current competence level and subsequently hampered the learning process.

Evidence from Classroom Studies

In mathematics classrooms, scaffolding has been shown to be necessary in early stages of learning particularly for students attempting to learn content just outside of their ZPD (Smit, van Eerde, & Bakker, 2013), but must be faded out successively in order to pass responsibility to the learner (Razzaq & Heffernan, 2006; Vorhölter, Kaiser, & Ferri, 2014). When scaffolds are not removed over time, combined with not taking into account the students' competence level, performance can suffer (Schwonke et al., 2011). As with laboratory work, the successful use of scaffolding in mathematics classrooms is based on individualized instruction through computer adaptive learning or one-on-one strategies (e.g., Salden, Aleven, Renkl, & Schwonke, 2009), rather than a one size fits all scaffolding approach.

Recommendations for Further Research

A lack of consensus on the definition of scaffolding has made assessing the effectiveness of this approach difficult. For example, proponents of Wood et al.'s (1976) perspective define scaffolding as support from a teacher or more competent peer to help students solve problems that are beyond their capacity, whereas van de Pol et al. (2010) require that scaffolding adjusts to the students' level of understanding, fading or successively removing scaffolds, and transferring responsibility from the teacher to the learner. Further, this definition has not yet been broadened to include scaffolds used in intelligent tutoring systems. In these computer-based contexts, the computer uses students' prior knowledge to determine the level of fading needed to adapt to the individual learner, making this activity less of a human-to-human interaction. As it stands, it is difficult to compare the effectiveness of this strategy across studies while also being aware that scaffolding does not look the same for every teacher-student interaction, and thus a general outline for what scaffolding is not may be an important part of reaching a consensus on what scaffolding includes (van de Pol et al., 2010).

Further, it seems clear that fading of support is necessary, but further work should investigate how and when it is best to fade for different mathematical content, not only by considering the learners' prior knowledge, but other cognitive characteristics (e.g., working memory, inhibitory control, and attention) that might improve or diminish learning if there are too few or too many scaffolds in place. Considering access to computer based learning, however, it may also be necessary to establish methods for providing faded support in traditional mathematics classrooms when personalized, computerized instruction is not in place, though this may not be feasible given the nature of scaffolds.

DISTRIBUTED PRACTICE EFFECT

Distributed practice is one of the most widely studied cognitive principles, whereby spreading out learning opportunities (spaced practice)—either within a single study session or across multiple sessions—leads to better long-term retention of studied material than does providing multiple learning opportunities one right after the other (massed practice) (Dunlosky et al., 2013). It is important to note that this principle does not refer to increasing the number of practice problems, but rather, the temporal distribution of those problems. As Rohrer (2009) clarifies,

> *For example, rather than work 10 problems on the same topic in one session, a student might divide the same 10 problems across two sessions separated by a week. Furthermore, spacing does not entail more recent practice, because in a well-designed spacing experiment, the interval between study and test, the test delay, is measured from the last practiced problem (Rohrer, 2009, p. 8).*

Research on distributed practice has yielded two types of effects, spacing, and lag effects, with implications for the distribution of both study time and of practice testing. A *spacing effect* refers to an advantage of spaced-out practice over massed, or blocked, practice (see Cepeda, Pashler, Vul, Wixted, & Rohrer, 2006 for a review). A *lag effect* refers to advantages of spacing with longer lags (i.e., longer time between practice sessions) versus spacing with shorter lags (i.e., shorter time between practice sessions) (Carpenter, Pashler, & Cepeda, 2009). If stimuli are being processed intentionally, spacing effects are larger than when processing is unintentional (Janiszewski, Noel, &

Sawyer, 2003); however, these effects can also be observed under conditions of incidental learning when deep processing occurs (Challis, 1993).

Evidence from Laboratory Studies

Within the domain of mathematics, laboratory studies are sparse, but available results indicate there are also benefits for spaced practice on tasks that require more than rote memorization. For instance, spaced practice is superior to massed practice for understanding abstract mathematical concepts, such as determining simple permutations (Rohrer & Taylor, 2006, 2007). Longer lags were also shown to be superior to shorter lags for learning basic arithmetic facts (Rickard, Lau, & Pashler, 2008).

Evidence from Classroom Studies

Within mathematics classrooms, benefits of spaced practice have been observed for fluent retrieval of both addition (Schutte et al., 2015) and multiplication facts (Rea & Modigliani, 1985). Some classroom studies have focused on more complex tasks, but have also often conflated testing of multiple cognitive principles. Thus, it is hard to isolate the benefit of spaced practice in mathematics for these more complex tasks. For example, Gay (1973) claimed to find greater benefit for spaced practice in algebra classrooms, but spacing was confounded with test delay in the study (Rohrer, 2009). A classroom study examining conceptual understanding in statistics, however, did confirm that longer lags between practice sessions were superior to shorter lags (Budé, Imbos, van de Wiel, & Berger, 2011)

Recommendations for Further Research

As noted by a number of researchers, there is a lack of studies exploring the effects of spacing of instruction or practice on mathematics learning. Spaced practice is not widely evident in mathematics instruction (Bahrick & Hall, 2005; Dempster, 1988; Mayfield & Chase, 2002; Rohrer, 2009; Willingham, 2002), and mathematics textbooks often promote blocked practice rather than spaced practice (Rohrer, 2009). Further study is needed to evaluate the effect of distributed practice on a variety of mathematics skills, especially those beyond rote fact memorization.

Another clear gap in the research is whether the effects of distributed practice vary with individual differences in learner characteristics. Several potential characteristics of interest have been suggested, including prior knowledge, and motivation (Dunlosky et al., 2013), and cognitive capacities (Delaney, Verkoeijen, & Spirgel, 2010). However, there have been no direct tests of whether the benefits of spacing and lag effects in mathematics learning are moderated by these individual differences variables.

Delaney et al. (2010) suggest that researchers must examine whether other classroom activities may be effectively spaced to achieve the same goals. For example, classroom discussions may similarly serve to remind students of the content, and distributing discussion of those topics over time may be even more effective than distributing practice on those ideas over time (Delaney et al., 2010).

FEEDBACK PRINCIPLE

Generally, feedback is considered to be any form of information whose purpose is to enlighten a recipient (Mory, 2004). The study of feedback—in the form of reinforcement—dates back to Skinner's assertion that it reinforces behavior (Kulhavy, 1977), but feedback itself is considered to be "instructionally powerful" (Cohen, 1985, p. 33) for many contexts, not just behavior. Within a learning context, feedback is typically given in response to an instructional question or task and is intended to improve the learners' problem solving accuracy or conceptual understanding of the topic (Carter, 1984; Kulhavy, 1977). The effectiveness of feedback has been widely studied (see Shute, 2008 for a review), though not all students interpret teacher feedback in the same way (Williams, 1997). Recent research has focused specifically on the nature and timing of the feedback given, and how those factors impact its effectiveness.

Evidence from Laboratory Studies

For mathematics learning, one laboratory study showed that feedback helps to promote the formation of strategies used to solve arithmetic problems (Alibali, 1999). Further, process-oriented feedback (i.e., individualized comments about a student's confidence, strengths, and weaknesses) is more helpful than grade-oriented feedback (i.e., assigning a score, based on student performance compared to a reference group) for improving both mathematics achievement and interest in 9th grade students (Harks, Rakoczy, Hattie, Besser, & Klieme, 2014).

Students with low prior ability or knowledge may benefit more strongly from either outcome or strategy based feedback than their higher ability or more knowledgeable peers (Fyfe, Rittle-Johnson, & DeCaro, 2012; Krause, Stark, & Mandl, 2009; Luwel, Foustana, Papadatos, & Verschaffel, 2011). A student's working memory capacity may also moderate the type of feedback that is most beneficial; for instance, outcome-based feedback may be more beneficial for students with a lower working memory capacity, as more complicated feedback may tax their already limited working memory resources, while strategy and outcome-based feedback may be equally helpful for students with a higher working memory capacity (Fyfe, DeCaro, & Rittle-Johnson, 2015).

Evidence from Classroom Studies

Classroom studies providing individualized feedback have confirmed that immediate feedback is better than delayed feedback (Kehrer, Kelly, & Heffernan, 2013; Singh et al., 2011) and that it can increase the development of expertise (Ellis, Klahr, & Siegler, 1993). Further, video feedback (e.g., an individual providing verbal explanation feedback while referring to a whiteboard illustration) may be even better than traditional textual feedback as it was found to slow down the pace of the learner, suggestively allowing more time for the student to internalize the concept (Ostrow & Heffernan, 2014).

Recommendations for Further Research

In classroom settings, information about the effectiveness of method of delivery and frequency of immediate, detailed, process-based feedback is limited. It remains to be determined how to best individualize feedback in classroom settings, and whether the types of feedback provided by computer assisted learning can be provided by teachers in classroom settings. In addition, more work is needed to test the effectiveness of the kinds of feedback that are typically given in the context of traditional classrooms that are not technology-based. Finally, more research is needed to assess the effects of negative feedback in terms of potential improvements in learning without undermining motivation.

WORKED EXAMPLE PRINCIPLE

The worked example principle suggests that having learners study examples of worked-out solutions to problems is more effective for learning than having them solve all of the problems themselves (Sweller & Cooper, 1985). Further learning benefits are found when the learners are asked to explain the examples (Renkl, Stark, Gruber, & Mandl, 1998; Hausmann & van Lehn, 2007). Several mechanisms have been invoked to explain this principle. First, studying worked examples is thought to reduce learners' cognitive load by reducing the attentional and working memory demands needed to remember all of the problem solving steps (see also Lee, Ng, & Bull, Chapter 9). Instead, they can focus their limited working memory capacity on understanding the reasoning behind the procedural steps taken in the example. Second, prompting learners to *explain* the steps in the example provides the additional benefits inherent to self-explanation, including making knowledge explicit, connecting new information to prior knowledge, and generating inferences to fill knowledge gaps (Roy & Chi, 2005). Studying and explaining worked examples may be most beneficial for novice than expert learners (Booth et al., 2015b; Kalyuga, Chandler, Tuovinen, & Sweller, 2001).

Evidence from Laboratory Studies

The worked example principle has been successfully applied in a number of laboratory studies within the domain of mathematics. These studies demonstrate that studying worked examples increases problem-solving performance or transfer for undergraduate students learning probability (, 1996, 1998) and business math (Hsiao, Hung, Lan, & Jeng, 2013), as well as high school students learning algebra (Cooper & Sweller, 1987; Sweller & Cooper, 1985), and geometry (Tarmizi & Sweller, 1988). However, these benefits may only be attained when the worked examples do not require learners to split their attention between two separate sources of information (Tarmizi & Sweller, 1988; see also Rittle-Johnson, Star, & Durkin, Chapter 12). Learners may also experience a decrease in the amount of time it takes them to learn mathematical content (Zhu & Simon, 1987) or to solve problems on their own (Sweller & Cooper, 1985).

Evidence from Classroom Studies

Most studies of the worked examples effect that have taken place in the context of real-world courses have relied on computer-assisted instruction. Studies conducted in computerized mathematics classrooms have revealed benefits including a decrease in the amount of time it takes high school students to learn in geometry (Schwonke et al., 2007), an increase in procedural performance for 5th graders learning about equivalent fractions (Lee & Chen, 2015), and increased performance and improved attitudes for middle school students learning basic mathematical concepts via worked example-based podcasts (Kay & Edwards, 2012). Benefits for conceptual understanding have also been found for middle and high-school algebra students explaining worked examples in computerized settings (Booth, Lange, Koedinger, & Newton, 2013; Reed, Corbett, Hoffman, Wagner, & MacLaren, 2013).

A growing number of worked examples studies have been conducted in more traditional classrooms. A study conducted in a high school algebra classroom found that students who studied worked examples made fewer problem solving errors, took less time to learn, and required less assistance from their teacher (Carroll, 1994). Longer-term classroom studies have shown a time-on-task benefit for Dutch 4th graders who studied worked examples while learning about subtraction (van Loon-Hillen, van Gog, & Brand-Gruwel, 2012) and benefits for conceptual and procedural performance as well as performance on standardized test items for middle and high school students who explained correct and incorrect examples while learning algebra (Booth et al., 2015a); however, benefits may be observed only for students with some prior knowledge who are not yet experts (Booth et al., 2015b). Finally, Retnowati, Ayres, and Sweller (2010) found that 7th graders improved their understanding of how to solve geometry problems after studying worked examples either individually or in group work; however, metacognitive training including comprehension, connection, and reflection questions while solving problems may be even more effective than studying worked examples for middle school students working in groups (Mevarech & Kramarski, 2003).

Recommendations for Further Research

Recent studies have tested the effectiveness of worked examples in computerized and real-world classrooms. While these studies typically demonstrate benefits of studying and explaining worked examples, there are published exceptions where the intervention does not work, or does not work for all students (e.g., Booth et al., 2015b). Research is needed to investigate the conditions under which worked examples are beneficial in educational settings. For instance, does the benefit of worked examples vary based on the particular mathematical content to be learned? Is feedback necessary when students self-explain examples? To what degree does the formatting of the example matter? And what level of detailed translational work is necessary to ensure that worked examples will be beneficial in real-world classrooms?

Also, while it has been established that prior knowledge impacts learning from worked examples, all studies to date have examined general achievement or prior knowledge of the particular topic the students are learning about. However, might the type of prior knowledge that is important for learning vary with different mathematical topics? For example, are there particular types of prerequisite knowledge or skills that might make students better prepared to learn a certain topic? If so, do students who lack that particular knowledge or skill benefit more from studying and explaining worked examples? Further investigation is needed to address these questions (see Rittle-Johnson, Star, & Durkin, Chapter 12).

INTERLEAVING PRINCIPLE

The interleaving principle suggests that when practice problems are alternated, with a problem on one concept followed by a problem on another concept, students learn better than if problems are blocked, or grouped by concept (Rohrer, 2012). Interleaving problem types is thought to be effective because it helps students to differentiate between problems of varying types and identify problem features that suggest different strategies for solution (Rohrer, 2012). It also may help students to build strong associations between problem types and appropriate solution strategies (Rohrer, Dedrick, & Burgess, 2014); in contrast, blocked practice allows students to solve problems by carrying out the same strategy on a group of problems without building such associations.

Evidence from Laboratory Studies

Mayfield and Chase (2002) are given credit for being the first to test the interleaving effect for mathematics learning. In their study, college students were given remediation on algebraic concepts, and those who received their problems in an interleaved format outscored the blocked group on application and problem-solving test items. However, the study confounded interleaving with spaced practice, as the interleaving group also had more space between practice problems of the same type. Subsequent work has investigated the effectiveness of interleaving independent of spacing. In two laboratory studies, college students were given interleaved practice for solving volume problems with different types of obscure solids and these students outscored students given blocked practice on the posttest (Rohrer & Taylor, 2007; LeBlanc & Simon, 2008). Taylor and Rohrer (2010) found that fourth grade students did not benefit immediately from interleaved problems on prisms, but outperformed peers who had experienced blocked practice at the follow-up test; blocked students often made errors by using strategies that were appropriate for a different problem type than they had been taught. Finally, using the SimStudent, an intelligent agent that is able to learn from demonstration and problem-solving experiences, Li, Cohen, and Koedinger (2012) found that interleaving allowed more opportunities for the simulated student to detect its errors and refine its knowledge on several different mathematical content areas (fractions, addition, and equation solving).

Evidence from Classroom Studies

In the last few years, several studies have tested the effectiveness of interleaved practice in real world mathematics classrooms. Ostrow, Heffernan, Heffernan, and Peterson (2015) recently conducted a study to directly replicate Rohrer and Taylor's (2007) results, but in a real-world instructional context. Using the ASSISTments system for homework assignments, Ostrow et al. (2015) found a positive interleaving effect for seventh grade students reviewing mathematics concepts, such as complementary and supplementary angles, surface area of pyramids, and probability of compound events. Rohrer, Dedrick, and Stershic (2015) also found that interleaving practice was more effective than blocked practice on immediate and delayed posttests, even when the blocked practice group was given an additional review period to control the delay between practice and tests.

Finally, other studies have replicated and extended laboratory findings in classroom settings, examining different features of problems that could be interleaved. For instance, one classroom study compared interleaving of task types with interleaving of the types of representations used to solve fraction problems; interleaving of task types was found to be more important than interleaving representations for 5th and 6th grade students learning about fractions (Rau, Aleven, & Rummel, 2013). Though most interleaving is conducted with problems that have some commonalities, Rohrer et al. (2014) found that interleaving problems was still effective even when problem types were completely dissimilar.

Recommendations for Further Research

Both laboratory and classroom studies have shown that interleaving problem types is beneficial for students of different ages and for learning about a wide variety of mathematics concepts. The benefit may emerge immediately or at a delay, and may be robust regardless of the similarities and differences of the problems being interleaved (Rohrer et al., 2014); further studies must be conducted to verify this robustness across varying mathematical topics.

In general, though results are positive, application of the interleaving effect to mathematics is relatively new and is being studied by only a few research groups. Perhaps because of this, classroom research is still relatively limited, and mathematics textbooks still support blocked rather than interleaved practice (Rohrer, 2012). Further research will thus clearly be required to make appreciable differences in mathematical practice. Future work must also investigate how blocked and interleaved practice might be effectively combined to enhance student learning.

ABSTRACT AND CONCRETE REPRESENTATIONS PRINCIPLES

Do students benefit more from a concrete representation of a concept, such as a physical object or contextualized story, or an abstract representation of that concept, such as a symbol or an equation? Though some early studies pitted the two types of representations against one another, and reported benefits of one type over the other, a larger body of work finds that each type of representation has unique advantages and students learn most effectively by making connections between those representations (Pashler et al., 2007). Concrete representations are thought to ground new information with prior knowledge, and abstract representations help learners to transfer this new knowledge flexibly to other situations (see Belenky & Schalk, 2014, for a review). Beyond just linking the two types of representations, *concreteness–fading* (i.e., progressively presenting students with concrete representations that become increasingly abstract) may support efficient learning (Goldstone & Son, 2005).

Evidence from Laboratory Studies

Within the domain of mathematics, laboratory studies across a wide variety of age groups suggest that concrete representations support initial understanding but may not lead to transfer (e.g., Kaminski, Sloutsky, & Heckler, 2005). In contrast, abstract representations can support transfer, but may not always promote conceptual understanding (De Bock, Deprez, Van Dooran, Roelens, & Verschaffel, 2011). Integrating concrete and abstract representations may provide synergistic benefits. For example, Uttal et al. (2013) found that for 7-year-old learning 2-digit arithmetic, highlighting the connections between manipulatives and symbolic, written problem representations help children both gain a sound conceptual understanding and transferable knowledge. Gestures can also be used to highlight similarities and differences between mathematics problems of different types (Richland & McDonough, 2010). In this study, students learned more about the relationship between permutation (e.g., the number of possible outcomes for gold and silver medals in a race with four runners) and combination problems (e.g., the number of ways four students could win two show tickets) when the instructor left both types of problem on the board and gestured back and forth between the two problems when explaining the similarities (the same number of participants, two winners) and differences (order of winning matters in the race, but not in winning tickets to a show).

In general, starting with concrete or grounded representations and fading to more abstract or formal representations seem to lead to the best learning and transfer (Braithwaite & Goldstone, 2013; McNeil & Fyfe, 2012). However, concrete representations with features irrelevant to the concept or task can be harmful; Uttal, Scudder, and DeLoache (1997) found that children have a hard time understanding that toy-like mathematics manipulatives could be both physical objects and representations of an abstract concept, and thus tended not to transfer learning to formal representations. Similarly, undergraduates working with abstract representations (i.e., generic shapes) learned more than students given concrete representations that included seductive details (i.e., representations of measuring cups of liquid when the learning objective was mathematical grouping) (Kaminski, Sloutsky, & Heckler, 2008). Even discrete objects that are individually relevant to the task may fail to improve learning if their *quantity* is irrelevant to the task (Kaminski & Sloutsky, 2013). Abstract representations may be more beneficial than concrete representations when problems are particularly complex; as shown for both elementary school students (McNeil & Uttal, 2009) and undergraduates (Koedinger, Alibali, & Nathan, 2008). Similarly, learners with low prior knowledge may benefit more from concrete, well-grounded materials (Braithwaite & Goldstone, 2015; Petersen & McNeil, 2013), while both older and higher prior knowledge students may be able to benefit from more abstract materials (Booth & Koedinger, 2012).

Evidence from Classroom Studies

Classroom studies in mathematics are more abundant for this principle than those discussed thus far, and are consistent with the idea that making connections between abstract and concrete representations is helpful for a wide range of age groups (e.g., Ainsworth, Bibby, & Wood, 2002; Bottge, 1999; Stephan & Akyuz, 2012). Starting with more concrete representations and fading towards more abstract ones was also found to be beneficial for learning in elementary school mathematics classrooms (Fyfe, McNeil, & Borjas, 2015). Overall, the effectiveness of concrete and abstract representations may vary with learner characteristics including age and prior knowledge, which are often confounded. Siler and Willows (2014) found that 8th grade students benefited from abstract representations, yet 6th grade students needed to see concrete representations in order to be successful with far transfer on the same problems. As they found that concrete representations were also beneficial for students with poor reasoning skills, it's possible that some age-related findings may stem from knowledge or skill differences.

Recommendations for Further Research

Future research on this principle should include more comprehensive examination of which type, combination, and ordering of representations are most beneficial for different learning goals and different learners. We need more precise and consistent ways of evaluating representations, learning goals, and learner characteristics. Apparently discrepant results regarding the effectiveness of interventions based on this principle could in fact be due to the operational definitions; for example, of *concrete* and *abstract* used in the studies, or the fact that prior knowledge affects whether a given representation is concrete or abstract for an individual learner. Prior knowledge and task familiarity may also underpin age- and grade-related trends that show differential benefits for concrete or abstract representations or from concreteness fading. Future research that operationalizes key factors related to the learner, the representations, the tasks, and learning goals, will yield more generalizable recommendations that inform practice and will enable practitioners to select the best representations or sequence of representations for their students and classroom objectives.

ERROR REFLECTION PRINCIPLE

The error reflection principle grew out of research on cognitive dissonance theory, originally proposed by Festinger (1957) to explain a series of findings from experiments on self-justification. The original theory proposed that a natural human motivation is to seek consistency amongst cognitions (i.e., thoughts, beliefs, or attitudes). *Cognitive dissonance* is thus a state of tension or discomfort that arises whenever one holds two cognitions that are inconsistent with one another (Festinger, 1957). While cognitive dissonance theory is often used to explain findings in social psychology on individuals' attitude change surrounding social topics (e.g., politics, stereotypes), one can see the great range of applicability of the theory to the study of cognitive change within learning environments. If experiencing cognitive dissonance can potentially motivate an individual to alter their cognitions to make them more consistent, then it follows that purposefully creating cognitive disequilibrium can effectively produce changes in students' thinking (Graesser, 2009). This idea can also be traced back to Piaget who argued that confronting discrepant ideas when in a state of disequilibrium is not only helpful but *essential* for knowledge growth (Piaget, 1980). Overoye and Storm (2015) review literature suggesting that learning tasks are particularly beneficial when they induce uncertainty in students and provoke them to attempt to resolve this uncertainty, such as presenting information contradictory to the learners' current knowledge.

There are a number of methods that can induce cognitive dissonance, including creating confusion (D'Mello & Graesser, 2014) and presenting students with discrepant events (Gorsky & Finegold, 1994). However, we focus here on a facet of cognitive dissonance that is particularly relevant to mathematics: reflection on errors—either one's own errors (Cherepinsky, 2011) or those of other learners as presented in the form of incorrect worked examples (Große & Renkl, 2007). Ohlsson's (1996) theory suggests that learning from errors can be particularly effective if learners are prompted to identify what features of the problem make the specific step taken incorrect, which can subsequently be used to correct faulty knowledge and fine-tune problem-solving. Studying errors is also thought to be beneficial because it provides exposure to multiple perspectives rather than just one's own perspective (Siegler & Chen, 2008).

Evidence from Laboratory Studies

One laboratory study within the domain of mathematics demonstrated that explaining *a combination* of correct and incorrect examples was superior for elementary students' learning about mathematical equality, compared to explaining correct examples alone (Siegler, 2002). However, Große and Renkl (2007) found that for undergraduate students learning about probability, those with lower prior knowledge were less able to benefit from incorrect examples, unless the segment of the solution that contained the error was highlighted.

Evidence from Classroom Studies

A growing number of classroom studies have demonstrated the effectiveness of reflecting on errors for mathematics learning. For instance, after being asked to correct errors in problems marked incorrect on their exams, undergraduate calculus students felt they gained a better understanding of the material (Cherepinsky, 2011). Having middle school algebra students explain incorrect examples led to improvement in their ability to encode algebraic equations (Booth et al., 2013); explaining a combination of correct and incorrect examples has also been shown to improve algebra students' conceptual and procedural knowledge (Booth et al., 2015a). A series of studies on students learning about decimal magnitudes have also demonstrated benefits for comparing correct and incorrect worked examples (Durkin & Rittle-Johnson, 2012), using incorrect worked examples within the context of a computerized tutoring intervention (Adams et al., 2014), and finding, explaining, and fixing errors within erroneous examples (McLaren, Adams, & Mayer, 2015); however, in each of these studies, benefits only emerged after a delay, not immediately after the activities.

Though many studies show the benefits of studying errors, there is less consensus on whether different groups of students experience differential benefits. For example, Heemsoth and Heinze (2014) found that students with high prior knowledge benefit more from identifying, explaining, and correcting errors than their less knowledgeable peers. In contrast, other studies have found that students with low prior knowledge benefit the most from studying and explaining errors (Barbieri & Booth, 2016; Booth et al., 2015b). Still others demonstrate that incorrect examples were equally effective for all students regardless of prior knowledge (Adams et al., 2014; Durkin & Rittle-Johnson, 2012). Thus, the degree of benefit from reflecting on errors may vary based on the particular activities required of the student, individual differences in prior knowledge, or perhaps even the particular mathematics domain. Though not previously explored, individual differences in cognitive competencies, such as working memory may also influence the effectiveness of reflecting on errors.

Recommendations for Further Research

The next step is to begin investigating and comparing particular features within error reflection activities. For example, do learners need to find, explain, and correct errors to achieve maximum benefit or would studying them be sufficient? Would comparing incorrect examples to correct ones be more effective than just studying and explaining incorrect examples alone? And does each of these types of error reflection activities have the potential to influence different types of knowledge or skills and at different time scales? Research must move from establishing the effectiveness of learning from errors in general to clarify how the two proposed mechanisms for benefits of error reflection map to particular task choices and resulting benefits for different groups of students.

ANALOGICAL COMPARISON PRINCIPLE

Analogical reasoning refers to the human ability to draw connections and notice contrasts between the relationships of two or more representations (see also DeWolf et al., Chapter 7). In the process of comparison, learners have to map between two relationships of representations based on alignments (or misalignments), leading to generalizable knowledge (Gentner, 1983; Gick & Holyoak, 1980). For an analogy to work, mental representations must share a common relation, but need not appear similar. For instance, two equations may be presented in a similar form (e.g. "3 + 4 = ?" and "5 + 3 = ?"), or may be presented in a dissimilar form (e.g., "3 + 4= ?" and "Johnny had five apples and Mary gave him three more"), however, the arithmetical relation between the two equations needs to correspond. The effectiveness of learning from analogies is known to vary with individual differences in executive function (EF); in particular, working memory and inhibitory control deficits make learning from comparisons difficult (Cho, Holyoak, & Cannon, 2007). Perhaps at least in part because prior knowledge offloads EF demands (Grossnickle, Dumas, Alexander, & Baggetta, 2016), prior knowledge also impacts students' ability to learn from comparisons (Gentner & Rattermann, 1991).

Evidence from Laboratory Studies

Within the domain of mathematics, laboratory research has confirmed that comparing multiple examples facilitates schema formation (see also Rittle-Johnson, Star, & Durkin, Chapter 12). This has been found for a variety of mathematical content areas including algebra (Novick & Holyoak, 1991), probability (Ross, 1989), number sense (Thompson & Opfer, 2010), ratio (Novick & Holyoak, 1991), statistics (Cummins, 1992), and equivalence (Hattikudur & Alibali, 2010). Further, providing instructional supports, such as visual alignment or linking gestures have been shown to facilitate learning from comparisons (Richland & McDonough, 2010).

Evidence from Classroom Studies

Classroom studies testing the analogical comparison principle have frequently shown benefits for conceptual understanding in a variety of mathematical content areas, including algebra (Rittle-Johnson & Star, 2009), computational estimation (Star & Rittle-Johnson, 2009), geometry (Guo & Pang, 2011), proportional reasoning (Begolli & Richland, 2016), numerical density (Vamvakoussi & Vosniadou, 2012), and division of natural and rational numbers (Richland & Hansen, 2013). For supporting conceptual understanding, Rittle-Johnson and Star (2009) found that comparing multiple solutions to a single problem was more beneficial than comparing multiple problem types using a single solution strategy, or comparing equivalent problem types with the same solution (Rittle-Johnson, Star, & Durkin, Chapter 12). Benefits of analogical comparison have also been found for procedural fluency and flexibility (Begolli & Richland, 2016; Star & Rittle-Johnson, 2009). Several types of instructional supports have been shown to facilitate learning from comparing multiple representations, including visually aligning the to-be-compared examples (Star & Rittle-Johnson, 2009), using linking gestures to help students notice critical features (Alibali & Nathan, 2007), and prompting exploration (Schwartz & Bransford, 1998) or providing conceptual instruction (Fyfe, DeCaro, & Rittle-Johnson, 2014) prior to studying contrasting cases.

Individual differences in effectiveness have also been found, as students with greater prior knowledge become more likely to notice critical features of the to-be-compared items (Rittle-Johnson, Star, & Durkin, 2009; Star & Rittle-Johnson, 2009). However, novices have also been found to benefit from comparison (Guo & Pang, 2011; Rittle-Johnson, Star, & Durkin, 2012). In particular, while students with high prior knowledge may benefit more from comparing solution methods, students with low prior knowledge may benefit more from sequential presentation or from comparing different problem types (Rittle-Johnson et al., 2009).

Recommendations for Further Research

Orchestrating instruction around comparisons remains a challenge for mathematics teachers (Ball, 1993; Stein, Engle, Smith, & Hughes, 2008). Teachers in the United States regularly use comparisons in their lessons, but they do not always provide necessary instructional supports (e.g., visually aligning source and target, using gestures) compared to teachers in countries with higher achieving students, such as Hong Kong and Japan (Richland et al., 2007). US teachers often try to control the analogies and only engage students in procedural aspects of the analogy, not requiring them to attend to the structural alignments between representations in order to participate in instructional discussions (Richland, Holyoak, & Stigler, 2004). This may lead to failed opportunities for students to learn the deep aspects of concepts and transferable knowledge that are typically found in laboratory studies. Thus, more translational research is needed to find ways to help teachers utilize analogical comparisons in their classrooms by examining year-long curriculum implementations of both interactional teaching strategies and instructional materials. A recent attempt suggests that a single 1-week professional development course seems to be insufficient for consistent uptake of instructional materials throughout a full curriculum, leading to small improvements in procedural knowledge (Star et al., 2015). Yet, these are important first steps towards testing and making instructional materials widely available for scalable use. Future scale-up studies need to systematically examine teacher-led discussions around instructional comparisons.

CONCLUSIONS AND FUTURE DIRECTIONS

Each of the principles reviewed in this chapter has a clear relevance to mathematics instruction and the potential to improve student learning. However, the principles vary widely in terms of the quantity and quality of the empirical support for their effectiveness for mathematics learning. For example, distributed practice and interleaving effects have been established in other content areas but have seldom been tested in mathematics contexts. These principles clearly need more attention, with studies conducted either in laboratory or classroom settings, to determine if and when they are effective for students' learning of different types of mathematical content. Other principles have been well-tested in laboratory studies, but need more investigations to determine when and how they are effective in real-world educational settings. For example, there is a growing body of evidence supporting error reflection, but less work detailing the benefits of different ways that errors can and should be used in the classroom. For certain principles, such as scaffolding and feedback, it is vital to focus on implementation and testing in more traditional mathematics classes where teachers cannot rely on aid of individualized technology. Finally, the effectiveness of many of the principles has been relatively well established in laboratory and classroom studies, including abstract/concrete representations, analogical comparison, and worked examples. Research on these principles needs to shift focus to understanding individual differences in the principle's effectiveness for learning different types of mathematics content, as well as nuances of implementation in real-world classrooms.

The principles also vary in terms of their effectiveness for simpler versus more complex mathematical content. For instance, distributed practice has been repeatedly shown to be effective for improving simple mathematics skills, such as memorizing arithmetic facts. However, research has not yet established whether and how distributed practice can be implemented to effectively increase learning of more complex mathematical concepts and skills. The distinction and connections between abstract and concrete representations may also vary depending on whether simple or complex content is to be learned. In particular, learning of complex content may require the inclusion of abstract representations, where simple content can be learned from concrete representations as long as they do not contain seductive details, such as decorative graphics. Still other principles have been shown to be effective for both simple and complex mathematical skills. For example, worked examples have been shown to be effective for simple content, such as basic mathematics skills as well as more complex content like probability and algebra. Similarly, scaffolding has been shown to enhance early stages of learning of both simple and complex content; feedback, interleaving, analogical comparison, and error reflection also have benefits for learning regardless of the type of content.

Finally, these principles also vary in terms of how difficult it is to implement them in real classrooms, with and without technology. For instance, concrete and abstract representations are already often incorporated into mathematics textbooks, though further efforts could be made in selecting appropriate graphics and determining how those components should be configured to minimize demands on cognitive load. Error reflection, analogical comparison, and interleaved practice can be incorporated into lessons via either interactive activities or in static, paper-based materials; though worked examples have most frequently been used in computer-based classrooms, methods for implementing them with paper and pencil in traditional classrooms have also been successful. However, distributed practice, scaffolding, and feedback may be particularly difficult to implement and test systematically in the context of noncomputerized classrooms. What do these principles even look like in traditional classrooms, when they cannot be provided with individualized computer programs? Can they be incorporated into written materials, or must they be actively implemented by teachers?

Recent work in the National Center for Cognition and Mathematics Instruction undertook the translation of several of these principles to redesign a popular middle school mathematics curriculum, Connected Mathematics 2 (Lappan,

Fey, Fitzgerald, Friel, & Phillips, 2006). In this project, worked examples were carefully designed to enhance the different kinds of practice problems inherent in the reform curriculum, and efforts were made to reduce graphics with seductive details and increase the mathematical relevance of all concrete representations included in the text. However, the other two included principles—distributed practice and feedback in terms of formative assessment—were trickier to implement and required supplemental guidebooks for teachers. Effective practice distribution in this case, without the aid of technology, may require a thorough understanding of what skills are being practiced in each problem type, how frequently they are revisited in different formats, and how to help students make connections between previously learned skills and ones that are upcoming (Davenport, Lepori, Hauk, Viviani, & Schneider, 2012). In this type of translational work, it is critical to understand what the enacted principle would look like in a classroom, which is not necessarily as straightforward as it seems from laboratory studies.

General Recommendations

Translational work is essential to ensure that relevant findings from basic science reach their potential to improve our education system (Koedinger et al., 2013). Publishing findings of cognitive principles and speculating about their implications may move research forward, but it rarely yields appreciable change in real-world classrooms. Few practitioners have access to academic journals, thus they are not frequently exposed to new research findings. Further, even knowing that something works in laboratory studies does not mean it will necessarily work in the classroom. Examples can be found in and outside of mathematics of laboratory-tested techniques failing to translate in classrooms (Davenport, Klahr, & Koedinger, 2007; Dynarski et al., 2007) or helping only a portion of the students in the class (Booth et al., 2015b). As classrooms are much less controlled than university laboratories, the work of translating laboratory findings to be applicable in the classroom requires intimate knowledge of both the principles and the education system, a willingness to develop lessons and materials iteratively, with many opportunities for revision as they are field-tested, and creative experimental design. Translational research may be best facilitated in collaboration with teachers and administrators from school districts to ensure that the development of materials and practices based on these principles yields classroom-ready products.

However, the best instruction is not one size fits all. Individual differences (e.g., prior knowledge, working memory capacity) may interact with the utility of different strategies. Differentiated instruction is heavily promoted in today's education system, and we need to be able to make good recommendations to practitioners about what techniques to use, when, and for whom. Studies on individual differences in effectiveness of instruction have mainly explored prior knowledge and cognitive factors; future research should further distinguish these factors and also investigate motivational characteristics that may impact whether a student benefits from a particular technique. We need to consider individual differences in demographic characteristics as well, especially given that factors like gender and race may influence the way in which instructional materials are viewed, interpreted, and accessed (Moreno & Flowerday, 2006). Continued investigation of individual differences is necessary in order to maximize the benefits of research findings for real-world educational settings.

ACKNOWLEDGMENTS

Funding for the writing of this chapter was provided by the Institute of Education Sciences and U.S. Department of Education through Grant R305B150014 to Temple University, Grant R305B130012 to the University of Delaware, Grant R305A100150 to the Strategic Education Research Partnership, and Grant R305C100024 to WestEd. The opinions expressed are those of the authors and do not represent views of the Institute or the U.S. Department of Education.

REFERENCES

Adams, D. M., McLaren, B. M., Durkin, K., Mayer, R. E., Rittle-Johnson, B., Isotani, S., & van Velsen, M. (2014). Using erroneous examples to improve mathematics learning with a web-based tutoring system. *Computers in Human Behavior, 36*, 401–411.

Ainsworth, S., Bibby, P., & Wood, D. (2002). Examining the effects of different multiple representational systems in learning primary mathematics. *The Journal of the Learning Sciences, 11*(1), 25–61.

Alibali, M. W. (1999). How children change their minds: strategy change can be gradual or abrupt. *Developmental Psychology, 35*(1), 127–145.

Alibali, M. W., & Nathan, M. J. (2007). Teachers' gestures as a means of scaffolding students' understanding: evidence from an early algebra lesson. *Video Research in the Learning Sciences*, 349–365.

Bahrick, H. P., & Hall, L. K. (2005). The importance of retrieval failures to long-term retention: a metacognitive explanation of the spacing effect. *Journal of Memory and Language, 52*, 566–577.

Ball, D. L. (1993). Halves, pieces, and twoths: constructing and using representational contexts in teaching fractions. *Rational Numbers: An Integration of Research*, 157–195.

Barbieri, C., & Booth, J. L. (2016). Support for struggling students in algebra: Contributions of incorrect worked examples. *Learning and Individual Differences, 48*, 36–44.

Begolli, K. N., & Richland, L. E. (2016). Teaching mathematics by comparison: analog visibility as a double-edged sword. *Journal of Educational Psychology, 108*(2), 194–213.

Belenky, D. M., & Schalk, L. (2014). The effects of idealized and grounded materials on learning, transfer, and interest: an organizing framework for categorizing external knowledge representations. *Educational Psychology Review, 26*(1), 27–50.

Belland, B. R., Walker, A. E., Olsen, M. W., & Leary, H. (2015). A pilot meta-analysis of computer-based scaffolding in STEM education. *Educational Technology & Society, 18*(1), 183–197.

Booth, J. L., & Koedinger, K. R. (2012). Are diagrams always helpful tools? Developmental and individual differences in the effect of presentation format on student problem solving. *British Journal of Educational Psychology, 82*(3), 492–511.

Booth, J. L., & Newton, K. J. (2012). Fractions: could they really be the gatekeeper's doorman? *Contemporary Educational Psychology, 37*, 247–253.

Booth, J. L., Lange, K. E., Koedinger, K. R., & Newton, K. J. (2013). Using example problems to improve student learning in algebra: differentiating between correct and incorrect examples. *Learning and Instruction, 25*, 24–34.

Booth, J. L., Cooper, L. A., Donovan, M. S., Huyghe, A., Koedinger, K. R., & Paré-Blagoev, E. J. (2015a). Design-based research within the constraints of practice: AlgebraByExample. *Journal of Education for Students Placed at Risk, 20*(1–2), 79–100.

Booth, J. L., Oyer, M. H., Paré-Blagoev, E. J., Elliot, A., Barbieri, C., Augustine, A. A., & Koedinger, K. R. (2015b). Learning algebra by example in real-world classrooms. *Journal of Research on Educational Effectiveness.*, *8*(4), 530–551.

Bottge, B. A. (1999). Effects of contextualized math instruction on problem solving of average and below-average achieving students. *Journal of Special Education, 33*, 81–92.

Braithwaite, D. W., & Goldstone, R. L. (2013). Integrating formal and grounded representations in combinatorics learning. *Journal of Educational Psychology, 105*(3), 666.

Braithwaite, D. W., & Goldstone, R. L. (2015). Effects of variation and prior knowledge on abstract concept learning. *Cognition and Instruction, 33*(3), 226–256.

Budé, L., Imbos, T., van de Wiel, M. W., & Berger, M. P. (2011). The effect of distributed practice on students' conceptual understanding of statistics. *Higher Education, 62*, 69–79.

Calder, N. (2015). Student wonderings: scaffolding student understanding within student-centered inquiry learning. *ZDM Mathematics Education, 47*, 1121–1131.

Carpenter, S. K., Pashler, H., & Cepeda, N. J. (2009). Using tests to enhance 8th grade students' retention of U.S. history facts. *Applied Cognitive Psychology, 23*, 760–771.

Carroll, W. M. (1994). Using worked examples as an instructional support in the algebra classroom. *Journal of Educational Psychology, 86*(3), 360–367.

Carter, J. (1984). Instructional learner feedback: a literature review with implications for software development. *Computing Teacher, 12*(2), 53–55.

Catrambone, R. (1996). Generalizing solution procedures learned from examples. *Journal of Experimental Psychology: Learning, Memory, and Cognition, 22*(4), 1020–1031.

Catrambone, R. (1998). The subgoal learning model: creating better examples so that students can solve novel problems. *Journal of Experimental Psychology: General, 127*(4), 355–376.

Common Core State Standards Initiative. (2010). *Common Core State Standards for Mathematics.* Availble from: http://www.corestandards.org/assets/CCSSI_Math%20Standards.pdf

Cepeda, N. J., Pashler, H., Vul, E., Wixted, J. T., & Rohrer, D. (2006). Distributed practice in verbal recall tasks: a review and quantitative synthesis. *Psychological Bulletin, 132*, 354–380.

Challis, B. H. (1993). Spacing effects on cued-memory tests depend on level of processing. *Journal of Experimental Psychology: Learning, Memory, and Cognition, 19*, 389–396.

Cherepinsky, V. (2011). Self-reflective grading: getting students to learn from their mistakes. *PRIMUS, 21*(3), 294–301.

Cho, S., Holyoak, K. J., & Cannon, T. D. (2007). Analogical reasoning in working memory: resources shared among relational integration, interference resolution, and maintenance. *Memory & Cognition, 35*(6), 1445–1455.

Cohen, V. B. (1985). A reexamination of feedback in computer-based instruction: implications for instructional design. *Educational Technology, 25*(1), 33–37.

Cooper, G., & Sweller, J. (1987). Effects of schema acquisition and rule automation on mathematical problem-solving transfer. *Journal of Educational Psychology, 79*(4), 347–362.

Cummins, D. D. (1992). Role of analogical reasoning in the induction of problem categories. *Journal of Experimental Psychology: Learning, Memory, and Cognition, 18*(5), 1103.

Davenport, J. L., Klahr, D., & Koedinger, K. R. (2007). *The influence of diagrams on chemistry learning.* Paper presented at the Biennial Conference of the European Association for Research on Learning and Instruction, Hungary, Budapest.

Davenport, J. L., Lepori, K. L., Hauk, S., Viviani, K., & Schneider, S. A. (2012). Integration of Cognitive Design Principles with the Connected Mathematics Project Curriculum. Paper presented at the 2012 Annual Meeting of the American Educational Research Association in Vancouver, B.C.

De Bock, D., Deprez, J., Van Dooren, W., Roelens, M., & Verschaffel, L. (2011). Abstract or concrete examples in learning mathematics? A replication and elaboration of Kaminski, Sloutsky, and Heckler's study. *Journal for Research in Mathematics Education, 42*(2), 109–126.

Delaney, P. F., Verkoeijen, P. P., & Spirgel, A. (2010). Spacing and testing effects: a deeply critical, lengthy, and at times discursive review of the literature. *Psychology of Learning and Motivation, 53*, 63–147.

Dempster, F. N. (1988). The spacing effect: a case study in the failure to apply the results of psychological research. *American Psychologist, 43*(8), 627.

Department of Education. (1997). *Mathematics equals opportunity.* White Paper prepared for U.S. Secretary of Education Richard W. Riley; Washington, DC.

D'Mello, S., & Graesser, A. (2014). Confusion and its dynamics during device comprehension with breakdown scenarios. *Acta Psychologica, 151*, 106–116.

Dunlosky, J., Rawson, K. A., Marsh, E. J., Nathan, M. J., & Willingham, D. T. (2013). Improving students' learning with effective learning techniques promising directions from cognitive and educational psychology. *Psychological Science in the Public Interest, 14*(1), 4–58.

Durkin, K., & Rittle-Johnson, B. (2012). The effectiveness of using incorrect examples to support learning about decimal magnitude. *Learning and Instruction, 22*, 206–214.

Dynarski, M., Agodini, R., Heaviside, S., Novak, T., Carey, N., Campuzano, L., Means, B., …, & Sussex, W. (2007). *Effectiveness of reading and mathematics software products: findings from the first student cohort.* Report provided to Congress by the National Center for Education Evaluation.

Ellis, S., Klahr, D., & Siegler, R. S. (1993). Effects of feedback and collaboration on changes in children's use of mathematical rules. In *A paper presented in Society for Research in Child Development.* New Orleans.

Festinger, L. (1957). *A theory of cognitive dissonance.* Standford, CA: Stanford University Press.

Fleischman, H. L., Hopstock, P. J., Pelczar, M. P., & Shelley, B. E. (2010). Highlights from PISA 2009: Performance of US 15-Year-Old Students in Reading, Mathematics, and Science Literacy in an International Context. NCES 2011-004. *National Center for Education Statistics.*

Fyfe, E. R., Rittle-Johnson, B., & DeCaro, M. S. (2012). The effects of feedback during exploratory mathematics problem solving: prior knowledge matters. *Journal of Educational Psychology, 104*(4), 1094–1108.

Fyfe, E. R., DeCaro, M. S., & Rittle-Johnson, B. (2014). An alternative time for telling: when conceptual instruction prior to problem solving improves mathematical knowledge. *British Journal of Educational Psychology*, 1–18.

Fyfe, E. R., DeCaro, M. S., & Rittle-Johnson, B. (2015a). When feedback is cognitively-demanding: the importance of working memory capacity. *Instructional Science, 43*(1), 73–91.

Fyfe, E. R., McNeil, N. M., & Borjas, S. (2015b). Benefits of "concreteness fading" for children's mathematics understanding. *Learning and Instruction, 35*, 104–120.

Gay, L. R. (1973). Temporal position of reviews and its effect on the retention of mathematical rules. *Journal of Educational Psychology, 64*(2), 171–182.

Gentner, D. (1983). Structure-mapping: a theoretical framework for analogy. *Cognitive Science, 7*(2), 155–170.

Gentner, D., & Rattermann, M. J. (1991). Language and the career of similarity. In *Perspectives on thought and language: Interrelations in development* (pp. 225–277).

Gick, M. L., & Holyoak, K. J. (1980). Analogical problem solving. *Cognitive Psychology, 12*(3), 306–355.

Goldstone, R. L., & Son, J. Y. (2005). The transfer of scientific principles using concrete and idealized simulations. *The Journal of the Learning Sciences, 14*(1), 69–110.

Gorsky, P., & Finegold, M. (1994). The role of anomaly and of cognitive dissonance in restructuring students' concepts of force. *Instructional Science, 22*, 75–90.

Graesser, A. C. (2009). Inaugural editorial for journal of educational psychology. *Journal of Educational Psychology, 101*(2), 259–261.

Große, C. S., & Renkl, A. (2007). Finding and fixing errors in worked examples: can this foster learning outcomes? *Learning and Instruction, 17*, 612–634.

Grossnickle, E. M., Dumas, D., Alexander, P. A., & Baggetta, P. (2016). Individual differences in the process of relational reasoning. *Learning and Instruction, 42*, 141–159.

Guo, J. P., & Pang, M. F. (2011). Learning a mathematical concept from comparing examples: the importance of variation and prior knowledge. *European Journal of Psychology of Education, 26*(4), 495–525.

Harks, B., Rakoczy, K., Hattie, J., Besser, M., & Klieme, E. (2014). The effects of feedback on achievement, interest and self-evaluation: the role of feedback's perceived usefulness. *Educational Psychology, 34*(3), 269–290.

Hattikudur, S., & Alibali, M. W. (2010). Learning about the equal sign: does comparing with inequality symbols help? *Journal of Experimental Child Psychology, 107*(1), 15–30.

Hausmann, R. G., & van Lehn, K. (2007). Explaining self-explaining: a contrast between content and generation. *Frontiers in Artificial Intelligence and Applications, 158*, 417.

Heemsoth, T., & Heinze, A. (2014). The impact of incorrect examples on learning fractions: a field experiment with 6th grade students. *Instructional Science, 42*(4), 639–657.

Hoven, J., & Garelick, B. (2007). Singapore Math: using the bar model approach, Singapore textbooks enable students to solve difficult math problems—and learn how to think symbolically. *Educational Leadership, 65*, 38-21.

Hsiao, J., Hung, C., Lan, Y., & Jeng, Y. (2013). Integrating worked examples into problem posing in a web-based learning environment. *Turkish Online Journal of Educational Technology, 12*(2), 166–176.

Janiszewski, C., Noel, H., & Sawyer, A. G. (2003). A meta-analysis of the spacing effect in verbal learning: implications for research on advertising repetition and consumer memory. *Journal of Consumer Research, 30*, 138–149.

Kalyuga, S., Chandler, P., Tuovinen, J., & Sweller, J. (2001). When problem solving is superior to studying worked examples. *Journal of Educational Psychology, 93*(3), 579–588.

Kaminski, J. A., & Sloutsky, V. M. (2013). Extraneous perceptual information interferes with children's acquisition of mathematical knowledge. *Journal of Educational Psychology, 105*(2), 351.

Kaminski, J. A., Sloutsky, V. M., & Heckler, A. F. (2005). Relevant concreteness and its effects on learning and transfer. In *Proceedings of the XXVII annual conference of the Cognitive Science Society* (pp. 1090–1095).

Kaminski, J. A., Sloutsky, V. M., & Heckler, A. F. (2008). The advantage of abstract examples in learning math. *Science*, 454–455.

Kay, R., & Edwards, J. (2012). Examining the use of worked example video podcasts in middle school mathematics classrooms: a formative analysis. *Canadian Journal of Learning and Technology, 38*(3), .

Kehrer, P., Kelly, K., & Heffernan, N. (2013). Does immediate feedback while doing homework improve learning. In Youngblood Boonthum-Denecke (Ed.), *Proceedings of the Twenty-Sixth International Florida Artificial Intelligence Research Society Conference* (pp. 542–545). St. Pete Beach, Florida: AAAI Press.

Kim, M. C., & Hannafin, M. J. (2011). Scaffolding problem solving in technology-enhanced learning environments (TELEs): Bridging research and theory with practice. *Computers & Education, 56*, 403–417.

Kim, D., Schneider, C., Engec, N., & Siskind, T. (2006). *South Carolina end-of-course examination program 2003–04 Algebra 1: Mathematics for the Technologies 2 state report.* Columbia, South Carolina: South Carolina Department of Education.

Koedinger, K. R., Alibali, M. W., & Nathan, M. J. (2008). Trade-offs between grounded and abstract representations: evidence from algebra problem solving. *Cognitive Science, 32*, 366–397.

Koedinger, K. R., Booth, J. L., & Klahr, D. (2013). Instructional complexity and the science to constrain it. *Science, 342*, 935–937.

Krause, U. M., Stark, R., & Mandl, H. (2009). The effects of cooperative learning and feedback on e-learning in statistics. *Learning and Instruction, 19*, 158–170.

Kulhavy, R. W. (1977). Feedback in written instruction. *Review of Educational Research, 47*(1), 211–232.

Kupers, E., van Dijk, M., & van Geert, P. (2015). Within-teacher differences in one-to-one teacher—student interactions in instrumental music lessons. *Learning and Individual Differences, 37*, 283–289.

Lannin, J., Townsend, B., & Barker, D. (2006). The reflective cycle of student error analysis. *For the Learning of Mathematics, 26*(3), 33–38.

Lappan, G., Fey, J. T., Fitzgerald, W. M., Friel, S. N., & Phillips, E. D. (2006). *Connected Mathematics 2*. Pearson.

LeBlanc, K., & Simon, D. (2008). Mixed practice enhances retention and JOL accuracy for mathematical skills. Paper presented at the 49th Annual Meeting of the Psychonomic Society, Chicago, IL.

Lee, C., & Chen, M. (2015). Effects of worked examples using manipulatives on fifth graders' learning performance and attitude toward mathematics. *Journal of Educational Technology & Society, 18*(1), 264–275.

Lehman, S., Schraw, G., McCrudden, M. T., & Hartley, K. (2007). Processing and recall of seductive details in scientific text. *Contemporary Educational Psychology, 32*(4), 569–587.

Li, N., Cohen, W. W., & Koedinger, K. R. (2012). Problem order implications for learning transfer. In *Intelligent Tutoring Systems* (pp. 185–194). Springer Berlin Heidelberg.

Luwel, K., Foustana, A., Papadatos, Y., & Verschaffel, L. (2011). The role of intelligence and feedback in children's strategy competence. *Journal of Experimental Child Psychology, 108*, 61–76.

Mayer, R., Sims, V. K., & Tajika, H. (1995). A comparison of how textbooks teach mathematical problem solving in Japan and the United States. *American Educational Research Journal, 32*, 443–460.

Mayfield, K. H., & Chase, P. N. (2002). The effects of cumulative practice on mathematics problem solving. *Journal of Applied Behavior Analysis, 35*(2), 105–123.

McLaren, B. M., Adams, D. M., & Mayer, R. E. (2015). Delayed learning effects with erroneous examples: a study of learning decimals with a web-based tutor. *International Journal of Artificial Intelligence in Education, 25*(4), 520–542.

McNeil, N. M., & Fyfe, E. R. (2012). "Concreteness fading" promotes transfer of mathematical knowledge. *Learning and Instruction, 22*(6), 440–448.

McNeil, N. M., & Uttal, D. H. (2009). Rethinking the use of concrete materials in learning: perspectives from development and education. *Child Development Perspectives, 3*(3), 137–139.

McNeil, N. M., Grandau, L., Knuth, E. J., Alibali, M. W., Stephens, A. S., Hattikudur, S., et al. (2006). Middle school students' understanding of the equal sign: the books they read can't help. *Cognition and Instruction*, *24*, 367–385.

Mevarech, Z. R., & Kramarski, B. (2003). The effects of metacognitive training versus worked-out examples on students' mathematical reasoning. *British Journal of Educational Psychology*, *73*(4), 449–471.

Moreno, R., & Flowerday, T. (2006). Students' choice of animated pedagogical agents in science learning: a test of the similarity-attraction hypothesis on gender and ethnicity. *Contemporary Educational Psychology*, *31*, 186–207.

Mory, E. H. (2004). Feedback research revisited. In D. Jonassen (Ed.), *Handbook of research on educational communications and technology: a project for the Association for Educational Communications and Technology* (2nd., pp. 745–783). Mahwah, NJ: Lawrence Erlbaum Associates.

Mullis, I. V., Martin, M. O., Foy, P., & Arora, A. (2012). TIMSS 2011 international results in mathematics. International Association for the Evaluation of Educational Achievement, Amsterdam, The Netherlands.

NCLB. (2002). No Child Left Behind (NCLB) Act of 2001 (pp. 107–110). US Department of Education. Washington, DC.

National Mathematics Advisory Panel. (2008). *Foundations for success: The Final Report of the National Mathematics Advisory Panel.* Washington, DC: U.S. Department of Education, 2008.

Novick, L. R., & Holyoak, K. J. (1991). Mathematical problem solving by analogy. *Journal of Experimental Psychology. Learning, Memory, and Cognition*, *17*(3), 398–415.

Ohlsson, S. (1996). Learning from performance errors. *Psychological Review*, *103*(2), 241.

Ostrow, K. S. & Heffernan, N. T. (2014). Testing the multimedia principle in the real world: a comparison of video vs. text feedback in authentic middle school math assignments. In J., Stamper, et al. (Eds.) Proceedings of the 7th International Conference on EDM. pp. 296–299.

Ostrow, K., Heffernan, N., Heffernan, C., & Peterson, Z. (2015). Blocking vs. interleaving: examining single-session effects within middle school math homework. In *Artificial Intelligence in Education* (pp. 338–347). Springer International Publishing.

Overoye, A. L., & Storm, B. C. (2015). Harnessing the power of uncertainty to enhance learning. *Translational Issues in Psychological Science*, *1*(2), 140–148.

Pashler, H., Bain, P., Bottge, B., Graesser, A., Koedinger, K., McDaniel, M., & Metcalfe, J. (2007). Organizing instruction and study to improve student learning: IES practice guide *(NCER 2007-2004)*. Washington, DC: National Center for Education Research.

Pennsylvania Department of Education (2011). 2011 Keystone Exams State Summary Report—Algebra I.

Petersen, L. A., & McNeil, N. M. (2013). Effects of perceptually rich manipulatives on preschoolers' counting performance: established knowledge counts. *Child Development*, *84*(3), 1020–1033.

Piaget, J. (1980). *Experiments in contradictions*. Chicago: University of Chicago Press.

Pratt, M. W., & Savoy-Levine, K. M. (1998). Contingent tutoring of long-divisions skills in fourth and fifth graders: experimental tests of some hypotheses about scaffolding. *Journal of Applied Developmental Psychology*, *19*, 287–304.

Rau, M. A., Aleven, V., & Rummel, N. (2013). Interleaved practice in multi-dimensional learning tasks: which dimension should we interleave? *Learning and Instruction*, *23*, 98–114.

Razzaq, L., & Heffernan, N. T. (2006). Scaffolding vs. hints in the assistment System. In *Intelligent Tutoring Systems* (pp. 635–644). Berlin, Heidelberg: Springer.

Rea, C. P., & Modigliani, V. (1985). The effect of expanded versus massed practice on the retention of multiplication facts and spelling lists. *Human Learning: Journal of Practical Research & Applications*, *4*(1), .

Reed, S. K., Corbett, A., Hoffman, B., Wagner, A., & MacLaren, B. (2013). Effect of worked examples and cognitive tutor training on constructing equations. *Instructional Science: An International Journal of The Learning Sciences, 41*(1), 1–24.

Renkl, A., Stark, R., Gruber, H., & Mandl, H. (1998). Learning from worked-out examples: the effects of example variability and elicited self-explanations. *Contemporary Educational Psychology, 23*(1), 90–108.

Retnowati, E., Ayres, P., & Sweller, J. (2010). Worked example effects in individual and group work settings. *Educational Psychology, 30*(3), 349–367.

Richland, L. E., & Hansen, J. (2013). Reducing cognitive load in learning by analogy. *International Journal of Psychological Studies, 5*(4), 69–80.

Richland, L. E., & McDonough, I. M. (2010). Learning by analogy: discriminating between potential analogs. *Contemporary Educational Psychology, 35*, 28–43.

Richland, L. E., Holyoak, K. J., & Stigler, J. W. (2004). Analogy use in eighth-grade mathematics classrooms. *Cognition and Instruction, 22*(1), 37–60.

Richland, L. E., Zur, O., & Holyoak, K. J. (2007). Cognitive supports for analogies in the mathematics classroom. *Science, 316*(5828), 1128.

Rickard, T. C., Lau, J. S. H., & Pashler, H. (2008). Spacing and the transition from calculation to retrieval. *Psychonomic Bulletin & Review, 15*(3), 656–661.

Rittle-Johnson, B., & Star, J. R. (2009). Compared with what? The effects of different comparisons on conceptual knowledge and procedural flexibility for equation solving. *Journal of Educational Psychology, 101*(3), 529–544.

Rittle-Johnson, B., Star, J. R., & Durkin, K. (2009). The importance of prior knowledge when comparing examples: influences on conceptual and procedural knowledge of equation solving. *Journal of Educational Psychology, 101*(4), 836–852.

Rittle-Johnson, B., Star, J. R., & Durkin, K. (2012). Developing procedural flexibility: are novices prepared to learn from comparing procedures? *The British Journal of Educational Psychology, 82*(Pt 3), 436–455.

Rohrer, D. (2009). The effects of spacing and mixing practice problems. *Journal for Research in Mathematics Education, 40*, 4–17.

Rohrer, D. (2012). Interleaving helps students distinguish among similar concepts. *Educational Psychology Review, 24*(3), 355–367.

Rohrer, D., & Taylor, K. (2006). The effects of overlearning and distributed practice on the retention of mathematics knowledge. *Applied Cognitive Psychology, 20*, 1209–1224.

Rohrer, D., & Taylor, K. (2007). The shuffling of mathematics practice problems boosts learning. *Instructional Science, 35*, 481–498.

Rohrer, D., Dedrick, R. F., & Burgess, K. (2014). The benefit of interleaved mathematics practice is not limited to superficially similar kinds of problems. *Psychonomic Bulletin & Review, 21*, 1323–1330.

Rohrer, D., Dedrick, R. F., & Stershic, S. (2015). Interleaved practice improves mathematics learning. *Journal of Educational Psychology, 107*(3), 900.

Ross, B. H. (1989). Distinguishing types of superficial similarities: different effects on the access and use of earlier problems. *Journal of Experimental Psychology: Learning, Memory, and Cognition, 15*(3), 456.

Roy, M., & Chi, M. T. (2005). The self-explanation principle in multimedia learning. In R. E. Mayer (Ed.), *The Cambridge handbook of multimedia learning* (pp. 271–286). New York: Cambridge University Press.

Salden, R. J. C. M., Aleven, V. A. W. M. M., Renkl, A., & Schwonke, R. (2009). Worked examples and tutored problem solving: redundant or synergistic forms of support? *Topics in Cognitive Science, 1*, 203–213.

Schutte, G. M., Duhon, G. J., Solomon, B. G., Poncy, B. C., Moore, K., & Story, B. (2015). A comparative analysis of massed vs. distributed practice on basic math fact fluency growth rates. *Journal of School Psychology*, *53*(2), 149–159.

Schwartz, D. L., & Bransford, J. D. (1998). A time for telling. *Cognition and Instruction*, *16*(4), 475–5223.

Schwonke, R., Wittwer, J., Aleven, V., Salden, R.J.C.M., Krieg, C., & Renkl, A. (2007). Can tutored problem solving benefit from faded worked-out examples? Paper presented at *The European Cognitive Science Conference*, Delphi, Greece.

Schwonke, R., Renkl, A., Salden, R., & Aleven, V. (2011). Effects of different ratios of worked solution steps and problem solving opportunities on cognitive load and learning outcomes. *Computers in Human Behavior*, *27*, 58–62.

Shute, V. J. (2008). Focus on formative feedback. *Review of Educational Research*, *78*(1), 153–189.

Siegler, R. S. (2002). Microgenetic studies of self-explanations. In N. Granott, & J. Parziale (Eds.), *Microdevelopment: Transition processes in development and learning* (pp. 31–58). New York: Cambridge University.

Siegler, R. S., & Chen, Z. (2008). Differentiation and integration: guiding principles for analyzing cognitive change. *Developmental Science*, *11*(4), 433–453.

Siler, S. A., & Willows, K. J. (2014). Individual differences in the effect of relevant concreteness on learning and transfer of a mathematical concept. *Learning and Instruction*, *33*, 170–181.

Singh, R., Saleem, M., Pradhan, P., Heffernan, C., Heffernan, N., Razzaq, L. Dailey, M. O'Connor, C. & Mulchay, C. (2011). Feedback during Web-Based Homework: The Role of Hints. In Biswas, et al., (Eds.), *Proceedings of the Artificial Intelligence in Education Conference* (pp. 328–336).

Smit, J., van Eerde, H. A. A., & Bakker, A. (2013). A conceptualisation of whole-class scaffolding. *British Educational Research Journal*, *39*, 817–834.

Star, J. R., & Rittle-Johnson, B. (2009). Making algebra work: instructional strategies that deepen student understanding, within and between algebraic representations. *ERS Spectrum*, *27*(2), 11–18.

Star, J. R., Pollack, C., Durkin, K., Rittle-Johnson, B., Lynch, K., Newton, K., & Gogolen, C. (2015). Learning from comparison in algebra. *Contemporary Educational Psychology*, *40*, 41–54.

Stein, M. K., Engle, R. A., Smith, M. S., & Hughes, E. K. (2008). Orchestrating productive mathematical discussions: five practices for helping teachers move beyond show and tell. *Mathematical Thinking and Learning*, *10*(4), 313–340.

Stephan, M., & Akyuz, D. (2012). A proposed instructional theory for integer addition and subtraction. *Journal for Research in Mathematics Education*, *43*(4), 428–464.

Stigler, J. W., & Hiebert, J. (1999). *The teaching gap: best ideas from the world's teachers for improving education in the classroom*. New York, NY: Free Press.

Sweller, J., & Cooper, G. (1985). The use of worked examples as a substitute for problem solving in learning algebra. *Cognition and Instruction*, *2*(1), 59–89.

Tarmizi, R. A., & Sweller, J. (1988). Guidance during mathematical problem solving. *Journal of Educational Psychology*, *80*(4), 424–436.

Taylor, K., & Rohrer, D. (2010). The effect of interleaving practice. *Applied Cognitive Psychology*, *24*, 837–848.

Thompson, C. A., & Opfer, J. E. (2010). How 15 hundred is like 15 cherries: effect of progressive alignment on representational changes in numerical cognition. *Child Development*, *81*(6), 1768–1786.

Uttal, D., Scudder, K., & DeLoache, J. (1997). Manipulatives as symbols: a new perspective on the use of concrete objects to teach mathematics. *Journal of Applied Developmental Psychology*, *19*, 37–54.

Uttal, D. H., Amaya, M., del Rosario Maita, M., Hand, L. L., Cohen, C. A., O'Doherty, K., & DeLoache, J. S. (2013). It works both ways: transfer difficulties between manipulatives and written subtraction solutions. *Child Development Research*, *2013*, Article ID 216367.

Vamvakoussi, X., & Vosniadou, S. (2012). Bridging the gap between the dense and the discrete: the number line and the "rubber line" bridging analogy. *Mathematical Thinking and Learning*, *14*, 265–284.

van de Pol, J., Volman, M., & Beishuizen, J. (2010). Scaffolding in teacher-student interaction: a decade of researcher. *Educational Psychology Review*, *22*, 271–296.

van Loon-Hillen, N., van Gog, T. V., & Brand-Gruwel, S. (2012). Effects of worked examples in a primary school mathematics curriculum. *Interactive Learning Environments*, *20*(1), 89–99.

Vorhölter, K., Kaiser, G., & Ferri, R. B. (2014). Modeling in mathematics classroom instruction: An innovative approach for transforming mathematics education. In *Transforming Mathematics Instruction* (pp. 21–36). Cham, Switzerland: Springer International Publishing.

Vygotsky, L. S. (1978). *Mind in society: The development of higher psychological processes*. Cambridge, MA: Harvard University Press.

Williams, S. E. (1997). Teachers' written comments and students' responses: A socially constructed interaction. *Paper presented at the Annual Meeting of the Conference on College Composition and Communication* (pp. 12–15). Phoenix, AZ.

Willingham, D. T. (2002). Ask the cognitive scientist allocating student study time "massed" versus "distributed" practice. *American Educator*, *26*(2), 37–39.

Wood, D., Bruner, J. S., & Ross, G. (1976). The role of tutoring in problem solving. *Journal of Child Psychology and Psychiatry*, *17*(2), 89–100.

Zhu, X., & Simon, H. (1987). Learning mathematics from examples and by doing. *Cognition and Instruction*, *4*(3), 137–166.

Index

D

M

S

T

FSC
www.fsc.org
MIX
Papier | Fördert gute Waldnutzung
FSC® C083411

Kolibri 360 GmbH
Ferdinand-Jühlke-Straße 7
99095 Erfurt, Deutschland
produktsicherheit@kolibri360.de